Thin Films and Heterostructures for Oxide Electronics

Multifunctional Thin Film Series

Editors:

**Orlando Auciello, Argonne National Laboratory,
Ramamoorthy Ramesh, University of California at Berkeley**

The basic applied science and applications to micro- and nano-devices
of multifunctional thin films span several fast-evolving
interdisciplinary fields of research and technological development
worldwide. A major driving force for the extensive research being
performed in many universities, industrial and national laboratories is
the promise of applications to a new generation of advanced micro-
and nano-devices. These applications can potentially revolutionize
current technologies and /or create new ones while simultaneously
creating multibillion dollar markets.) Multifunctional thin films cover
a wide range of materials from metals to insulators to organics,
including novel nanostructured metals, semiconductors, oxides,
widebandgap, polymers, self-assembled organic layers and many
other materials.

Series titles:

Nanoscale Phenomena in Ferroelectric Thin Films
Seungbum Hong

Graded Ferroelectrics, Transpacitors, and Transponents
Joseph V. Mantese and S. Pamir Alpay

Thin Films and Heterostructures for Oxide Electronics
Satischandra B. Ogale

Thin Films and Heterostructures for Oxide Electronics

Satischandra B. Ogale
The University of Maryland
U.S.A.

 Springer

Library of Congress Cataloging-in-Publication Data

Ogale, Satischandra B.
    Thin films and heterostructures for oxide electronics / Satischandra B. Ogale.
    p. cm. – (Multifunctional thin film series.)
    Includes bibliographical references and index.
ISBN 10: 0-387-25802-7   ISBN 10: 0-387- 26089-7 (e-book)
ISBN 13: 9780387258027   ISBN 13: 9780387260891 (e-book)
    1.   Electronics—Materials. 2. Thinfilms. 3. Metallic oxides. I. Title. II. Series

TK7871.O32 2005
621.381—dc22

                                                                    2005049016

Printed in the United States of America.

9 8 7 6 5 4 3 2 1   SPIN 11054252

springeronline.com

CONTENTS

vi

Preface

It has been a pleasure to edit this book covering a variety of topics in the rapidly developing field of oxide electronics. This book has thirteen excellent chapters written by researchers who have made significant contributions to the field over the past several years. This book came about through the active interest of Mr. Greg Franklin (Senior Editor, Materials Science, Springer) in this effort and the valuable recommendations and suggestions by Prof. R. Ramesh, University of California, Berkeley, and Dr. Orlando Auciello, Argonne National Laboratory, who are the co-editors of this series. I take this opportunity to thank them for their strong support and patience. I also wish to express my sincere thanks to all the contributors who agreed to participate in this effort and made it possible to complete the project fairly on time in spite of their active research and teaching commitments.

With the ever-shrinking dimensions and ever-increasing speeds of conventional semiconductor-based micro and opto-electronic devices, scientific limitations are beginning to be felt by the technology sector in keeping pace with the newer device demands of the modern times. Novel approaches involving new materials such as functional oxides are being explored to tackle this problem. This research is already beginning to pave way for the next generation device concepts and configurations for advanced oxide electronics, encompassing magneto-optics and spintronics. This book, presented in the form of comprehensive chapters written by active researchers in the field of oxide films and heterostructures, captures the excitement of this emerging frontier of immense scientific and technological interest. These well designed chapters highlight the materials, physics, device and even key fabrication issues. Many interesting material systems such as magnetic, ferroelectric and superconducting oxides, wide band-gap semiconducting oxides and high k dielectric oxides are discussed. With many references to the vast resource of recently published literature on the subject, this book intends to serve as a significant and insightful source of valuable information pertaining to the ongoing scientific debates, current state of understanding and future directions. As such, it should be very useful to university students, post docs, professors and research scientists in industries and national laboratories.

I take this opportunity to thank my collaborators at the University of Maryland Prof. T. Venkatesan, Prof. R. Ramesh (now at Berkeley), Prof. R. L. Greene, Prof. H. D. Drew, Prof. S. Das Sarma, as well as Prof. A. J. Millis (Columbia) and Prof. S. -W. Cheong (Rutgers) for many interesting discussions on the interesting subject matter covered in this book. I also wish to thank Prof. Ellen Williams, Director, NSF-MRSEC at the University of Maryland, College Park, for her support. I further wish to thank my collaborators Dr. Darshan C. Kundaliya and Dr. Sanjay R. Shinde at the University of Maryland and Ms. Carol Day at Springer for their immense help in the editorial process. Without their kind and active help it would have been very difficult to deliver the book on time. Finally I am happy to mention the constant support and encouragement by my wife Anjali, my parents, youngsters Abhijit, Deepti and Neeti.

I greatly appreciate the interest and support of Springer Publishers in this effort.

Satishchandra Ogale
Department of Materials Science
Centre for Superconductivity Research, Department of Physics
University of Maryland, College Park, MD 20742.

Ferroelectrics, Nano-scale phenomena, High k Dielectrics

CHAPTER 1

NANOSCALE PHENOMENA IN FERROELECTRIC THIN FILMS

V. Nagarajan*, T. Zhao, H. Zheng and R. Ramesh

Department of Materials Science and Engineering and Department of Physics, University of California, Berkeley, CA 94720
* Institute for Solid State Physics and CNI, Forschungzentrum Juelich, D 52425 Germany

1.1 INTRODUCTION

At present there is considerable interest in ferroelectric thin films as a medium for non-volatile data storage.[1] In particular much attention is focused on investigating high-density giga-bit data storage using scanning probe techniques.[2] As ferroelectric random access memory (FeRAM) devices scale down to the Gbit density regimes, it becomes imperative to understand physical phenomena that affect device performance at the nanoscale.

This is elucidated in figure 1. Figure 1(a) is the fundamental oxide perovskite heterostructure. It shows three unit cells; beginning from the bottom is the substrate, then the oxide electrode and finally the ferroelectric. Clearly at each interface there is wealth of materials information to be explored. One can now take these layers, and when grown in thin film form, the ferroelectric can then self-assemble itself in a mesoscopic pattern of domains (this will be explained in detail later) shown in figure 1(b). Several of such domains can form in a single grain (figure 1(c), and several such grains are found in a single capacitor (figure(1(d)) and several such capacitors form the basis of a FeRAM memory device, such as a smart card(1(e). In this paper we deal with the same evolution of the hierarchical level of complexity. We begin with a model system where the microstructure can be systematically changed. In this case we begin by investigating the key fundamental ferroelectric

4

phenomenon with the atomic force microscope. Table 1 shows each of the fundamental materials issues and the physical phenomenon associated with it.

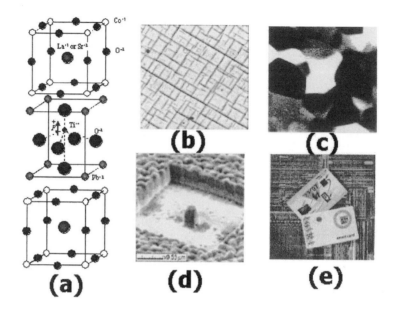

Figure 1: The hierarchical evolution of materials in FeRAM Technology. (a) Shows the basic perovskite structure of the substrate, oxide electrode and the ferroelectric.(b) Mesoscopic domain structure (c) A polycrystalline film with several grains. Each of these grains could have several mesoscopic domains, depending on the grain size. (d) A nanoscale capacitor fabricated via focused ion beam milling. (e) A smart card using several capacitors.

Table 1: Fundamental materials issues and physical phenomenon in thin film ferroelectrics for nonvolatile memory technology.

Fatigue	Dynamic interactions of domain walls with defects
Imprint	Static interaction of domain walls with defects
Retention	Polarization relaxation
Switching Speed	Domain dynamics under applied field
Stress Effects	Interplay between domain walls and stress fields
Leakage	Transport of charged species
Polarization	Domain dynamics, defect interactions

For example, the ferroelectric element in these storage media is subject to progressive loss of polarization, commonly referred to as "polarization relaxation". Polarization relaxation shows a very complex interplay between various factors such as defects, domain wall dynamics, crystallographic structure, etc and hence it has been a subject of intense scientific research. All of the above mentioned factors have characteristic dimensions in the range of a few hundred to a few thousand angstroms, which really means that in order to fully appreciate their role one must have a approach to study phenomena at those scales. With the advent of Piezo response Scanning Force Microscopy (Piezoresponse SFM), investigation of nanoscale ferroelectric phenomena has not only become viable, but also a necessity. It is a unique non-destructive characterization tool coupled with a relatively easy method of sample preparation, which can provide critical information on key nanoscale processes. In this paper we show in the first part, how this tool has been so effective in observing domain relaxation. In this section we focus on imaging nanoscale phenomena, particularly polarization relaxation. These studies are focused on understanding the role of "defects" such as domain walls as nucleation centers and how defect sites act as pinning centers to inhibit complete reversal. The physical nature of this problem and studies on the microscopic origins of polarization reversal using scanning force microscope techniques[3-9], show that they have the potential for quantifying the physical origin of the relaxation kinetics. The Johnson-Mehl-Avrami-Kolmorogov (JMAK) theory is a useful approach for systems exhibiting nucleation and growth based phase change kinetics[10]. In this paper we review in detail a study to apply this theory to the nanoscale investigation of polarization reversal in ferroelectric thin films via the atomic force microscope. In addition the same technique can be used to image the domain structure of epitaxial ferroelectric thin films. We also show, by using careful interpretation of domain images, one can reconstruct the 3-dimensional polarization picture in these thin films. The second section on our observations shall concentrate on obtaining quantitative information from such nanoscale capacitors. Particularly the dependence of the piezoelectric behavior of the capacitor on orientation, size and constraint has been investigated. Since the probe itself is only several tens of nanometers, it gives us the advantage of probing piezoelectric and ferroelectric properties of sub-micron capacitors. Consequently, we have been able to show piezoelectricity in capacitors as small as 70x70 nm^2 and direct hysteresis loops in sub-micron capacitors. We summarize this review with some results direct hysteresis and fatigue measurements on nanoscale capacitors. Finally, some areas for future research are also highlighted.

1.2 THIN FILM MATERIALS AND CHARACTERIZATION

In order to create films with systematically varying crystalline quality, two approaches were employed. This first approach was to use SrTiO3 /Si substrates where the thickness of STO is 15 nm. Details of the STO deposition process are described elsewhere).[11] On such substrates the PZT was deposited by sol-gel process, which were (001) oriented. The second approach used SrTiO$_3$ oxide substrates to obtain highly epitaxial structures. On such substrates the films were deposited via pulsed laser deposition.[12] For both type of films about 70nm thick LSCO top and bottom electrodes were deposited by RF-sputtering and crystallized in oxygen for 1 hour at 600-650°C. The sol-gel PZT was annealed at 450 °C. Details of the process can be found elsewhere.[13] X-ray diffraction (XRD) and transmission electron microscopy (TEM) were employed to probe the phase formation and microstructure of the samples. Top electrodes (Pt and LSCO) were made by photolithography and sputtering so that the capacitors for electrical measurements show a symmetric configuration. The ferroelectric measurement was done using a RT-66A tester.

A significant fraction of our domain imaging studies use epitaxial PZT films as model systems. In such systems strains associated with lattice mismatch and phase transformation from the cubic to the tetragonal (ferroelectric) phase are relaxed via domain formation.14 Figure 2 shows an example of such a model system. In this case highly tetragonal films of

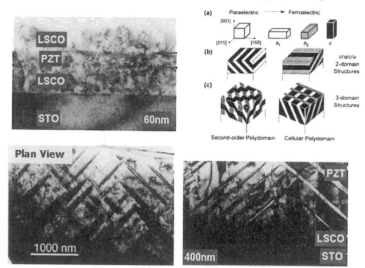

Figure 2: Polydomain formation in highly tetragonal PZT thin films. Figure 2(a) shows a 60 nm film where there is no polydomain formation observed. Figure 2(b) shows the schematic structure of a polydomain film 2(c) is the plan view of a 400 nm thick film and 2(d) is the cross section. Clearly in 2(c) and (d) the polydomains are seen as needle shaped structures.

composition PbZr$_{0.2}$Ti$_{0.8}$O$_3$ have been grown via pulsed laser deposition. By systematically changing the thickness one can systematically change the relaxation of strain and hence the domain fractions. This now allows us to insert artificial interfaces and size constraints on the system. Such 90° domains have been shown to have a strong impact on the hysteretic polarization behavior of ferroelectric films.[15] The quantitative piezoresponse measurements involve films with systematically different microstructures (epitaxial to polycrystalline) as well as a variety of substrates.

Voltage-modulated scanning force microscopy[16] has been employed to study the piezoelectric effect and to image and modify the domain structure in thin film samples. The experimental set up is shown in figure 3. Piezo-SFM is based on the detection of the local electromechanical vibration of the ferroelectric sample caused by an external ac voltage. The voltage is applied through the probing tip, which is used as a movable top electrode.

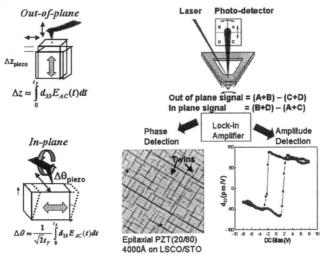

Figure 3: Experimental set up for Piezoresponse Imaging.

The external driving voltage with frequency ω generates a sample surface vibration with the same frequency due to the converse piezoelectric effect. This surface displacement can be approximated by :

$$\Delta z(t) = \int_{0}^{t_f}{}^{thin\_film} d_{33} E_{AC}(t,z)dz \qquad ....(1)$$

where E_{ac} is the ac electric field in function of time t and displacement z, and t_F is the thickness of the ferroelectric layer. The modulated deflection signal from the cantilever, which oscillates together with the sample, is detected using the lock-in technique. The amplitude of the first harmonic signal from the lock in amplifier is a function of the piezoelectric displacement and phase shift between the ac field and the cantilever displacement. Mathematically it is given as

$$H_\varpi = k_d d_{33}^{thin - film} V_{AC} \cos\theta \qquad \ldots.(2)$$

The first 2 terms in the above equation represent the cantilever displacement, which is coupled to magnitude of the polarization. The last term($\cos\theta$) represents the phase of the lock in, which (through some lock in techniques) can be made to assume values of 1, 0 or −1. This means that regions with opposite orientation of polarization, vibrating in counter phase with respect to each other under the applied ac field, should appear as regions of opposite contrast in the piezoresponse image. Since this technique depends on the piezoresponse of the ferroelectric it is also termed as piezo response microscopy.

For our experiments, we used a commercial Digital Instruments Nanoscope IIIA Multimode scanning probe microscope equipped with standard silicon tips coated with Pt/Ir alloy for electrical conduction. The typical force constant of these tips was 5 N/m and the apex radius was approximately 20 nm. The contact force was estimated to be around 70 to 100 nN.

1.3 NANOSCALE DOMAIN IMAGING IN FERROELECTRIC THIN FILMS

Figure 4 shows AFM images of a typical epitaxial PZT(0/20/80) film with : (a) the topography and (b) being the piezo-response image. The

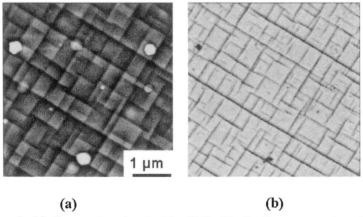

(a)   (b)

Figure 4: (a) Topography of epitaxial $PbZr_{0.2}Ti_{0.8}O_3$ film grown via pulsed laser deposition. (b) Piezoresponse images of the same film. The 90° domain structure is clearly seen in both images as a two dimensional grid.

topography clearly shows the 2 dimensional 90° domain structure. The topographical relief in the structure originates from the spontaneous strain in the ferroelectric and tilting of *a*-domains away from the normal, which has been discussed in detail elsewhere.[17,18] The piezoresponse image compares where very well to the topographical image. Piezoresponse images of the surface of the film show the presence of long needle-like orthogonal structures which have been identified to be 90° twinned domains. The image contrast in between the needle like regions is uniformly bright suggesting that the entire region is pre-poled in a specific direction. This was confirmed by measurements of the piezo-hysteresis loop from a local region,[19] which showed a marked asymmetry indicating a built-in internal field. This as grown sample was switched into the opposite state by scanning the surface of the film with the AFM tip biased at −10V (scan speed 1Hz) leading to a strong change in the image contrast from bright to dark as will be shown later. We observed this switched state (in the *c*-axis regions of the film) begin to relax back to the original bright state.[20]

We first discuss the use of this technique to map out the polarization orientation in a given ferroelectric thin film. This technique was first applied by Eng *et al.*[21] to image the 3 dimensional polarization map in BaTiO3 single crystal and later Roelofs *et al.*[22] demonstrated 3D polarization imaging polycrystalline PZT thin films. In order to map the polarization vectors in these *a*-domains, the lateral differential signal from the photodiode was tracked. This signal effectively represents torsion deformation of the AFM cantilever. As the tip scans over the *a*-domain regions, the electrical field through the tip produces the shearing deformation mentioned earlier, giving rise to the torsion signal in the photodiode.

This is schematically illustrated in Figure 5(a) and (b) for oppositely oriented *a*-domains. The phase of this torsion motion of the tip with respect to the input sinusoidal electric field provides information about the orientation of polarization in the domain. It should be pointed out that the scan direction of the tip is chosen to be such that the orthogonal *a*-domains contribute equally to the tip motion. The tip, therefore, scans at 45° to the a_1- and a_2-domains.

Figure 6(a) shows typical out of plane and in-plane piezoresponse images of the same region in the PZT film. In an earlier publication we showed that it was possible to find out the inclination of the *a*-domain with respect to the substrate by obtaining a line scan of the piezoresponse signal across the domain.[23-24]By combining this information with the

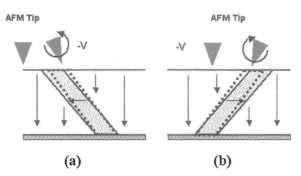

Figure 5: Interaction of the field with the tip to give torsion signal. Depending upon the output (a) or (b) , the orientation of the in-plane image can be mapped out.

Knowledge about the polarization direction within the *a*-domain, it is possible to see that the head to tail configuration of the polarization vectors is maintained across the 90° domain wall. Thus, domains aligned at 45° and 135° to the substrate are phase shifted by 180° and therefore lead to different contrast levels in the in-plane image. This can be seen in the figure 6(b). Here when traversing from point A to B, we travel from a sharp interface to a diffuse interface and hence along a domain which inclined at 135°. The in-plane domain image for the same region shows that domain contrast is now black, compared to the white contrast for the earlier case.

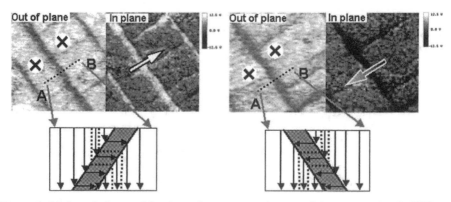

Figure 6: (a) Out-of-plane and In-plane piezo response images of the same region in PZT film for domain inclined at 45°. (b) Out-of-plane and In-plane images for a domain inclined at 135°. Notice the difference in contrast for the in-plane regions, although the out–of-plane contrast is the same. This change in in-plane orientation is necessary to maintain a head to tail domain configuration.

The absence of a top electrode in our samples leads to a favorable state of polarization along the *c*-axis as seen in Figure 4(b). This stable state of polarization can be switched into the opposite unstable state as illustrated in Figure 7(a). This unstable state decays over time and through nucleation and growth of the original stable state converts the switched region back into the as grown state of polarization. The next section will focus on the underlying concepts and materials phenomena that govern "polarization relaxation".

For the relaxation experiments, a 5μm×5μm area of the sample was switched from its original state (bright contrast) to the reversed state (dark contrast) by scanning the film with the AFM tip biased at −5V. Piezoresponse images were recorded from the same region as a function of time from a few minutes to several days. The polarization spontaneously reversed its direction as illustrated in the sequence of piezoresponse images in Figs 7(b-d). However it is evident in 7(d) that the complete reversal of polarization does not occur, rather there are some small regions of dark contrast, which seem to be pinned.

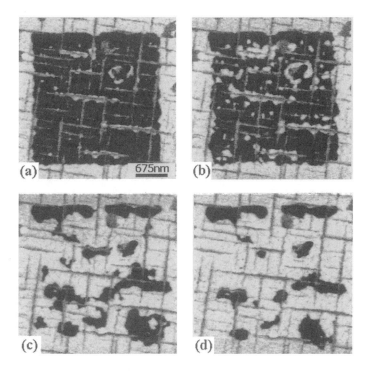

Figure 7: (a) A 5μm by 5μm region is switched into oposite polarity by applying a DC volatge through the tip. (b), (c) and (d) show the relaxation of the switched state after times of 39040, 75520 and 322740 seconds respectively.

The key aspects of the reversal experiments are as follows. The internal field present due to the asymmetry of the electrode structure drives the reversal. Nucleation of the reverse domains was found to occur preferentially at the twin boundaries. The bright circle in figure 8(a) highlights this. White incipient nuclei are clearly seen followed by the growth of the reversed regions to cover an increasing area of the film. These observations were found to be very repeatable in several experiments conducted over various parts of the sample. The thermodynamic driving force for relaxation is a result of the built-in field due to charge redistribution within the electrode. In the as-grown state this built-in field is compensated by the film polarization, so that the net electrostatic field in the film is close to zero. The polarization charge is considerably less that P_s due to the effect of internal screening near the interface. Since this poled state can exist for a long time and can be repeatedly obtained after reheating the sample, it is to be naturally assumed that this is close to the 'stable' state of the film and that the electrostatic field in the film is absent (or close to zero) for this state. After "writing" or switching the polarization vector under an external DC field, an unstable

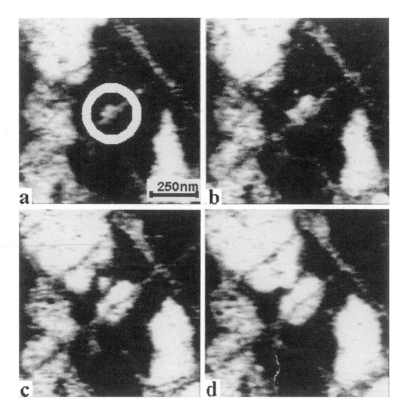

Figure 8: (a)-(d) : This figure shows the role of 90o domains in the reversal of polarization. In (a) the bright circle highlights the incipient reverse nuclei, in white contrast, growing along the edge of the 90° domains. This shows the role of heterogeneous nucleation sites. Figure (b) shows that the reverse domain now grows towards the edges of the "cell". (c) Now incipient nuclei begin to show up at the center of the cell. (d) Shows the merging of two domain walls advancing towards each other, each having nucleated at opposite 90° domain walls.

state of polarization is created which then relaxes toward the as-grown stable state. (Gruverman et al., similarly observed suppression of relaxation, in a recent publication 25).

Our original interpretation of these relaxation results followed an empirical fit of the experimental data which yielded a stretched exponential-type behavior. Such a stretched exponential dependence, of course, is very well known in condensed matter physics[26] and is a consequence of either a broad distribution of pinning energies and/or activation times for pinning sites with the same pinning energy.

An alternative approach to interpret this data uses a classical nucleation and growth argument. In such a model, we use the Johnson Mehl Avrami Kolmogorov(JMAK) model to understand this time dependence of fraction of reversed regions. Consider the mesh of c/a/c/a domains as shown schematically in Fig 6(a). The spacing between the a-domains is random. Nucleation is modeled to occur midway along each of the a-domain lines at time t=0.

From JMAK theory we can relate the fractional volume transformed to the extended volume through

$$f = 1 - e^{-V_{ext}} \qquad \text{....(3)}$$

where f is the actual transformed volume and V_{ext} is the extended volume. To calculate V_{ext} consider the growth of a cylindrical stable (white) domain surrounded by an unstable (black) region (Fig. 7(b)). For a small change in the radius dR, we can write the change in the free energy $d\Delta F$:

$$d\Delta F = -2\pi R h_f P_s E(1-f)dR + 2\pi R h_f P_s \alpha E dR + 2\pi \Gamma h_f dR \qquad \text{....(4)}$$

where ΔF is the change in free energy, R the instantaneous radius of the growing nucleus, P_s the spontaneous polarization of the ferroelectric thin film ($0.6 C/cm^2$), f is the fraction of reversed domains, E is the imprint driving field in the sample, αE is a reverse driving field and Γ is the surface energy of the 180° domain wall. From the piezoresponse images at early times ($t < 10^4 s$), it can be seen that domain wall is rough. This implies that we can preclude 2D-nucleation assisted domain wall motion. Therefore we can write the rate equation as:

$$\frac{dR}{dt} = -\frac{L}{2\pi R}\frac{d\Delta F}{dR} \qquad \text{....(5)}$$

where the rate constant L can be expressed through[27]:

$$L = \upsilon v_0 \frac{exp\left(\dfrac{-U_a}{kT}\right)}{kT} \qquad \qquad ....(6)$$

where U_a is the activation energy, υ the volume of critical nucleus and v_0 the jump frequency (10^{13} Hz).

The critical radius can be estimated from $\left.\dfrac{d\Delta F}{dR}\right|_{R=R_{crit}} = 0$:

$$R_{crit} = \left.\frac{\Gamma}{P_s E(1-f-\alpha)}\right|_{f\to 0} = \frac{\Gamma}{P_s E(1-\alpha)} \qquad \qquad ....(7)$$

Therefore

$$\frac{dR}{dt} = -\frac{L}{2\pi R}\frac{d\Delta F}{dR} = P_s ELh_f(1-f-\alpha)\left(1-\frac{R_{crit}}{R}\right) \qquad ....(8)$$

R_{crit} has been calculated to be of the order of 40Å, which is much smaller than the resolution limit of our imaging technique. Thus we can ignore the term involving R_{crit} in equation (6). This leads to an estimate for V_{ext} as

$$V_{ext} = \int_0^t N(\xi)\pi(R(t-\xi))^2 h_f d\xi = N_0\pi h_f^3 P_s^2 E^2 L^2\left\{\int_0^t(1-f-\alpha)dt\right\}^2$$

$$....(9)$$

Finally, we can write the fraction-reversed f as

$$f = 1 - exp(-V_{ext}) = 1 - exp\left\{-N_0\pi h_f^3 P_s^2 E^2 L^2\left(\int_0^t(1-f-\alpha)dt\right)^2\right\}$$

$$....(10)$$

This equation can be plotted numerically to fit the pre-exponent parameters $C = N_0\pi h_f^3 P_s^2 E^2 L^2$ and α. When the numerical solution to eqn. 10 is fitted to the experimental data, the values of L and α are obtained as 5.95×10^{-8} m^3/Js and 0.1 respectively (see Table 2). This gives an estimate of 0.962eV for U_a. It is enticing to correlate this activation energy to the activation energy for the diffusion of specific defects, such as oxygen vacancies. However, detailed understanding of the atomic scale origins of this activation process needs considerable further study.

An important question that arises is this: is this polarization reversal accompanied by reversal within the a-domain? Figures 9(a) and (b) are in-plane and out of plane domain images of the PZT film before writing (as

Table 2: Values of N_o, α and U_a obtained after fitting the data to equation 10 for various DC write voltages.

Voltage	N_0	α	L	U_a
-8.0	8.89×10^{12}	0.35	1.428×10^{-17}	0.9076
-10.0	7.77×10^{12}	0.40	2.381×10^{-18}	0.9522
-12.0	1.89×10^{12}	0.82	7.142×10^{-18}	0.9249

(a) (b)

(c) (d)

Figure 9: In plane domain image and out of plane image for the same region.(a) Shows the as-grown area. (b) The center of the as-grown area is now written with DC bias of -8V. The center appears in dark contrast and corresponding in-plane images show the in plane polarization also changes contrast to avoid charge build up at the interface. (c) As the written region relaxes we see that the corresponding in –plane regions also change contrast. (d) the switched region has shrunk to almost 50% of it's original size, and we can see the in-plane polarization also return to their original state.

grown). The out-of-plane polarization state within the region scanned was switched into the opposite state by scanning with the tip biased at - 8 V. The in-plane polarization within the a-domains was seen to switch at the same time, such that no charged interfaces could be identified. Images (c) and (d), (e) and (f), and (g) and (h) were acquired after wait times of 1.8×10^4 s, 2.6×10^4 s and 5.9×10^4 s, respectively. The reversed c-domains in (c) are seen to reach the c/a domain interface AB but no reversal of the in-plane polarization within the a-domain is observed. This interface therefore is likely to be charged as it represents a head-head arrangement of dipoles. The polarization within the a-domain is seen to reverse only after stable c-

16

domains have nucleated on the opposite side of AB as seen in (e) and grow to meet the interface AB (in (g)) thereby causing electrostatic pressure to be applied on both sides of the *a*-domain. It is possible that the sluggish nature of in-plane reversal under zero external field (relaxation) could be attributed to the small dimension (width) of the *a*-domain.

Until now, we have discussed the polarization phenomenon on a more or less macro-scale. It becomes pertinent to investigate this physical phenomenon on a very local scale. Figure 10(a-d) show the transformation in a single "cell" 500nm wide. The domain wall is seen to meander across from the nucleation site inside one of the sides towards the opposite *a*-domain wall till it gets pinned by defect sites as observed in Fig. 10(c). In

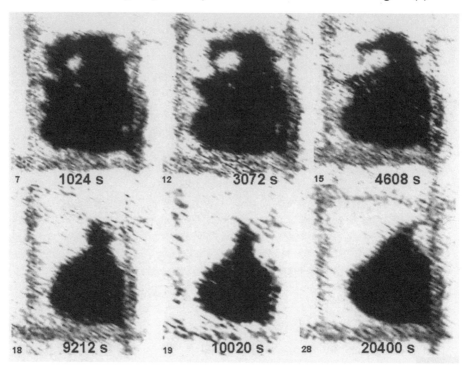

Figure 10: (a) – (f) Relaxation in a "single cell". Data illustrates the role of pinning in governing the local velocity of the 180° domain wall. In (b) pinning is clearly observed inside the circle. The moving domain wall breaks away from the site in (c), moving in the direction shown by the arrows and gets pinned at another site(d). Figure 10(e) and (f) show how faceting of the domain wall surface occurs, accompanied by a dramatic reduction in wall velocity.

the early stages there is enough driving force for the domain wall to overcome the pinning sites as can be seen in the series of images Figs. 10(b) – (c). However, as the relaxation proceeds to later stages, the driving force is exhausted and the wall, unable to break free from the pinning sites, bows out. The difficulty of estimating this small driving field at the end of the relaxation prevents an accurate estimation of the wall energy. The direct observation of pinned walls at small fields and their non-crystallographic orientation allows us to suggest that the main controlling mechanism of domain wall mobility is the overcoming of pinning points in the field. The domain wall movement should then require 2D-nucleation and this diminishes the mobility dramatically. In addition to these features we have also reported faceting of the domain walls, which result in metastable shapes with clear crystallographic orientations.28

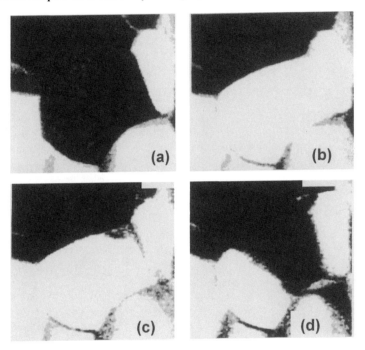

Figure 11:(a) − (d) Relaxation in a "single grain". Data illustrates the role of grain boundaries in the relaxation process. In (c) polarization reversal is clearly observed adjacent to the grain boundary.

Figure 11 (a) to (c) show relaxation observed for a polycrystalline film. It is clear that grain boundaries play a crucial role in the relaxation behavior. Figure 11(a) is a single grain,which shows the polarization with a dark contrast. Figure 11(b) is the image of the same grain after switching with a DC bias of 6V.

Figure 11(c) and 11(d) are images taken after 60 seconds and 10^6 seconds respectively. In figure 11(c) we observe that the relaxation process proceeds from the grain boundaries. In this way, it is analogous to the role of 90° domain boundaries in the nucleation of reverse polarization nuclei. Secondly it is seen that even after 10^6 seconds the polarization is not completely reversed back. In fact it is pinned by almost 50% of its original value. These set of images reveal clearly the role of grain boundaries on polarization relaxation. It is important to note that there is very little rigorous understanding of the local electrodynamics, especially at structural interfaces such as grain boundaries. The fact that nucleation of reversal occurs at grain boundaries should not be surprising ; indeed, this is pervasive in solid state phase transformations. However, it is imperative to understand these phenomena in terms of the structure of the grain boundary, the misorientation of polarization direction across the grain boundary (which leads to local polarization gradients) as well as the possible role of possible second phases at grain boundaries.

1.4 NANOSCALE PIEZOELECTRIC AND FERROELECTRIC BEHAVIOR

This section focuses in detail our recent observations and findings on the quantification of ferroelectric and piezoelectric phenomenon in sub micron capacitors. The SFM probe tip is used to make contact with ferroelectric through a top electrode, typically LSCO/Pt, and hence SFM can now be used as a "nano-probe" station. Our motivation for such investigation is clear from figure 12. Figure 12(a) shows a polycrystalline film where the grain marked with a cross is poled into the opposite state. It is seen clearly that the surrounding grains also change polarity. Therefore we would like to study individual capacitors which are free each other, and compare their properties to continuous devices.

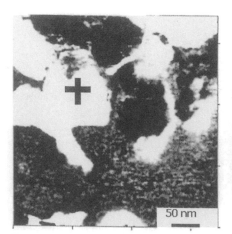

Figure 12: (a) Cross-talk between grains in a polycrystalline film. The region marked with a cross is switched into opposite polarity. It is seen that the surrounding regions change contrast also.

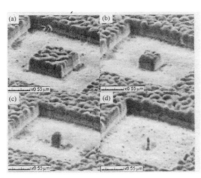

Figure 13: Nanoscale capacitors fabricated via focused ion beam milling. (a) 0.5 by 0.5µm
(b) 0.25 by 0.25 µm (c) 0.1 by 0.1 µm(d) 0.07 by 0.07 µm

One such approach to modify the interface is to change the mechanical boundary conditions on the film by "cutting" it into small dimensions (of the order of 1micron or smaller). This has been successfully achieved via focused ion beam milling.[29,30] Figure13(a)-(d) show such capacitors from 1 µm by µm to 0.07 µm by µm that have been fabricated via focused ion beam milling. As a result, on a nanoscale, we now have and effectively freestanding ferroelectric thin film.

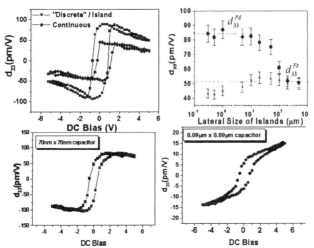

Figure14: Piezoelectric measurements for nanoscale capacitors. (a) d_{33} for a constrained film vs. a free capacitor. The free standing capacitor shows a 50% increase in d_{33}.(b) d_{33} as a function of device size cut capacitor and continuous capacitors. (c) d_{33} for 70 nm by 70nm PNZT capacitor (d) d_{33} for 90 by 90nm SBT capacitor.

Figure 14(a) plots the measured d_{33} for a constrained capacitor and a free-standing capacitor which is 0.01 µm² in area. It is clear that the free-standing capacitor shows a higher value of d_{33} than the continuous film. The measured value of d_{33} is 50% more than the clamped value and hence in good agreement with the Lefki and Dormans model.[31] Figure 14(b) plots the piezoresponse from various such capacitors that have been fabricated by FIB from 100 x 100 μm^2 down to 0.1 X 0.1 μm^2. a clear increase in the d_{33} is seen when the capacitors are milled to 10 microns or below. However when one mills only the top platinum, it is seen that the piezo-response is clamped, does not change much with capacitor size, and is plotted as solid triangles in the same figure.

Figure 14(c) and (d) is piezo-response from a 70nm by 70nm capacitor for PNZT and SBT respectively, which is clearly piezoelectric, and hence ferroelectric. To date, this is the smallest capacitor stack, whose piezoelectric coefficients have been reported.

Figure 15 shows the static hysteresis loops that are measured directly via the scanning force microscope with a conductive Pt/Ir coated tip as the probe for various capacitors from 100μm^2 to 0.01 μm^2. The cantilever was connected to a commercial TF Analyzer 2000 from aixACCT Systems to

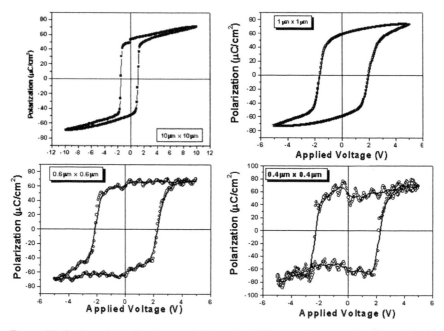

Figure 15: Static polarization hysteresis loops for PZT capacitors as a function of device size.(a) 10 by 10μm(b) 5 by 5μm(c) 1by 1μm(d)0.4 by 0.4 μm

apply the excitation signal and record the current response of the device.[32] The response obtained from the 0.16μm^2 capacitor is shown after subtracting the contribution from the parasitic capacitances and is compared with the response from 100μm^2 cap. It is clearly seen that there is no loss of polarization even when scaled down to the sub micron sized ferroelectric capacitors. Another question of significant technological interest, is what fatigue behavior do sub-micron capacitors show? We have performed fatigue on LSCO/PZT/LSCO and Pt/PZT/Pt capacitors before and it is well known that oxide electrodes prevent fatigue. However in nanoscale regimes, again such an investigation becomes imperative. Figure 16 plots fatigue for two such capacitors, one with LSCO electrodes and the other with Pt electrodes.

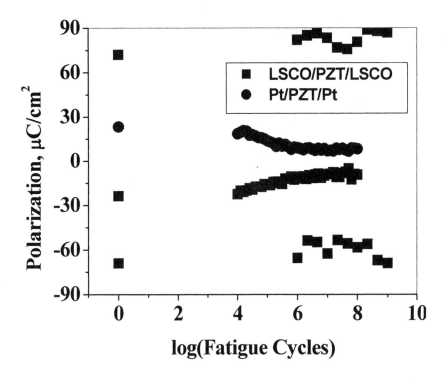

Figure 16: Nanoscale fatigue: Figure shows that Pt/PZT/Pt capacitors show polarization fatigue while LSCO/PZT/LSCO capacitors are fatigue free.

It is clear that for the capacitor with LSCO electrodes no loss of polarization is observed while the capacitor with PT electrodes shows a marked decrease in switched polarization.

1.5 NEW MATERIALS : "GREEN FERROELECTRICS"

Fabricating fatigue and imprint-free capacitors using PZT as the ferroelectric requires the use of a conducting oxide contact electrode; perovksite conducting oxides such as strontium ruthenate and lanthanum cobaltite have emerged as attractive candidates, along with simpler oxides such as iridium oxide. The SBT system does not require the use of such conducting oxide electrodes, although the processing temperatures are higher (750C) compared to the PZT family (450-650C). In the ideal case,

22

one would like to have the intrinsic "green-ness" of the SBT system (the individual elements are relatively chemically "benign" compared to the "toxicity" of lead in the PZT system) with the low temperature processability of the cubic perovskites.

Recent work on lead-free ferroelectric/multiferroics, Fig.17, such as Bismuth ferrite (BFO), has opened up the attractive possibility of using such perovskites as ferroelectric storage elements. This is primarily driven by the fact that these compounds posses a very high ferroelectric Curie temperature (~800C), a large spontaneous polarization of 90microC/cm2, in conjunction with a dielectric constant that is much smaller than the corresponding PZT system (80-140 compared to 350). These studies have

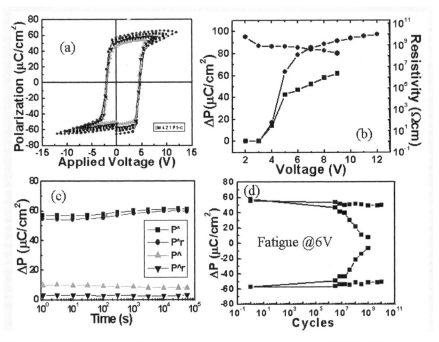

Figure 17: a) Ferroelectric hysteresis measured at a frequency of 15kHz. The remnant polarization is ~45μC/cm². b) Pulsed polarization, ΔP vs. electric field, measured with electrical pulses of 100μsec width and resistivity vs. electric field measured with 1sec pulse. c) Polarization retention shows no change after several days. d) fatigue data with oxide electrode and Pt.

also demonstrated the feasibility of processing these materials at a temperature as low as 450C, making this system very amenable to full scale integration with Si-CMOS to make prototype memory devices. Using a novel epitaxial template approach, we have demonstrated a pathway to grow these ferroelectrics epitaxially on Si substrates. These initial results lay the foundation for an extensive program that will enable the integration of such "environmentally-friendly" materials with Si-CMOS to create prototype memory elements.

1.6 NEW MATERIALS: "Ferroelectric/Ferrimagnetic Nanostructures"

Materials with coexistence of ferroelectric and ferromagnetic properties, which are called "multiferroics", have caused intensive recent research interest both from the device design and fundamental science of view. It was proposed that composites of ferroelectric (piezoelectric) and ferromagnetic (magnetostrictive) phases can be electromagnetically coupled via a stress mediation[33]. In a film-on-substrate geometry, such composites can be created in two extreme forms, one is the multilayered structure and the other is the vertically aligned nanostructures.

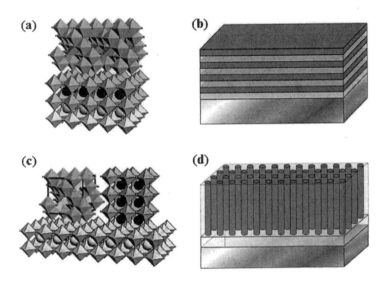

Figure 18 (a) Superlattice of a spinel (top) and a perovskite (middle) on a perovskite substrate (bottom). (b) Schematic illustration of a multilayer structure on a substrate. (c) Epitaxial alignment of a spinel (top left) and a perovskite (top right) on a perovskite substrate (bottom). (d) Schematic illustration of a self-assembled nanostructured thin film formed on the substrate.

Fig. 18((a), (b)) show a "multilayer" geometry consisting of alternating layers of the ferroelectric phase (e.g., perovskite $BaTiO_3$) and the ferro/ferrimagnetic phase (e.g., spinel $CoFe_2O_4$). When the magnetoelectric coupling is purely through elastic interactions, the effect in a multilayer structure will be negligible due to the clamping effect of the substrate[34]. There was a recent report[35] on creating and analyzing a vertically aligned

$BaTiO_3$-$CoFe_2O_4$ structure (Fig. 18 (c), (d)). The approach was based on two concepts[36]. The first is the intrinsic similarity in crystal chemistry between perovskites and spinels, both of which are based on an octahedral oxygen coordination. This leads to crystal lattice parameters that are reasonably commensurate. The second key aspect is the fact that although many of these complex oxides can accommodate considerable cationic solid solution solubility, the perovskite/spinel system behaves like line compounds with little solid solubility into each other. These two aspects then present an interesting opportunity to create perovskite-spinel nanostructures through a spontaneous phase separation process.

Figure 19 (a) AFM topography image of the films. (b) Cross-section dark field TEM image taken with g = [260] of $CoFe_2O_4$

Atomic force microscopy (AFM) image of the films, Fig.19 (a) shows an interesting morphological pattern which consists of local hexagonal features. Transmission electron microscopy (TEM) studies resolve these patterns to be arrays of 20-30nm in diameter size of $CoFe_2O_4$ pillars embedded in a $BaTiO_3$ matrix, see cross section TEM image in Fig. 19 (b). $CoFe_2O_4$ pillars are in three-dimensional epitaxy with $BaTiO_3$ matrix and $SrTiO_3$ single crystal substrate. The size of the pillars increases with the increase of the growth temperature, which suggests that the growth of the nanostructures is controlled by diffusion.

Ferroelectric measurements of the thin film nanostructures demonstrate well-defined ferroelectric hysteresis, as shown in Fig. 20 (a). The polarization values were normalized to the volume fraction of $BaTiO_3$ (~65%) yielding a saturation polarization (Ps) of ~23 $\mu C/cm^2$. Piezoelectric measurements reveal a clear hysteresis loop (Fig. 20 (b)) with a maximum value of ~50 pm/V (as compared to the value of ~130 pm/V for single crystal $BaTiO_3$). This decrease is primarily due to clamping effects from both substrate and the $CoFe_2O_4$ nanopillars.

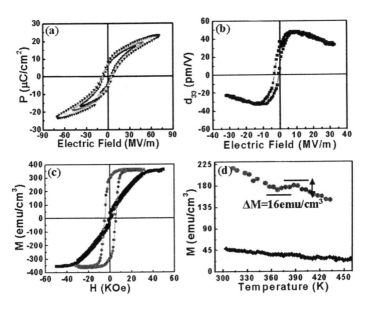

Figure 20 (a) Polarization-electric field hysteresis loop showing that the film is ferroelectric with a saturation polarization $P_s \sim 23$ $\mu C/cm^2$. (b) Small signal piezoelectric d_{33} hysteresis loop for a 50 μm diameter capacitor. (d) Out of plane (gray) and in plane (black) magnetic hysteresis loops depicting the large uniaxial anisotropy. (d) Magnetization vs temperature curve measured at H = 100 Oe, which shows a distinct drop in magnetization at the ferroelectric Curie temperature for the vertically self-assembled nanostructure (gray curve) while the multilayered nanostructure (black curve) shows negligible change in magnetization

The nanostructure thin films show a saturation magnetization of ~350emu/cm^3 (normalized to the volume fraction of $CoFe_2O_4$, ~35%), see Fig. 20 (c), with an easy axis normal to the film plane. The results show a strong anisotropy between the out-of-plane ([001]) and in-plane [100] directions with a uniaxial anisotropy field of ~35 kOe. It was calculated that the compressive stress along the $CoFe_2O_4$ pillars is the main contribution to the large anisotropy field.

Results of temperature-dependent magnetization measurements (gray curve in Fig. 20 (d)) show coupling between the electric and magnetic order parameters in the self-assembled nanostructure. This is manifested as a distinct drop in the magnetization of ~16 emu/cm^3 (~5% of magnetization at 100 Oe external field) around the ferroelectric Curie temperature (Tc~390 K). At temperatures higher than Tc the $CoFe_2O_4$ is compressed due to the lattice mismatch with $BaTiO_3$. For T < Tc, the tetragonal distortion in the $BaTiO_3$ lattice decreases this compression in the $CoFe_2O_4$. Since $CoFe_2O_4$ has a negative magnetostriction, it results in a reduction of the moment, as observed in our experiments. For comparison, the temperature dependence of magnetization at 100 Oe (black curve in Fig. 20 (d)) for a $CoFe_2O_4$-$BaTiO_3$ multilayer sample with a layer thickness of ~30 nm shows negligible change around the ferroelectric Curie temperature. This can be understood as a consequence of the fact that the in-plane piezo-deformation in the multilayer structure is clamped by the substrate, thus precluding any deformation in the magnetic layer. This also proves that the coupling is dominated by elastic interactions in two-phase nanostructures.

1.7 CONCLUSIONS

Scanning piezo-force microscopy is evolving to be a powerful tool that has enabled investigations of a variety of phenomena in ferroelectric thin films at the nanoscale. Our work has focused on understanding size scaling issues, dynamics of both 180° and 90° domain walls in constrained as well as constraint-relieved nanostructures, and a fundamental understanding of polarization relaxation behavior. This surface probe, in conjunction with nano-fabrication techniques such as focused ion beam milling or e-beam lithography provides an ideal "playground" for nanoscale science of ferroelectric materials.

Over the past few years, our studies on polarization relaxation behavior indicates that relaxation occurs through the nucleation of reverse domains at heterogeneities in the The JMAK model for phase change kinetics has been applied to describe the observed transformation-time data for 180° domain reversal under constantly decaying internal driving field. The effect of local curvature on driving the domain relaxation kinetics has been studied and it has been observed that the curvature of growing domains decreases. Pinning and bowing of 180° domain walls is observed in the late stages of relaxation.

Future work on size effects undoubtedly will throw light on the origins of polarization suppression (if there is any) as a function of lateral and thickness dimensions. This paper did not overtly discuss such size effects, especially as a function of film thickness. Indeed, this is quite a controversial aspect of nanoscale effects in ferroelectrics. Another exciting area that is emerging is lead-free perovskite ferroelectrics (such as BiFeO3). This particular material has demonstrated very large polarization values (figure 17) along with the coexistence of antiferromagnetic order., thus making it a multiferroic material. Undoubtedly, interest in these types of materials will continue to climb as their unique combination of properties are explored in detail.

ACKNOWLEDGEMENTS

We have been fortunate to be able to collaborate with several colleagues both within Maryland as well as around the world. We wish to gratefully acknowledge their past and on-going contributions to the central theme of our research effort. Several outstanding postdoctoral researchers, graduate students and undergraduates have contributed and continue to contribute to this effort. Specifically, we wish to acknowledge the work of Dr. A. Stanishefsky on focused ion beam milling, Professors A. Roytburd and M. Wuttig on theoretical modeling and piezo-mechanics, Professors Ellen Williams, Rainer Waser (RWTH, Aachen) and Lukas Eng (TU Dresden) on SFM and fundamental studies of dynamics ; Dr. Orlando Auciello (Argonne NL) on scaling physics ; This work is supported by NSF under grants DMR 00-80008 and DMR 99-03279; and partially from a DOE S&P Center through Argonne National Laboratory, a grant from PNNL and by a grant from Motorola.

Our research on Crystalline Oxides on Semiconductors is collaboration between Maryland, Penn. State (Professor Schlom), U. of Michigan (Professor X. Pan), University of Wisconsin (Professor Eom) and Motorola/Freescale Semiconductor (R. Droopad). We gratefully acknowledge their contributions to and support of this blossoming project.

REFERENCES

[1] J F Scott, Ferroelectric Memories, Vol.3 of the Springer series on "Advanced Microelectronics", (Springer, Heidelberg, April 2000); D Damjanovic, *Rep. Prog. Phys.* **61**, 1267 (1998).

[2] T. Hidaka, T. Mayurama, M. Saitoh, N. Mikoshiba, M. Shimizu, T. Shiosaki, L. A. Wills, R. Hiskes, S. A. Dicarolis and J. Amano, *Appl. Phys. Lett.* **68**, 2358 (1996).

[3] C. H. Ahn, T. Tybell, L. Antognazza, K. Char, R. H. Hammond, M. R. Beasley, Ø. Fisher, and J.-M. Triscone, *Science* **276**, 1100 (1997).

[4] A. Gruverman, H. Tokumoto, A. S. Prakash, S. Aggarwal, B. Yang, M. Wuttig, R. Ramesh, O. Auciello and T. Venkatesan, *Appl. Phys. Lett.* **71**, 3492 (1997).

[5] Hong S, Woo J, Shin H, Jeon JU, Pak YE, Colla EL, Setter N, Kim E, Noh K, *J. Appl. Phys.* **89**, 1377 (2001).

[6] S. Hong, E.L. Colla, E. Kim, D.V. Taylor, A.K. Tagantsev, P. Muralt, K. No, N. Setter, *J. Appl. Phys.* **86,** 607 (1999).

[7] J.W. Hong, K.H. Noh, S. Park, S.I. Kwun, Z.G. Khim, *Phys. Rev. B* **58**, 5078 (1998).

[8] S.V. Kalinin and D.A. Bonnell, *Appl. Phys. Lett.* **78,**1116 (2001).

[9] S.V. Kalinin and D.A. Bonnell, *Phys. Rev. B* **63** , 125411 (2001).

[10] J. D. Axe and Y. Yamada, *Phys. Rev. B* **34**, 1599 (1986). K. Sekimoto, *Physica A* **125**, 261 (1984); **128**, 132 (1984); **135**, 328 (1986). S. Ohta, T. Ohta and K. Kawasaki, *Physica A* **140**, 478 (1987). R. M. Bradly and P. N. Strenski, *Phys. Rev. B* **40**, 8967 (1989). K. Sekimoto, *Intl. J. Mod. Phys. B* **5**, 1843 (1991). Y. A. Andrienko, N. V. Brilliantov, and P. V. Krapivsky, *Phys. Rev. A* **45**, 2263 (1992).

[11] K. Eisenbeiser, J. M. Finder, Z. Yu, J. Ramdani, J. A. Curless, J. A. Hallmark, R. Droopad, W. J. Ooms, L. Salem, S. Bradshaw, and C. D. Overgaard, *Appl. Phys. Lett.* **76**, 1324 (2000); R. A. McKee, F. J. Walker, and M. F. Chisholm, *Science* **293**, 468-471,2001; R. McKee, F. Walker, and M. Chisholm, *Phys. Rev. Lett.* **81**, 3014 (1998).

[12] R. Ramesh, H. Gilchrist, T. Sands, V.G. Keramidas, R. Haakenaasen, D.K. Fork, *Appl. Phys Lett.* **63**, 3592-3594 (1993).

[13] K. Maki, N. Soyama, S. Mori, and K. Ogi, Integr. Ferroelectr. 30, 193 (2000); B.T. Liu, K. Maki, V.Nagarajan and R. Ramesh, *submitted ,Appl. Phys Lett.*

[14] A L Roitburd, *Phys. Status Solidi A* **37**, 329 (1976).

[15] V. Nagarajan, I. G. Jenkins, S. P. Alpay, H. Li, S. Aggarwal, L. Salamanca-Riba, A. L. Roytburd and R. Ramesh, *J. Appl. Phys.* **86**, 595 (1999).

[16] P Güthner and K Dransfeld, *Appl. Phys. Lett.* **61**, 1137 (1992); A Gruverman, O Auciello, and H Tokumoto, *Appl. Phys. Lett.* **69**, 3191 (1996); T Tybell, C H Ahn, and J -M Triscone, *Appl. Phys. Lett.* **72**, 1454 (1998); O Auciello, A Gruverman, H Tokumoto, S A Prakash, S Aggarwal and R Ramesh, *Mater. Res. Bull.* **23**, 33 (1998).

[17] C. M. Foster, Z. Li, M. Buckett, D. Miller, P. M. Baldo, L. E. Rehn, G. R. Bai, D. Guo, H. You, and K. L. Merkle, *J. Appl. Phys.* **78**, 2607 (1995).

[18] B. S. Kwak, A. Erbil, J. D. Budai, M. F. Chrisholm, L. A. Boatner, and B. J. Wilkens, *Phys. Rev. B* **49**, 14865 (1994).

[19] G. E. Pike, W. L. Warren, D. Dimos, B. A. Tuttle, R. Ramesh, J. Lee, V. G. Keramidas and J. T. Evans, Jr., *Appl. Phys. Lett.* **66**, 484 (1995).

[20] C. S. Ganpule, V. Nagarajan, H. Li, A. S. Ogale, D. E. Steinhauer, S. Aggarwal, E. Williams and R. Ramesh, *Appl. Phys. Lett.* **77**, 292 (2000).

[21] L. M. Eng, M. Abplanalp, and P. Günter, *Appl. Phys. A: Mater. Sci. Process.* **66A**, S679 (1998).

[22] A. Roelofs, U. Böttger, R. Waser, F. Schlaphof, S. Trogisch, and L. M. Eng, *Appl. Phys. Lett.* **77**, 3444 (2000).

[23] C. S. Ganpule, V. Nagarajan, S. B. Ogale, A. L. Roytburd, E. D. Williams, and R. Ramesh, *Appl. Phys. Lett.* **77**, 3275 (2000)

[24] C. S. Ganpule, V. Nagarajan, B. K. Hill, A. L. Roytburd, E. D. Williams, R. Ramesh, S. P. Alpay, A. Roelofs, R. Waser, and L. M. Eng, *J. Appl. Phys.* **91**, 1477 (2002)

[25] A. Gruverman and M. Tanaka, *J. Appl. Phys.* **89**, 1836 (2001)

[26] J. Kakilios, R. A. Street, and W. B. Jackson, *Phys. Rev. Lett.* **59**, 1037 (1987); R. G. Palmer, D. L. Stein, E. Abrahams, and P. W. Anderson, *ibid.* **53**, 958 (1984); R. V. Chamberlin, G. Mozurkewich, and R. Orbach, *ibid.* **52**, 867 (1984); D. K. Lottis, R. M. White, and E. Dan Dahlberg, *ibid.* **67**, 362 (1991).

[27] Paul G. Shewmon, *Transformations in Metals*, McGraw-Hill Book Company, 1969; B. Y. Lyubov, *Kinetic Theory of Phase Transformation*, National Bureau of Standards, New Delhi, 1978.

[28] C. S. Ganpule, A. L. Roytburd, V. Nagarajan, B. K. Hill, S. B. Ogale, E. D. Williams, R. Ramesh, and J. F. Scott *Phys. Rev. B* **65**, 014101 (2002)

[29] C. S. Ganpule, A. Stanishevsky, S. Aggarwal, J. Melngailis, E. Williams, R. Ramesh, V. Joshi, and Carlos Paz de Araujo, *Appl. Phys. Lett.* **75**, 3874 (1999)

[30] C. S. Ganpule, A. Stanishevsky, Q. Su, S. Aggarwal, J. Melngailis, E. Williams, and R. Ramesh, *Appl. Phys. Lett.* **75**, 409 (1999)

[31] K. Lefki and G. J. M. Dormans, *J. Appl. Phys.* **76**, 1764 (1994)

[32] S. Tiedke, T. Schmitz, K. Prume, A. Roelofs, T. Schneller, U. Kall, R. Waser, C. S. Ganpule, V. Nagarajan, A. Stanishevsky, and R. Ramesh.

[33] J. Van Suchtelen, *Philips Res. Rep.* **27**, 28 (1972).

[34] K. Lefki, G. J. M. Dormans, *J. Appl. Phys.* **76**, 1764 (1994).

[35] H. Zheng, J. Wang, S. E. Lofland, Z. Ma, L. Mohaddes-Ardabili, T. Zhao, L. Salamanca-Riba, S. R. Shinde, S. B. Ogale, F. Bai, D. Viehland, Y. Jia, D. G. Schlom, M. Wuttig, A. Roytburd, and R. Ramesh, *Science* **303**, 661 (2004).

[36]   H. Zheng, J. Wang, L. Mohaddes-Ardabili, D. G. Schlom, M. Wuttig, L. Salamanca-Riba, R. Ramesh, *Appl. Phys. Lett.* **85**, 2035 (2004).

CHAPTER 2

HIGH-*K* CANDIDATES FOR USE AS THE GATE DIELECTRIC IN SILICON MOSFETS

D.G. Schlom,[1a] C.A. Billman,[1] J.H. Haeni,[1a] J. Lettieri,[1] P.H. Tan,[1b] R.R.M. Held,[2] S. Völk,[2] and K.J. Hubbard[1c]

[1] *Department of Materials Science and Engineering, Penn State University, University Park, PA 16803-6602, U.S.A.Fax: +1 (814) 863-0618, e-mail: schlom@ems.psu.edu*
[2] *Experimentalphysik VI, Center for Electronic Correlations and Magnetism, Institute of Physics, Augsburg University, 86135 Augsburg, Germany*
[a] *On sabbatical at Experimentalphysik VI, Center for Electronic Correlations and Magnetism, Institute of Physics, Augsburg University, 86135 Augsburg, Germany*
[b] *Present address : TPC (Malaysia, 11900 Penang, Malaysia*
[c] *Present address: Kagan Binder, Stillwater, MN 55082-5021, U.S.A.*

2.1 INTRODUCTION

An "immediate grand challenge" identified by the international semiconductor industry is the identification and development of an alternative gate dielectric for future metal-oxide-semiconductor field-effect transistors (MOSFETs) on silicon.[1] This alternative gate dielectric, which needs to be fully implemented in production by 2007 in order for integrated circuits to continue to follow Moore's Law, needs (1) to have a significantly higher dielectric constant (K) than amorphous SiO_2 ($K = 3.9$), (2) to be stable in contact with silicon (capable of withstanding a ~900 °C annealing step)[2], (3) to have a bandgap high enough (4-5 eV) to provide sufficiently low gate leakage, (4) to have a low density of electrically-active defects at the dielectric / silicon interface (D_{it}), and (5) to be the electrical equivalent (with a much lower leakage current) of an SiO_2 layer with a physical thickness, t_{ox}, of <10 Å.[1,3,4] SiO_2 itself is no longer suitable because at such thinnesses the gate leakage, due to the quantum-mechanical tunneling of electrons between the semiconductor channel and gate electrode though the SiO_2, exceeds 1 A/cm² and increases exponentially with decreasing thickness of the SiO_2 dielectric (about ten fold per Å). Here we report the results of our research focused on identifying candidate materials that satisfy the first two of these requirements: high K and stability in direct contact with silicon.

Until recently, the high-K materials being investigated most widely as possible alternative gate dielectrics were closely related to the high-K materials that had been identified as promising in dynamic random access memory (DRAM) research and development: [3] tantalum pentoxide (Ta_2O_5),[5-9] titanium dioxide (TiO_2),[10,11] and barium strontium titanate $((Ba,Sr)TiO_3)$.[12] Unfortunately, all of these materials are thermodynamically *unstable* in direct contact with silicon as is evident from the chemical reactions below. In DRAMs a metallic barrier layer can be used between the silicon and the dielectric to avoid the reaction that accompanies the placement of these dielectrics in direct contact with silicon.[3] In MOSFETs, however, such a barrier layer cannot be tolerated as it would screen the applied electric field from reaching the silicon channel.

$$\frac{13}{2}\,Si + Ta_2O_5 \xrightarrow{\Delta G^{\circ}_{1000\,K} = -413.332\,\frac{kJ}{mol}} 2\,TaSi_2 + \frac{5}{2}\,SiO_2, \tag{1}$$

$$3\,Si + TiO_2 \xrightarrow{\Delta G^{\circ}_{1000\,K} = -95.357\,\frac{kJ}{mol}} TiSi_2 + SiO_2, \tag{2}$$

$$3\,Si + SrTiO_3 \xrightarrow{\Delta G^{\circ}_{1000\,K} = -87.782\,\frac{kJ}{mol}} TiSi_2 + SrSiO_3, \tag{3}$$

and

$$3\,Si + BaTiO_3 \xrightarrow{\Delta G^{\circ}_{1000\,K} = -101.802\,\frac{kJ}{mol}} TiSi_2 + BaSiO_3. \tag{4}$$

For each of these reactions $\Delta G^{\circ}_{1000\,K}$ is the free energy change of the system when the reaction between reactants and products, all taken to be in their standard state (the meaning of the $^{\circ}$ superscript), proceeds in the direction indicated at a temperature of 1000 K.[13] Note that all of the above reactions are energetically favorable ($\Delta G < 0$). This is true not only at 1000 K, but at all temperatures between room temperature and the ~900 °C needed for MOSFET processing. Further, the reaction products all involve dielectrics with low K, e.g., SiO_2. $SrSiO_3$ has a dielectric constant of about 7.[14] $BaSiO_3$ is estimated to have a dielectric constant of about 8 using the method of Shannon[15] described below. Such low-K reaction layers in series with the desired high-K dielectrics rapidly nullify the benefits of the high-K dielectrics when a capacitance corresponding to a total SiO_2

physical thickness of <10 Å is needed for future MOSFETs. Experimental observations are consistent with the above thermodynamic expectation of interfacial reaction between silicon and Ta_2O_5, TiO_2, and $(Sr,Ba)TiO_3$ layers. 5·8[-10,16-20]

Thermodynamic instability is a necessary, but insufficient criterion to conclude that an interface will react. It indicates the presence of a driving force for reaction, but it is possible for a kinetic barrier to limit the extent of reaction or even prevent a thermodynamically-unstable interface from reacting at all. However, at the relatively high temperature considered, ~900 °C during the annealing step in which the implanted dopants are activated,[2] such kinetic barriers must be significant to prevent interaction.

One example of where low growth temperature has been used to prevent interfacial reaction between a high-K dielectric that is unstable with silicon, is in the growth of CeO_2 on silicon. Although Ce_2O_3 is thermodynamically stable in contact with silicon, CeO_2 (which has a K in the range 16.6 [15] to 26 [21]) is not.[22] Nonetheless epitaxial CeO_2 films have been grown at room temperature on (111) Si and high-resolution cross-sectional transmission electron microscopy (TEM) has revealed an atomically-abrupt interface free of any reaction layer.[23] This is an example of how at sufficiently low temperature, kinetics can minimize the extent of the interfacial reaction despite there being a driving force for reaction. However, when CeO_2 films are grown or annealed at higher temperatures, e.g., above 700 °C, the kinetic barrier is overcome and significant interfacial reactions are seen by TEM.[24,25] Even extremely thin kinetically-limited reaction layers rapidly use up the <10 Å SiO_2 equivalent thickness budget of the dielectric layer for future MOSFETs.

An obvious way to avoid reactions between a high-K dielectric and silicon is to select a dielectric that is thermodynamically stable in contact with silicon. Alternatively, a silicon-incompatible material could be used as a gate oxide, provided that it is physically separated from the underlying silicon by a silicon-compatible buffer layer. Examples of the latter approach include the use of a SrO buffer layer between $SrTiO_3$ and silicon[30] or a Si_3N_4 buffer layer between Ta_2O_5 or TiO_2 and silicon.[9,31] The benefit of the high-K material, however, is rapidly diluted by the existence of the lower-K buffer layer because the capacitance of the buffer layer is in series with that of the desired high-K layer. The capacitance of the buffer layer can be minimized by making it thin, but concomitant with such thinning is the increased likelihood of reaction between the high-K layer and the silicon either via diffusion through the extremely thin buffer layer or by the increased likelihood of pinholes in the buffer layer as it is made thinner. Regardless if one is looking for a buffer layer for a silicon-incompatible dielectric or wishes to have the entire dielectric compatible with silicon, identifying silicon-compatible high-K dielectrics is an important first step.

In this paper we describe a comprehensive search to identify alternative gate dielectrics (or buffer layers for alternative gate dielectrics). We begin by reviewing the thermodynamic stability of all binary[†] oxides and all binary nitrides in contact with silicon. Ranking the resulting silicon-compatible binary dielectrics by K reveals many more binary oxides with a high K than binary nitrides with a high K. Concentrating on the oxides, we then consider potentially stable multicomponent oxides comprised of silicon-compatible binary oxides and use the Clausius-Mossotti equation and Shannon's dielectric polarizabilities[15] of the constituent ions to estimate the K of these multicomponent oxides. The result is a ranked list of high-K, potentially silicon-compatible, gate dielectric candidate materials. A preliminary overview of this work was published previously. [32,33]

2.2 METHOD

2.2.1 Thermodynamic stability of binary oxides (MO_x) and binary nitrides (MN_x)

The thermodynamic stability of all binary oxides (MO_x) in contact with silicon was investigated by Hubbard and Schlom.[22] More recently, Tan and Schlom[34] investigated the thermodynamic stability of all binary nitrides (MN_x) in contact with silicon. The same method[35] was used in both of these studies and only its salient aspects will be repeated here. All binary oxides and nitrides that are solid at 1000 K and not radioactive were considered. The key concept was that any reaction between silicon and the binary oxide or nitride under consideration that lowered the Gibbs free energy of the system resulted in the elimination of the binary oxide or nitride from further consideration. The reaction did not need to be the most favorable, as *any* reaction with $\Delta G < 0$ implies that the interface between silicon and the binary oxide or nitride under consideration is thermodynamically unstable. Even though there are many possible reactions, several key reactions were identified by considering the metal-(oxygen or nitrogen)-silicon (M-O/N-Si) phase diagrams. Initially phase diagrams with only one binary oxide or nitride and no ternary phases were considered. For the binary oxide or nitride to be stable in contact with silicon a tie line must exist between the binary oxide or nitride and silicon. To determine if a tie line exists, several reactions were considered. If the

---

[†]   A binary compound contains two elements. In this work "binary oxides" are taken to have the general formula MO_x, where one of the elements is M and the other is oxygen. Similarly for binary nitrides the general formula is MN_x, where one of the elements is M and the other is nitrogen.

system passed these reactions, that is $\Delta G > 0$ for each reaction, then additional reactions were checked if the system contained more than one binary oxide or nitride or contained ternary oxides or nitrides, to the extent that thermodynamic data existed.

It is important to realize the presence of two simplifying assumptions in this thermodynamic analysis method. First, it only includes volume free energies; interfacial energies are neglected. Second, reactions involving gaseous species are not considered, as the idea is to test the stability of the interface between two solids (silicon and the dielectric), and there is no free volume in which a gas can exist. For very thin gate dielectrics, however, both of these assumptions come into question. As the gate dielectric layer becomes thinner, the interfacial free energies become more important, and conceivably they could alter the sign of ΔG for reactions in which its magnitude is close to zero. Also as the dielectric gets thinner, it is possible for gaseous species (e.g., SiO) to diffuse through a thin dielectric layer at high temperature and escape from the dielectric / silicon interface.[36-38] Notwithstanding these assumptions, this method serves to cull unsuitable dielectrics from consideration and identify the best candidates for detailed study.

The results of these comprehensive studies on the thermodynamic stability of binary oxides[22] and binary nitrides[34] in contact with silicon are graphically summarized on the periodic tables in Figs. 1 and 2, respectively. Elements M having no binary oxide (MO_x) or no binary nitride (MN_x) that is stable or potentially stable in contact with silicon are crossed out by diagonal lines in the respective periodic tables (Figs. 1 and 2), and the reason for their elimination is given. An element M is only crossed out when *all* of its binary oxides or nitrides have been eliminated. For example, iron ($M = $ Fe) has three binary oxides that are solid at 1000 K: FeO, Fe_3O_4, and Fe_2O_3.[39] But all of these fail ($\Delta G < 0$) balanced reactions of the type $Si + MO_x \rightarrow M + SiO_2$, as shown by the following specific reactions:

$$\frac{1}{2} Si + FeO \xrightarrow{\Delta G^{\circ}_{1000 K} = -158.138 \frac{kJ}{mol}} Fe + \frac{1}{2} SiO_2, \qquad (5)$$

$$2 Si + Fe_3O_4 \xrightarrow{\Delta G^{\circ}_{1000 K} = -668.326 \frac{kJ}{mol}} 3 Fe + 2 SiO_2, \qquad (6)$$

36

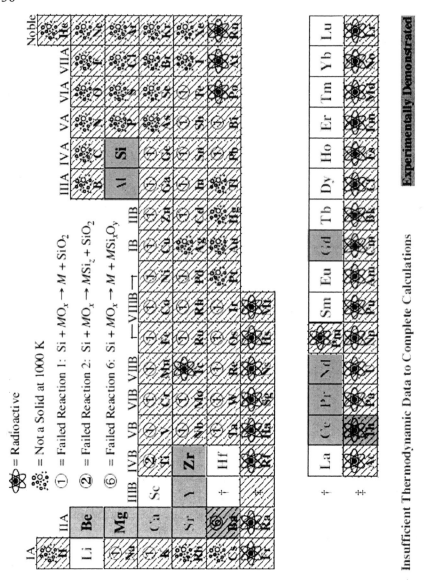

Figure 1: Summary of which elements M have an oxide (MO_x) that *may* be thermodynamically stable in contact with silicon at 1000 K, based on Ref. 22. Elements M having no thermodynamically-stable or potentially thermodynamically-stable oxide (MO_x) are crossed out by diagonal lines and the reason for their elimination is given. The elements M for which sufficient thermodynamic data exist to complete all necessary calculations to establish the thermodynamic stability of MO_x in contact with silicon appear in bold. The elements M having an oxide (MO_x) that has been experimentally demonstrated to be stable in direct contact with silicon are underlined. Performing the thermodynamic analysis over the full range of temperatures for which relevant thermodynamic data are available (as much as 300-1600 K as shown in Table 1) does not alter the conclusions shown.

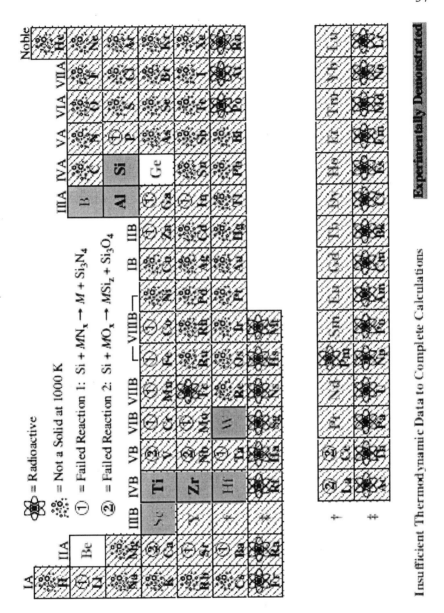

Figure 2: Summary of which elements M have a nitride (MN_x) that *may* be thermodynamically stable in contact with silicon at 1000 K, based on Ref. 34. Elements M having no thermodynamically-stable or potentially thermodynamically-stable nitride (MN_x) are crossed out by diagonal lines and the reason for their elimination is given. The elements M for which sufficient thermodynamic data exist to complete all necessary calculations to establish the thermodynamic stability of MN_x in contact with silicon appear in bold. The elements M having a nitride (MN_x) that has been experimentally demonstrated to be stable in direct contact with silicon are underlined. Performing the thermodynamic analysis at temperatures between room temperature and 1000 K does not alter the conclusions shown.

and

$$\frac{3}{2}\,\text{Si} + \text{Fe}_2\text{O}_3 \xrightarrow{\Delta G^{\circ}_{1000\,\text{K}}\ =\ -534.693\,\frac{\text{kJ}}{\text{mol}}} 2\,\text{Fe} + \frac{3}{2}\,\text{SiO}_2, \qquad (7)$$

where $\Delta G^{\circ}_{1000\,\text{K}}$ is the free energy change of the system for each reaction when it proceeds in the direction indicated at a temperature of 1000 K.[13] This indicates that all of these binary oxides are thermodynamically unstable in contact with silicon and thus Fe is crossed out by diagonal lines on the periodic table in Fig. 1.

An example of a binary oxide for which there were sufficient thermodynamic data to complete our calculations and demonstrate its thermodynamic stability in contact with silicon is ZrO_2, as shown below:

$$\text{Si} + \text{ZrO}_2 \xrightarrow{\Delta G^{\circ}_{1000\,\text{K}}\ =\ +177.684\,\frac{\text{kJ}}{\text{mol}}} \text{Zr} + \text{SiO}_2, \qquad (8)$$

$$3\,\text{Si} + \text{ZrO}_2 \xrightarrow{\Delta G^{\circ}_{1000\,\text{K}}\ =\ +24.742\,\frac{\text{kJ}}{\text{mol}}} \text{ZrSi}_2 + \text{SiO}_2, \qquad (9)$$

$$\frac{3}{2}\,\text{Si} + \text{ZrO}_2 \xrightarrow{\Delta G^{\circ}_{1000\,\text{K}}\ =\ +9.129\,\frac{\text{kJ}}{\text{mol}}} \frac{1}{2}\,\text{ZrSi}_2 + \frac{1}{2}\,\text{ZrSiO}_4, \qquad (10)$$

and

$$\frac{1}{2}\,\text{Si} + \text{ZrO}_2 \xrightarrow{\Delta G^{\circ}_{1000\,\text{K}}\ =\ +85.600\,\frac{\text{kJ}}{\text{mol}}} \frac{1}{2}\,\text{Zr} + \frac{1}{2}\,\text{ZrSiO}_4. \qquad (11)$$

$\Delta G^{\circ}_{1000\,\text{K}}$ is the free energy change of the system for each reaction when it proceeds in the direction indicated at a temperature of 1000 K.[13] As we were able to complete all necessary calculations to establish the thermodynamic stability of ZrO_2 in contact with silicon, Zr appears in bold on the periodic table in Fig. 1.

Because of the more up-to-date thermodynamic data compilation[13] that has become available since the publication of Ref. 22, we have redone the summary table of all those binary oxides that may be thermodynamically stable in contact with silicon (Table VII in Ref. 22) and present the updated $\Delta G^{\circ}_{1000\,\text{K}}$ values in Table 1. References 40 and 41 were used for those

Table 1: Summary of the only binary oxides that *may* be thermodynamically stable in contact with silicon from the analysis of Ref. 22. All ΔG°_{1000} have been recalculated with the thermodynamic data from Refs. 13, 40, and 41. Due to the lack of free energy data for relevant silicides, ternary phases, or both, only for a limited number of binary oxides (shown in bold) could thermodynamic stability with silicon be concluded. BaO is also included since it has been experimentally demonstrated to be stable in epitaxial form in contact with silicon at low temperature (see Sec. 3.1).

Oxide	ΔG°_{1000} per MO_x for Reaction 1 [a] (kJ/mol)	ΔG°_{1000} per MO_x for Reaction 2 [a] (kJ/mol)	ΔG°_{1000} per MO_x for Reaction 5 [a] (kJ/mol)	ΔG°_{1000} per MO_x for Reaction 6 [a] (kJ/mol)	Temperature Range (K) over which ΔG°_T Calculated and found Positive
Li_2O	+101.279	b	b	−2.419[g,h]	300—800
BeO	+144.947	c	c	+69.307[f]	300—1600
MgO	+127.730	+93.351	+31.071[f]	+48.260[f]	300—1300
CaO	+165.959	+43.987	−7.873[g]	+40.916[f]	ℓ
SrO	+126.104	b	b	+9.331[e,i]	300—1200
BaO	+93.214	b	b	−19.249[j]	m
Sc_2O_3	+518.182	+372.982[n]	d	d	300—1600
Y_2O_3	+519.121	b	b,d	d	300—1600
La_2O_3	+412.282	+270.882[o]	d	d	300—1600
Ce_2O_3	+417.775	b	b,d	d	300—1200
Pr_2O_3	+425.766	b	b,d	d	300—1600
Nd_2O_3	+429.654	+291.853[p]	d	d	300—1600
Sm_2O_3	+437.633	b	b,d	d	300—1600
EuO	+139.066	b	b,d	d	300—1600
Gd_2O_3	+444.039	+325.438[q]	d	d	300—1600
Tb_2O_3	+483.140	b	b,d	d	300—1600
Dy_2O_3	+473.011	b	b,d	d	300—1600
Ho_2O_3	+496.279	b	b,d	d	300—1500
Er_2O_3	+511.389	b	b,d	d	300—1600
Tm_2O_3	+488.732	b	b,d	d	300—1600
Yb_2O_3	+434.960	b	b,d	d	300—1400
Lu_2O_3	+489.970	b	b,d	d	300—1600
ThO_2[r]	+307.246	+142.469	d	d	300—1600
UO_2[r]	+181.636	+47.721	d	d	300—1600
ZrO_2	+177.684	+24.742	+9.129[g,s]	+85.600[f]	300—1500
HfO_2	+230.879	b	b,d	d	300—1600
Al_2O_3	+265.947	c	c	+79.997[e,k]	300—1600

[a] Reactions from Ref.22 (balanced appropriately)

1: $Si + MO_x \rightarrow M + SiO_2$

2: $Si + MO_x \rightarrow MSi_z + SiO_2$

5: $Si + MO_x \rightarrow MSi_z + MSi_xO_y$

6: $Si + MO_x \rightarrow M + MSi_xO_y$

[b] Free energy data for the silicide(s) were unavailable in Refs. 13, 40, and 41.

[c] No known silicides.

[d] Free energy data for the relevant ternary phase(s) were unavailable in Refs. 13, 40, and 41.

[e] Free energy data for the relevant ternary phase(s) were not available. The free energy change at 1000 K (ΔG°_{1000}) for each of the phases for which data were available was positive. The ΔG°_{1000} value given is for the reaction involving a ternary phase that was the least positive.

[f] Free energy data for the relevant ternary phase(s) were available and reactions 5 and 6 from Ref. 22 were positive in all cases.

[g] The magnitude of this reaction is close to the approximate uncertainty of the thermodynamic data, making its result inconclusive.

[h] Reaction 6 from Ref. 22 with $MSi_xO_y = Li_4SiO_4$.

[i] Reaction 6 from Ref. 22 with $MSi_xO_y = Sr_2SiO_4$.

[j] Reaction 6 from Ref. 22 with $MSi_xO_y = Ba_2SiO_4$.

[k] Reaction 6 from Ref. 22 with $MSi_xO_y = Al_6Si_2O_{13}$.

[l] Reaction 5 from Ref. 22 with $MSi_xO_y = Ca_3SiO_5$ remained negative over the entire temperature range tested (300—1200 K).

[m] Reaction 6 from Ref. 22 with $MSi_xO_y = Ba_2SiO_4$ remained negative over the entire temperature range tested (300—1600 K).

[n] The thermodynamic data for the relevant silicide, ScSi, from Ref. 40 are only valid over the temperature range 825—1045 K.

[o] The thermodynamic data for the relevant silicide, $LaSi_{1.69}$, from Ref. 40 are only valid over the temperature range 960—1060 K.

[p] The thermodynamic data for the relevant silicide, $NdSi_{1.8}$, from Ref. 40 are only valid over the temperature range 920—1020 K.

[q] The thermodynamic data for the relevant silicide, $GdSi_{1.89}$, from Ref. 40 are only valid over the temperature range 830—960 K.

[r] Radioactive, but still considered since thermodynamic data exist.

[s] If $\Delta H^\circ_{f,298\,K}$ from Ref. 42 is used to adjust the free energy of $ZrSi_2$ (together with the C_P and S° values of $ZrSi_2$ from Ref. 13), ΔG°_{1000} becomes –1.616 kJ/mol for reaction 5 (Eqn. 10).

phases not present in Ref. 13. Note that BaO is also included in Table 1 as it is relevant to the discussion in Sec. 3.1. The changes in the thermodynamic data set[13] did not affect the sign of $\Delta G^\circ_{1000\,K}$ of any of the reactions and thus did not affect any of the conclusions of Ref. 22. In addition, we also redid the entire thermodynamic analysis at temperatures from 300 K to 1600 K (i.e., not just 1000 K as was done in Ref. 22) using the latest thermodynamic data.[13] Even over this broad temperature range, the conclusions of Ref. 22 are unchanged. The last column of Table 1 gives the temperature range over which the ΔG°_T of all of the relevant chemical reactions for which data are available remain positive. This is the temperature range over which the particular binary oxide is either thermodynamically stable (when sufficient thermodynamic data exist to complete all calculations) or possibly stable (when insufficient thermodynamic data were available). A temperature range smaller than 300—1600 K does not indicate that ΔG_T° became negative for any of the reactions; rather thermodynamic data were only available over a limited part of this temperature range, allowing the thermodynamic stability to only be assessed over this limited range. Similarly, our analysis of the thermodynamic stability of binary nitrides in contact with silicon was not found to change at all over the temperature range from 300 K to 1000 K. [34] Additional details, including all of the thermodynamic reactions considered, are given elsewhere.[22,34]

Use of the recently reported value of $\Delta H_{f,298\,K}^\circ$ for $ZrSi_2$,[42] which differs by 21 kJ/mol (13%) from the value in Ref. 13, could, however, affect our conclusion that ZrO_2 is thermodynamically stable in contact with silicon. [43] Unfortunately only $\Delta H_{f,298\,K}^\circ$ was reported in this recent study of $ZrSi_2$ and not its Gibbs free energy. But if we use the heat capacity and entropy values of $ZrSi_2$ from Ref. 13 in combination with the revised value of $\Delta H_{f,298\,K}^\circ$ from Ref. 42, the value of $\Delta G_{1000\,K}^\circ$ for Eqn. 10 becomes $\Delta G_{1000\,K}^\circ = -1.616$ kJ/mol. Although negative (which would imply that ZrO_2 is not stable in direct contact with silicon[43]), the magnitude of $\Delta G_{1000\,K}^\circ$ is within the approximate uncertainty of the thermodynamic data,[22] making the result of this calculation inconclusive.

2.2.2 Experimental determination of the K of nine new materials

The K of single crystals of yttria-stabilized cubic HfO_2,[44] single crystals of the perovskites $LaLuO_3$,[45] $NdScO_3$,[46] $SmScO_3$,[46] $GdScO_3$,[46] $DyScO_3$,[46] $SrZrO_3$,[47] and $SrHfO_3$[47] and a fully-dense, but polycrystalline $CaZrO_3$ sample[48] were measured at room temperature using an HP 4284A LCR Meter at 1 MHz, a 2 V_{rms} oscillation level, an HP 16034E test fixture, and open and short corrections. For each material except $CaZrO_3$, two rectangular plates were sliced from single crystals whose crystallinity had been confirmed by Laue patterns. Evaporated gold electrodes were applied to both sides of the parallel plate samples that were typically 1 mm thick and had electrode areas of about 100 mm^2 for $GdScO_3$, $DyScO_3$, and the yttria-stabilized cubic HfO_2 and about 5 mm^2 for $LaLuO_3$, $NdScO_3$, $SmScO_3$, $CaZrO_3$, $SrZrO_3$, and $SrHfO_3$. No edge capacitance corrections (e.g., that of Subramanian et al. [49]) were employed because of the large ratio of the electrode area to the thickness of the samples. The dielectric loss, tan δ, was less than 10^{-4} for all samples except cubic HfO_2 for which tan δ was 0.0039. Note that other than $GdScO_3$ and $DyScO_3$, for which the full dielectric constant tensor was established,[46] the plates were not crystallographically oriented and additional measurements are needed to establish the dielectric constant tensor (a second-rank polar tensor)[50] of $LaLuO_3$, $NdScO_3$, $SmScO_3$, $CaZrO_3$, $SrZrO_3$, and $SrHfO_3$. These perovskites are all orthorhombic at room temperature, so their dielectric tensor contains three independent coefficients. Yttria-stabilized cubic HfO_2 is cubic, so its dielectric constant tensor contains only one independent coefficient.

2.2.3 Estimation of the K of nearly 2000 multicomponent oxides

To identify all ~silicon-compatible multicomponent oxides, i.e., those composed entirely of binary oxides that may be compatible with silicon, a comprehensive analysis of all known crystalline phases was performed using the NIST-ICDD Crystal Data database which contains about 150,000 inorganic entries.[51] As most of these oxides do not have measured K-values, we applied a technique developed by Shannon[15] to estimate the K of the nearly 2,000 inorganic structures composed entirely of the binary oxides listed in Table 1. This method uses the approximate relationship between the dielectric polarizability of a material and its K given by the Clausius-Mossotti equation:[15,52-54]

$$K = \frac{3\,V_m + 8\pi\,\alpha_D^T}{3\,V_m - 4\pi\,\alpha_D^T}, \qquad (12)$$

where V_m is the molar volume and α_D^T is the total dielectric polarizability of the material. Electronic, ionic, dipolar, and space charge mechanisms of polarizability all contribute, in general, to α_D^T.[55]

Although the Clausius-Mossotti equation is exact for some simple structures with cubic symmetry, it has been found by Shannon to provide a reasonably accurate (2-3% accuracy) general means of estimating the average K of oxides, so long as they are not conductive, ferroelectric, nor piezoelectric, do not contain "rattling" or "compressed" cations, and do not contain dipolar impurities.[15] These conditions amount to materials for which ionic and electronic contributions to polarizability are the sole contributors to α_D^T and for which the space in which each ion resides is typical for that ion, i.e., close to its ionic radius. For MOSFETs, only gate dielectrics with ionic and electronic contributions to polarizability are desired. At the high frequencies at which they operate, space charge and dipolar contributions to polarization will no longer contribute to α_D^T,[55] making it desirable to screen candidates based only on their ionic and electronic components of α_D^T, just as Shannon's method does. Because the multicomponent oxides that are stable or potentially stable in contact with silicon do not contain ions that often give rise to dipolar or space charge contributions to α_D^T, we expected Shannon's method to be an excellent way to narrow the list of high-K candidate materials for use as the gate dielectric in silicon MOSFETs. It is nonetheless possible that some of the structures may contain "compressed" cations,[56] but other than this caveat Shannon's method seemed ideally suited to the MOSFET gate oxide problem.

The parameters in Eqn. 12 are readily calculated for each material considered. The molar volume, V_m, is calculated from the unit cell parameters contained in the NIST-ICDD Crystal Data database using the equation [57]

$$V_m = \frac{a\,b\,c\sqrt{1 - \cos^2\alpha - \cos^2\beta - \cos^2\gamma + 2\cos\alpha \cdot \cos\beta \cdot \cos\gamma}}{Z},$$

(13)

where a, b, and c are the lengths of the unit cell axes, α, β, and γ are the interaxial angles, and Z is the number of formula units per unit cell. To calculate the total dielectric polarizability the additivity rule was used, which states that the total dielectric polarizability of a multicomponent material can be broken up into the sum of the individual dielectric polarizabilities of all the ions in the formula unit [15,54], i.e.,

$$\alpha_D^T = \sum_{i=1}^{n} v_i\, \alpha_D(i),$$

(14)

where n is the total number of types of ions in the formula unit of the substance and v_i is the number of ions of type i in the formula unit. Shannon[15] extracted a list of dielectric polarizabilities for 61 ions for general use in the Clausius-Mossotti equation by performing a best fit analysis on accurate K measurements made on 154 different oxides and fluorides, most of which were single crystals.

Of the ions comprising the binary oxides that we found to be stable or potentially stable in contact with silicon, only one was not on Shannon's list, Hf^{4+}. We estimated the dielectric polarizability of Hf^{4+} from that of Zr^{4+} based on the observation that the dielectric polarizability of ions of the same valence is roughly proportional to r^3,[15,56,58] where r is the ionic radius. Both Zr^{4+} and Hf^{4+} have nearly identical ionic radii[59], so a value of 3.25 Å^3 (the same as that for Zr^{4+}) was used for the dielectric polarizability of Hf^{4+}.

This dielectric polarizability for Hf^{4+} is also consistent with the K that we measured for a yttria-stabilized cubic HfO_2 single crystal: as the measured K and V_m of yttria-stabilized cubic ZrO_2 and yttria-stabilized cubic HfO_2 are comparable, the dielectric polarizabilities of these ions must also be comparable. The dielectric polarizabilities of all the ions used in our comprehensive estimation of the K of multicomponent oxides are listed in Table 2, along with the approximate uncertainties [60] in these values. The comparative magnitudes of the dielectric polarizabilities of the relevant ions for gate oxide applications are shown graphically on the periodic table in Fig. 3.

For relatively few multicomponent oxides, the molar volume per oxygen,

$$V_{ox} = \frac{V_m}{(\text{\# of oxygens per formula unit})}, \qquad (15)$$

exceeded 25 $Å^3$. For these oxides, α_D values from the first column of α_D values in Table III of Ref. 15 and a dielectric polarization of the O^{2-} ion with units $Å^3$ given by[*]

$$\alpha_D(O^{2-}) = 1.68 + 5 \times 10^{-4} \, V_{ox}^2, \qquad (16)$$

were used in accordance with Shannon's method.[15]

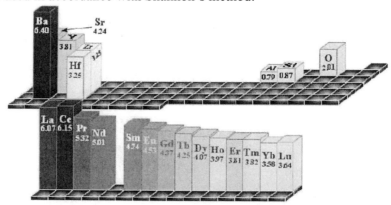

Figure 3: The dielectric polarizabilities (α_D with units of $Å^3$ from Ref. 15) of the ions whose binary oxides (that are solid at 1000 K and not radioactive) *may* be thermodynamically stable in contact with silicon. Ba^{2+} is also included since BaO has been experimentally demonstrated to be stable in epitaxial form in contact with silicon at low temperature (see Sec. 3.1).

---

[*] Equation 16 differs from that given in Ref. 15. There appears to be a typographical error, however, in the formula given in Ref. 15, as the formula given is inconsistent with all discussion related to it. Equation 16 was deduced from the discussion in Ref. 15.

Achieving high K is synonymous with driving the denominator of Eqn. 12 toward zero, i.e., making the total dielectric polarizability (α_D^T) large and the molar volume (V_m) small. When this occurs, small uncertainties in the dielectric polarizabilities of constituent ions (or in principle V_m, although in practice the uncertainty in α_D^T dominates) can lead to large uncertainties in the estimated average K of a material. Thus, keeping track of the uncertainty in the calculated K values is important.

Applying standard propagation of error methods [61] to Eqn. 12, the uncertainty in the calculated average K is given for cubic crystals by

$$\Delta K = 36\pi \sqrt{\frac{a^4 \, Z^2 \left(9 \left(\alpha_D^T\right)^2 (\Delta a)^2 + a^2 \left(\Delta \alpha_D^T\right)^2\right)}{\left(3 \, a^3 - 4\pi \, Z \, \alpha_D^T\right)^4}}, \qquad (17)$$

for hexagonal crystals by

$$\Delta K = 72\pi\sqrt{3} \sqrt{\frac{a^2 \, Z^2 \left(a^2 \left(\alpha_D^T\right)^2 (\Delta c)^2 + c^2 \left(4 \left(\alpha_D^T\right)^2 (\Delta a)^2 + a^2 \left(\Delta \alpha_D^T\right)^2\right)\right)}{\left(3\sqrt{3} \, a^2 \, c - 8\pi \, Z \, \alpha_D^T\right)^4}} \qquad (18)$$

for tetragonal crystals by

$$\Delta K = 36\pi \sqrt{\frac{a^2 \, Z^2 \left(a^2 \left(\alpha_D^T\right)^2 (\Delta c)^2 + c^2 \left(4 \left(\alpha_D^T\right)^2 (\Delta a)^2 + a^2 \left(\Delta \alpha_D^T\right)^2\right)\right)}{\left(3 \, a^2 \, c - 4\pi \, Z \, \alpha_D^T\right)^4}} \qquad (19)$$

,

and for orthorhombic crystals by

$$\Delta K = 36\pi \sqrt{\frac{Z^2 \left(a^2 b^2 \left(\alpha_D^T\right)^2 (\Delta c)^2 + a^2 c^2 \left(\alpha_D^T\right)^2 (\Delta b)^2 + b^2 c^2 \left(\alpha_D^T\right)^2 (\Delta a)^2 + a^2 b^2 c^2 \left(\Delta \alpha_D^T\right)^2\right)}{\left(3abc - 4\pi Z \alpha_D^T\right)^4}} \qquad (20)$$

,

where Δa, Δb, and Δc are the uncertainty in the lattice constants, and $\Delta \alpha_D^T$ is the uncertainty in the total dielectric polarizability of the material given by

$$\Delta \alpha_D^T = \sqrt{\sum_{i=1}^{n} v_i^2 \left(\Delta \alpha_D(i)\right)^2} . \qquad (21)$$

Table 2: Dielectric polarizabilities (α_D) of all ions in multicomponent oxides likely to be stable in contact with silicon and the uncertainty ($\Delta\alpha_D$) in each α_D.

Ion	α_D (Å^3) from Ref. 15[a]	$\Delta\alpha_D$ (Å^3) from Ref. 60
Li$^+$	1.20	0.1
Be^{2+}	0.19	0.05
Mg^{2+}	1.32	0.1
Ca^{2+}	3.16	0.1
Sr^{2+}	4.24	0.3
Ba^{2+}	6.40	0.3
Sc^{3+}	2.81	0.1
Y^{3+}	3.81	0.1
La^{3+}	6.07	0.3
Ce^{3+}	6.15	0.3
Pr^{3+}	5.32	0.3
Nd^{3+}	5.01	0.3
Sm^{3+}	4.74	0.3
Eu^{2+}	4.83	0.3
Gd^{3+}	4.37	0.3
Tb^{3+}	4.25	0.3
Dy^{3+}	4.07	0.3
Ho^{3+}	3.97	0.2
Er^{3+}	3.81	0.2
Tm^{3+}	3.82	0.2
Yb^{3+}	3.58	0.2
Lu^{3+}	3.64	0.2
Zr^{4+}	3.25	0.2
Hf^{4+}	3.25[b]	0.2
Al^{3+}	0.79	0.05
Si^{4+}	0.87	0.05
O^{2-}	2.01	0.1

[a]   When $V_{ox} > 25$ Å^3, α_D values from the first column of α_D values in Table III of Ref. 15 and a $\alpha_D(O^{2-})$ that depends on V_{ox}, i.e., Eqn. 16, were used in accordance with Shannon's method[15].

[b]   Estimated (α_D is related to the cube of the ionic radius, r^3 [15,54,56]. Since $r_{Hf}4+ \cong r_{Zr}4+$, $\alpha_D(Hf^{4+}) \cong \alpha_D(Zr^{4+})$).

Table 3: All binary oxides (that are solid at 1000 K and not radioactive) that *may* be stable in contact with silicon, ranked by measured K. Due to the lack of free energy data for relevant silicides, ternary phases, or both, only for four of the binary oxides (shown in bold) could thermodynamic stability with silicon be concluded [22]. BaO is also included since it has been experimentally demonstrated to be stable in epitaxial form in contact with silicon at low temperature.

Binary Oxide (MO_x)	$K_{measured}$ (at room temperature)[a]
BaO[b],[116] (water soluble)	31.1[15]—37.4[119],[120]
EuO (too conductive)	23.9[15]
HfO$_2$ (stabilized with 15% yttria)	22[44] ± 1
ZrO$_2$	22[15]
La$_2$O$_3$	20.8[c],[121]
Sm$_2$O$_3$	18.4[c],[121]
Nd$_2$O$_3$	16[15]—19.7[c],[121]
Pr$_2$O$_3$	14.9[c],[15]
SrO (water soluble)	14.5[15]—14.6[120],[122]
Tb$_2$O$_3$	13.3[c],[15]
Sc$_2$O$_3$	13[15]
Er$_2$O$_3$	13.0[c],[15]—12.5[c],[121]
Ho$_2$O$_3$	13.1[c],[15]—12.3[c],[121]
Lu$_2$O$_3$	12.5[c],[15]—12.9[c],[121]
Dy$_2$O$_3$	13.1[c],[15]—12.1[c],[121]
Tm$_2$O$_3$	12.6[c],[15],[c],[121]
Y$_2$O$_3$	11.3[15]—14.0[c],[121]
Gd$_2$O$_3$	13.6[c],[15]—11.4[c],[121]
Yb$_2$O$_3$	11.8[15]—12.6[c],[121]
CaO (water soluble)	12.0[15]—12.2[120],[123]
Al$_2$O$_3$[b],[117]	9.4 (c)[15], 11.6 (a)[15]
MgO[b],[118]	9.8[15],[124]
Li$_2$O	8.1[15]—8.8[c],[125]
BeO	6.9 (a)[15],[126], 7.7 (c)[15],[126]
Ce$_2$O$_3$	7.0[c],[21]
SiO$_2$	4.5 (a)[15],[127], 4.6 (c)[15],[127]

a   All measurements on single crystals unless otherwise noted.

b   Epitaxial growth at room temperature demonstrated.

c K measured on a polycrystalline solid.

Although it is straightforward to derive the general form of this equation for the remaining crystal system, triclinic, all of the high-K candidates identified in this study (Table 7) fall into the above crystal systems where the above formulas are applicable (and simpler). The error estimates are included in all of the K values calculated (Tables 7 and 8). For materials with the highest calculated K values, the ability of the denominator of Eqn. 12 to go to zero in combination with the large magnitude of $\Delta \alpha_D^T$ causes the ΔK given by the above formulas to lose its physical meaning. For example, for LaAlO$_3$ Eqn. 12 yields $K = 344$ and Eqn. 18 yields $\Delta K = 1315$. In such cases, a physically more meaningful value of ΔK is arrived at by considering the effect of $+\Delta \alpha_D^T$ and $-\Delta \alpha_D^T$ separately in Eqn. 12 and ignoring the insignificant effects of Δa, Δb, and Δc. In such a way, the asymmetric error caused by $\pm \Delta \alpha_D^T$ on K is seen to be $K = 344 {}_{-274}^{+\infty}$. The later method of calculating the uncertainty in K was used for materials with the highest K (specifically, $K > 20$).

2.3 RESULTS AND DISCUSSION

2.3.1 Comparison between thermodynamic stability analyses and experimental observations

The results of our thermodynamic stability analyses,[22,34] which are summarized in Tables 1, 3, and 4 and Figs. 1 and 2, are consistent with experimentally-demonstrated stable [62,82] and unstable 5-8,[10,24,62,83-87] binary oxide / silicon interfaces (including those in which silicon-compatible binary oxide or nitride buffer layers are used between silicon and a binary oxide material that would otherwise react with silicon [931]) and stable [88-106] and unstable [97,107-109] binary nitride / silicon interfaces, with the notable exception of the BaO / silicon interface.[28,68,110,111] BaO / silicon interfaces free of reaction layers have been experimentally demonstrated.[28] These BaO / silicon interfaces, however, were grown at low temperature.[28,112] BaO is predicted to react with silicon by the reaction

$$\frac{1}{4}\,Si + BaO \xrightarrow{\Delta G^{\circ}_{1000\,K}\ =\ -19.249\,\frac{kJ}{mol}} \frac{1}{2}\,Ba + \frac{1}{4}\,Ba_2SiO_4, \qquad (22)$$

where $\Delta G^{\circ}_{1000\,K}$ is the free energy change of the system when the

Table 4: All binary nitrides (that are solid at 1000 K and not radioactive) that *may* be stable in contact with silicon, ranked by measured K. Due to the lack of free energy data for relevant silicides, ternary phases, or both, only for two of the binary nitrides (shown in bold) could thermodynamic stability with silicon be concluded [34].

Binary Nitride (MN_x)	$K_{measured}$ (at room temperature)[a]
AlN	9.1[128]
Si_3N_4	7
BN	cubic: 7.1[128]
	hexagonal: 6.8 (a)[128], 5.1 (c)[128]
Ge_3N_4	5.5—6.9[129]
Be_3N_2	$(E_g \approx 12$ eV$)$[b],[130]
ScN	$(E_g \approx 2.1$ eV$)$[b],[131]
TiN	$(\rho \approx 2.2 \times 10^{-5}$ $\Omega \cdot$cm$)$[d],[132]
ZrN	$(\rho \approx 1.4 \times 10^{-5}$ $\Omega \cdot$cm$)$[d],[132]
HfN	$(\rho \approx 3.3 \times 10^{-5}$ $\Omega \cdot$cm$)$[d],[132]
δ-WN	
β-W_2N	$(\rho \approx 7.1 \times 10^{-5}$ $\Omega \cdot$cm$)$[d],[133]

[a]   All measurements on single crystals unless otherwise noted.
[b]   Dielectric constant unknown.
[c] K measured on a polycrystalline solid.
[d]   Resistivity at room temperature.

reaction proceeds in the direction indicated at a temperature of 1000 K.[13] Although small, the above driving force significantly exceeds the ± 3 kJ/mol[22] estimated uncertainty in the above $\Delta G^{\circ}_{1000\ K}$ value calculated from the thermodynamic data. An important aspect of bulk thermodynamic calculations is that they do not include interfacial free energies. In cases where the volume free energy change is small, the film is thin, and the interfacial areas are large, as is the case for the BaO / silicon system, the interfacial free energy difference could overcome the volume free energy difference. The lattice mismatch between (100) BaO and (100) silicon is low (1.6% at room temperature), making it possible for the interfacial free energy to be lower for the well lattice-matched epitaxial BaO / silicon interface than the Ba_2SiO_4 / silicon interface. At temperatures in the 500-750 °C range, however, BaO is observed to react with silicon to form barium silicate[112,113] in agreement with thermodynamic expectations (Eqn. 22). Thus, the differences in interfacial free energy are insufficient to overcome the volume free energy differences (Eqn. 22).

As was mentioned in Sec. 2.1, if the recently reported value of $\Delta H^{\circ}_{f,298\,K}$ for ZrSi$_2$ is used in Eqn. 10, the stability of silicon in contact with ZrO$_2$ comes into question.[43] Several experimental studies confirm the stability of ZrO$_2$ in direct contact with silicon to high temperatures.[114,115] In these studies, the ZrO$_2$ / silicon interface was created under conditions in which the ZrO$_2$ is not reduced, e.g., by physical vapor deposition (PVD). If deposition conditions are used in which the ZrO$_2$ may be reduced, e.g., a chemical vapor deposition (CVD) process, then reaction is seen at the interface.[43,115] The lack of reaction at ZrO$_2$ / silicon interfaces deposited by PVD is a sufficient test of the intrinsic stability of this interface. Materials that are thermodynamically stable in direct contact with silicon are only stable for a limited range of processing conditions as described in the Sec. 3.3. In contrast, dielectrics that are thermodynamically unstable in contact with silicon are unstable under all processing conditions.

2.3.2 K of silicon-compatible binary oxides (MO_x) and binary nitrides (MN_x)

The dielectric constants of all the binary oxides[22] (including BaO since it has been experimentally demonstrated to be stable in epitaxial form in contact with silicon at low temperature) and binary nitrides[34] that are stable or potentially stable in contact with silicon are shown in Tables 3 and 4, respectively. Note that there is a significant disparity in the values of K reported in the literature for these oxides and nitrides. The values listed in Tables 3 and 4 are believed to be the most accurate reported values conveying the intrinsic K of these binary materials and are from critically-reviewed compilations or measurements on single crystals wherever possible.[15,21,119-129]

Comparing Figs. 1 and 2 reveals that fewer nitrides than oxides are stable or potentially stable in contact with silicon. Of these nitrides, many are conductors. While conductors that are stable in contact with silicon are not of interest as gate dielectrics, they are of interest as electrodes in direct contact with silicon, e.g., for use in DRAM. Several of the conducting nitrides (and their alloys) listed in Table 4 are used for this application. Of those that are insulators, the dielectric constants range from 5 to 9.

In contrast to the silicon-compatible binary nitrides, all of the binary oxides that are stable or potentially stable in contact with silicon are insulators. Their dielectric constants range from 4 to about 24. As these values are significantly higher than those of the binary nitrides, oxides appear more promising than nitrides as potential replacement gate insulator materials.

BaO has the highest K of the oxides listed in Table 3. It has the disadvantage, however, of being water soluble and reacts quickly on exposure to air to form carbonates and hydroxides. It also reacts with silicon at high temperatures (in the 500-750 °C range),[112,113] in agreement with thermodynamic expectations (Eqn. 22).

The binary oxide with the next highest K, EuO, is also unstable in air and has an electrical conductivity that depends sensitively on stoichiometry,[134] making it undesirable as a gate dielectric. Its bandgap (1.12 eV between the $4f$ and $5d$ states [135]) is far too low for an alternative gate dielectric for silicon-based MOSFETs. Nonetheless, EuO is unique in that it is the only binary oxide that is both ferromagnetic and possibly stable in direct contact with silicon. The fact that it is approximately 100% spin-polarized [136] and may be epitaxially grown on silicon [137] makes it of significant interest for spintronics applications.

Next come HfO_2 and ZrO_2. The dielectric constant of ZrO_2 is well established from measurements on single crystals to be 22.[15]
. Measurements on HfO_2 single crystals have not been reported and values for HfO_2 thin films in the literature range from 18 [138] to 50.[139] We have measured the K of yttria-stabilized cubic halfnia $((Y_2O_3)_{0.15}$—$(HfO_2)_{0.85})$ single crystals. The result, $K = 22 \pm 1$, is comparable to that of Y_2O_3—ZrO_2 single crystals and is consistent with Zr^{4+} and Hf^{4+} ions having similar dielectric polarizabilities, α_D. Far UV spectroscopic ellipsometry and visible-near UV optical transmission measurements on these same $(Y_2O_3)_{0.15}$—$(HfO_2)_{0.85}$ single crystals indicate a d-state bandgap of about 5.8 eV. [140]

Next come the rare-earth oxides, Re_2O_3, where Re is a rare-earth element (with the exception of Eu since Eu_2O_3 is *not* stable in contact with silicon[22]), and the related oxides Sc_2O_3, and Y_2O_3, with K ranging from 21 (La_2O_3) to 12 (Yb_2O_3). Like BaO, SrO, and CaO, several of these Re_2O_3 materials react with water vapor. In addition, the cubic Re_2O_3 compounds, i.e., Re = Gd, Tb, Dy, Ho, Er, Tm, Yb, and Lu, Y_2O_3, and Sc_2O_3 [13,141] have relatively high oxygen ion diffusion coefficients due to one-fourth of the potential oxygen sites being vacant in this oxygen-deficient fluorite structure type known as bixbyite. For example, at 900 °C the oxygen diffusion coefficient in Y_2O_3 is more than five orders of magnitude higher than it is in SiO_2.[142]

2.3.3 Limited processing window for gate oxides stable in contact with silicon

Oxides with low permeability to oxygen are desired for gate oxides for MOSFETs because even for gate oxide materials that are thermodynamically stable in contact with silicon there is a driving force for SiO_2 to form at the interface between silicon and the gate oxide if excess oxygen is transported through the gate oxide during sample processing. The driving force for this is illustrated by the reaction below:

$$Si + \text{Stable Gate Oxide} + O_2 \xrightarrow{\Delta G^{\circ}_{1000\,K} = -730.256\,\frac{kJ}{mol}} SiO_2 + \text{Stable Gate Oxide},$$

$$(23)$$

where $\Delta G^{\circ}_{1000\,K}$ is the free energy change of the system when the reaction proceeds in the direction indicated at a temperature of 1000 K.[13] Such reactions are common, even in the growth of thermodynamically-stable oxides on silicon [36,70,143-146], because growth is typically performed with excess oxygen in order to fully oxidize the gate dielectric. The amount of SiO_2 formed depends on how much oxygen is transported through the gate oxide and will be highest for materials with high oxygen permeability, when the oxide is thin (i.e., at the early stages of gate oxide growth), and for high growth temperatures.

Fortunately, the ~900 °C high temperature annealing step is not performed in an intentionally oxidizing environment, so it is during growth rather than this implant-activation step that oxidation due to diffusivity through the gate oxide is a of greatest concern. Nonetheless, Eqn. 23 can occur during the ~900 °C implant-activation step if the annealing atmosphere contains a sufficient partial pressure of oxygen (e.g., from oxygen impurities in the gas), which typically is the case.[36] Although the reaction described by Eqn. 23 is ideally not a concern for a silicon-compatible gate oxide material during the high-temperature annealing step, if a reducing environment is used in this implant-activation step, it could result in a reduction or decomposition reaction of the gate oxide material. And in contrast to the gate oxide material, the products of a decomposition reaction could well be unstable in contact with silicon.[37,115] Thus, unwanted reactions with silicon can occur even when using gate oxide materials that are thermodynamically stable in contact with silicon if either (1) processing occurs in an excess oxygen environment where Eqn. 23 can take place or (2) processing occurs in an environment (e.g., reducing) where the gate oxide material is reduced or decomposes and the reduced or decomposed reaction product then reacts with silicon. Although the dielectric / silicon interface is stable only under a limited range of processing conditions for a dielectric that is thermodynamically stable with silicon, a dielectric that is thermodynamically unstable in contact with silicon will be unstable under *all* processing conditions.

The minimum partial pressure of oxygen needed to stabilize each of the binary oxides listed in Table 3 is shown in Fig. 4 for temperatures relevant to the implant-activation step. These Ellingham diagrams of each binary oxide were calculated from thermodynamic data.[13,40] At oxygen partial pressures above each line, the binary oxide is thermodynamically stable. At oxygen partial pressures below each line the binary oxide is unstable and a driving force exists for it to decompose into its elemental form, which would then be free to react with silicon. Note that at oxygen partial pressures below that necessary to make SiO_2 stable, there is no driving force for the oxidation of the underlying silicon substrate even if the gate oxide material is fully permeable to oxygen. This makes all of the binary oxides listed in Table 3 special in that a window of oxygen partial pressures exists in which these gate oxide materials are stable and yet there is no driving force to oxidize the silicon.

This processing window is explicitly shown in Fig. 4. Accessing this processing window directly with oxygen partial pressure is not feasible. Instead, it is common to use H_2/H_2O mixtures or CO/CO_2 mixtures to attain such low equivalent partial pressures of oxygen.[147] At the high

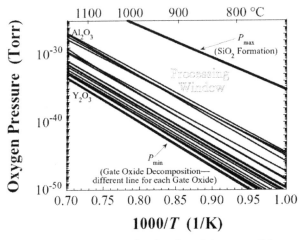

Figure 4: Thermodynamic stability lines showing the minimum partial pressure of oxygen as a function of reciprocal temperature needed to sustain the binary oxides listed in Table 1. Each binary oxide is thermodynamically stable at an oxygen partial pressure above its stability line and is thermodynamically unstable at oxygen pressures below its stability line. The region labeled "processing window" is a region in which all of the binary oxides in Table 1 are thermodynamically stable, yet SiO_2 is thermodynamically unstable. The order of the binary oxides between Y_2O_3 and Al_2O_3 at the left edge of the figure is Y_2O_3, Sc_2O_3, Er_2O_3, CaO, Ho_2O_3, Lu_2O_3, Tm_2O_3, Tb_2O_3, Dy_2O_3, ThO_2, Gd_2O_3, Sm_2O_3, Yb_2O_3, BeO, Nd_2O_3, Pr_2O_3, EuO, Ce_2O_3, La_2O_3, MgO, SrO, HfO_2, Li_2O, BaO, UO_2, ZrO_2, and Al_2O_3. Based on thermodynamic data from Ref. 13.

temperatures shown in Fig. 4, the volatility of the binary oxides (and their suboxides) should also be considered. Although Fig. 4 may indicate that a particular binary oxide is stable, it will evaporate if it has sufficient vapor

54

pressure. For example, even at oxygen partial pressures well above the SiO_2 line shown in Fig. 4, a silicon surface will remain free of oxide if the temperature is high enough to desorb the SiO_2 (e.g., via SiO loss in vacuum). [148-150]

2.3.4 *K* of multicomponent oxides

Silicon-compatible multicomponent oxides could offer performance advantages over binary oxides for gate dielectric applications. They could have higher *K* or have other characteristics, e.g., water insolubility, high bandgap, favorable band offsets, low D_{it}, or low oxygen permeability, that would make them more suitable than the binary oxides identified in Table 3. Although there are insufficient thermodynamic data to evaluate the thermodynamic stability of most multicomponent oxides in contact with silicon,* all reported silicon-compatible multicomponent oxides (Table 5)[62,63,68-71,74,76,79,143,151-161] are composed of silicon-compatible binary oxide constituents. This fact led us to propose a rule of thumb for selecting potential multicomponent oxide dielectrics for MOSFETs: to choose those made of combinations of binary oxides that are all thermodynamically compatible with silicon.[22]

This rule of thumb is equivalent to a bond strength argument. If bond strengths were assigned to all of the relevant *M*—O bonds, where *M* is each of the ions listed in Table 2, and we know that it is not favorable for *M*—O bonds to react with silicon when they are part of a binary oxide (MO_x), then it will also not be favorable for them to react with silicon when they are part of a multicomponent oxide. This is not a proof, just a way of restating the assumption since bond strength arguments do not depend on the crystalline or amorphous phase in which the bonds reside. This approximation may well overlook some compounds that contain small amounts of ions not listed in Table 2, but are nonetheless thermodynamically-stable in contact with silicon, or may select others that are not thermodynamically stable in contact with silicon. It is a simple first-order method for filtering the list of possible multicomponent materials and applies equally well to amorphous gate dielectrics.

---

* Since the free energy of the ternary or higher, multicomponent oxide must be lower than its binary constituents in order for it to form, and the binary constituents selected are individually compatible with silicon, the only potential reactions with silicon (i.e., where $\Delta G < 0$) involve ternary or higher, multicomponent products (e.g., MM'_xO_z, MM'_xSi_y, or $MM'_xSi_yO_z$ phases). Although this greatly reduces the number of reactions that need to be tested to establish thermodynamic stability, thermodynamic data for nearly all of the relevant multicomponent materials do not exist.

When multicomponent oxides are considered, it is no longer necessary to cull binary oxide constituents that are not solids at 1000 K, one of the

Table 5: Multicomponent oxides experimentally demonstrated to be compatible in contact with silicon, ranked by measured K. All measurements were made on single crystals.

Material	$K_{measured}$ (at room temperature)
Y_2O_3—ZrO_2 [69-71,76,143,151-157]	25—29.7[15,191,192]
$Ba_xSr_{1-x}O$ [68,158]	37.4[120] ($x = 1$) to 14.6[120] ($x = 0$)
$LaAlO_3$ [62,74]	24 (a)[185,186], 24—25 (c)[185,186]
$ZrSiO_4$ [62,74]	11.5—12.6 (a)[15], 11.5—12.8 (c)[15]
$MgAl_2O_4$ [62,79,159-161]	8.3[15]
$BeAl_2O_4$ [63]	9.4 (a)[15], 9.1 (b)[15], 8.3 (c)[15]

steps (see Table I in Ref. 22) leading to Table 1. Such low melting point oxides could conceivably be components of a multicomponent oxide that is both solid at high temperature and thermodynamically stable in contact with silicon. Evaluating reactions of the type

$$Si + MO_x \rightarrow M + SiO_2, \qquad (24)$$

(balanced appropriately) for all such systems for which thermodynamic data exist, revealed that at 300 K all such binary oxides for which thermodynamic data exist are thermodynamically unstable ($\Delta G < 0$) in contact with silicon. The results of these calculations are shown in Table 6. Equation 24 was evaluated at 300 K and all temperatures over which thermodynamic data of the condensed phases exist. The last column of Table 6 gives the temperature range over which the ΔG_T° of Eqn. 24 could be assessed. In all cases it remained negative, meaning that no additional ions other than those listed in Table 2 are relevant to the rule of thumb used to identify promising multicomponent oxides for alternative gate dielectrics.

Using the Clausius-Mossotti equation,[15,52-54] the additivity rule, Shannon's dielectric polarizability values for the relevant ions (Table 2),[15] and the NIST-ICDD Crystal Data database,[51] we estimated the average K of all known multicomponent oxide compounds that are composed of ~silicon-compatible binary oxides constituents. Compounds that are metastable at room temperature and pressure were eliminated from

Table 6: Gibbs free energy change at 300 K (ΔG°_{300}) for the reaction $Si + MO_x \rightarrow M + SiO_2$ (balanced appropriately) for each of the binary oxides for which thermodynamic data exist that were eliminated from consideration in Ref. 22 because they were not solids at 1000 K. Data from Ref. 13.

Binary Oxide (MO_x)	ΔG°_{300} per MO_x for $Si + MO_x \rightarrow M + SiO_2$ (balanced appropriately) (kJ/mol)	Temperature Range (K) over which ΔG°_T Calculated and found Negative ΔG°_{300} (balanced appropriately) (kcal/mol)
Rb_2O	−128.237	300—900
Cs_2O	−119.880	300—700
Ag_2O	−416.992	300—500
Au_2O_3	−1362.526	300—500
HgO	−369.725	300—600
Tl_2O	−284.745	300—1000
B_2O_3	−91.851	300—1000
P_2O_5	−809.461	300—800
As_2O_3	−708.002	300—700
As_2O_5	−1359.060	300—600
Sb_2O_3	−650.349	300—1000
Sb_2O_5	−1312.007	300—700
SeO_2	−684.970	300—600

the list of highest K candidates, which included several materials for which Eqn. 12 indicated a negative K—a sign that the denominator had become negative. Although in principle metastable phases with high K could be stabilized in thin films, e.g., via epitaxial stabilization,[162-164] our desire is to identify candidate materials that are not only stable in contact with silicon, but are also stable themselves and thus have the best chance of surviving a ~900 °C anneal in a MOSFET. For some of these compounds slightly more accurate lattice constants are available than are in the NIST-ICDD Crystal Data database,[51] and in such cases the most accurate values were used.[165,166] The resulting ranked list of the top 20 high-K phases that are

~silicon-compatible and stable is given in Table 7. The calculated K of several additional compounds, including those that have been identified by other researchers as promising materials for gate oxides,[3,4,167] are listed in Table 8. Those few compounds for which the K has been reported are also listed in these tables for comparison, as is the uncertainty in the calculated K values (see Sec. 2.3).

From a qualitative perspective, the K measurements that are available confirm that the top ranked candidates in Table 7 do for the most part have high K. The dominant structural family of the top-ranked multicomponent oxides in Table 7 is the perovskite family. $CeAlO_3$, $LaAlO_3$, $PrAlO_3$, $LaScO_3$, $NdAlO_3$, $SmAlO_3$, $PrScO_3$, $LaLuO_3$, $LaMg_{1/2}Zr_{1/2}O_3$, $BaHfO_3$, $NdScO_3$, and $SmScO_3$ are all perovskites. Several perovskite-related phases are also present. $SrLa_2Al_2O_7$, $SrCeAlO_4$, and $SrLaAlO_4$ are $n = 2$, $n = 1$, and $n = 1$ phases, respectively, in a homologous series of phases with general formula $Sr(La,Ce)_nAl_nO_{3n+1}$. This structural series, known as the Ruddlesden-Popper homologous series,[170-173] is shown in Fig. 5. These compounds may be thought of as layered rare-earth aluminate perovskites, $ReAlO_3$, in which extra SrO layers have been

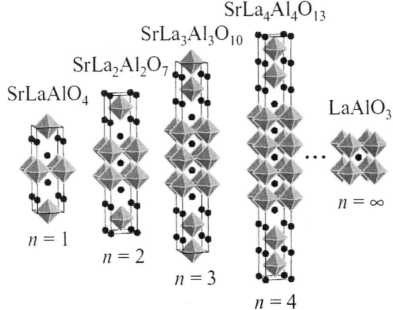

Figure 5: Schematic of the crystal structure of a unit cell of the $n = 1$ ($SrLaAlO_4$), $n = 2$ ($SrLa_2Al_2O_7$), $n = 3$ ($SrLa_3Al_3O_{10}$), $n = 4$ ($SrLa_4Al_4O_{13}$), and $n = \infty$ ($LaAlO_3$) members of the $SrLa_nAl_nO_{3n+1}$ Ruddlesden-Popper homologous series. The strontium and lanthanum atoms (which are both statistically distributed over these sites) are represented by spheres. Aluminum atoms reside at the center of the octahedra (coordination polyhedra), with oxygen atoms at the apices of the octahedra.

Table 7: Top 20 multicomponent oxides comprised of only the ions in Table 2, ranked by calculated K. The uncertainty (ΔK) in the calculated K is given. The measured K of single crystals (unless otherwise noted) is also listed for comparison. \

Material	$K_{calculated}{}^a \pm \Delta K^b$	$K_{measured}$
CeAlO$_3$	-282 (131 to ∞) $+\infty$	$24\,(a)^{[185,186]},\ 24\text{—}25\,(c)^{[185,186]}$
LaAlO$_3$	$344\ \begin{smallmatrix}\\ -274\end{smallmatrix}$	
Ce$_2$O$_3{}^d$	$67\ \begin{smallmatrix}+303\\ -\ 31\end{smallmatrix}$	$7^{c,[21]}$
PrAlO$_3$	$64\ \begin{smallmatrix}+191\\ -\ 28\end{smallmatrix}$	$25^{[15]}$
LaScO$_3$	$47\ \begin{smallmatrix}+40\\ -15\end{smallmatrix}$	$30^{c,[168]}\text{—}68^{c,[169]}$
NdAlO$_3$	$47\ \begin{smallmatrix}+59\\ -18\end{smallmatrix}$	$22.5^{[15]}$
SrLa$_2$Al$_2$O$_7$	$42\ \begin{smallmatrix}+31\\ -13\end{smallmatrix}$	
La$_2$O$_3{}^d$	$40\ \begin{smallmatrix}+37\\ -14\end{smallmatrix}$	$20.8^{c,[121]}$
SmAlO$_3$	$38\ \begin{smallmatrix}+36\\ -13\end{smallmatrix}$	$19^{[15]}$
La$_2$Be$_2$O$_5$	$35\ \begin{smallmatrix}+24\\ -11\end{smallmatrix}$	$38.7\,(a)^{[15]},\ 25.9\,(b)^{[15]},\ 22.6\,(c)^{[15]}$
PrScO$_3$	$32\ \begin{smallmatrix}+16\\ -\ 8\end{smallmatrix}$	
LaLuO$_3$	$31\ \begin{smallmatrix}+13\\ -\ 8\end{smallmatrix}$	$22 \pm 2^{[45]}$
SrCeAlO$_4$	$30\ \begin{smallmatrix}+14\\ -7\end{smallmatrix}$	
LaMg$_{\frac{1}{2}}$Zr$_{\frac{1}{2}}$O$_3$	$30\ \begin{smallmatrix}+14\\ -7\end{smallmatrix}$	
LaLiO$_2{}^d$	$30\ \begin{smallmatrix}+15\\ -8\end{smallmatrix}$	
BaHfO$_3$	$30\ \begin{smallmatrix}+12\\ -7\end{smallmatrix}$	
NdScO$_3$	$29\ \begin{smallmatrix}+12\\ -7\end{smallmatrix}$	$23 \pm 2^{[46]},\ 27^{c,[168]}$
SrLaAlO$_4$	$28\ \begin{smallmatrix}+12\\ -6\end{smallmatrix}$	$16.8\,(a)^{[15]},\ 20.0\,(c)^{[15]}$
Pr$_2$O$_3{}^d$	$28\ \begin{smallmatrix}+18\\ -\ 8\end{smallmatrix}$	$14.9^{c,[15]}$
SmScO$_3$	$28\ \begin{smallmatrix}+11\\ -7\end{smallmatrix}$	$18 \pm 2^{[46]}$

a Calculated using Eqn. 12.
b Calculated using the $\Delta \alpha_D$ values in Table 2 [60].
c K measured on a polycrystalline solid.
d V_{ox} exceeds 25 Å^3, so Eqn. 16 was used to calculate $\alpha_D(O^{2-})$.

Table 8: Calculated K of selected ~silicon-compatible materials of current interest. The uncertainty (ΔK) in the calculated K is given. The measured K of single crystals (unless otherwise noted) is also listed for comparison.

Material	$K_{calculated}{}^{a} \pm \Delta K^{b}$	$K_{measured}$
$BaZrO_3$	$26 {}^{+8}_{-6}$	$43^{c,[168]}$
BaO^{d}	$24 {}^{+10}_{-5}$	$31.1^{[15]}$—$37.4^{[119],[120]}$
$DyScO_3$	$24 {}^{+9}_{-5}$	$22\ (a)^{[46]},\ 19\ (b)^{[46]},\ 35\ (c)^{[46]}$
$GdScO_3$	$24 {}^{+9}_{-5}$	$23\ (a)^{[46]},\ 19\ (b)^{[46]},\ 29\ (c)^{[46]}$
HfO_2	$23 {}^{+10}_{-5}$	$22\pm 1^{[44]}$
ZrO_2	$21 {}^{+8}_{-5}$	$22^{[15]}$
$SrZrO_3$	$15 {}^{+3}_{-3}$	$30 \pm 2^{[47]}$
$SrHfO_3$	$17 {}^{+3}_{-3}$	$28 \pm 2^{[47]}$
$CaZrO_3$	$14 {}^{+2}_{-2}$	$23 \pm 2^{[48]}$
$HfSiO_4$	$12 {}^{+2}_{-2}$	
$ZrSiO_4$	$11 {}^{+2}_{-2}$	11.5—$12.6\ (a)^{[15]},\ 11.5$—$12.8\ (c)^{[15]}$

a   Calculated using Eqn. 12.
b   Calculated using the $\Delta\alpha_D$ values in Table 2 [60].
c K measured on a polycrystalline solid.
d V_{ox} exceeds 25 Å^3, so Eqn. 16 was used to calculate $\alpha_D(O^{2-})$.

interspersed (and the Re^{3+} and Sr^{2+} ions are statistically distributed over the combination of their resulting sites [174]). When $n = \infty$, this is the perovskite structure. Although, in principle Ruddlesden-Popper $SrRe_nAl_nO_{3n+1}$ structures with n intermediate between 2 and ∞ could exist, none have ever been synthesized in single-phase form and thus do not exist in the the NIST-ICDD Crystal Data database. If they did, they too would have appeared in Table 7 because their calculated K, from Eqn. 12, would be intermediate between $n = 1$ $SrLaAlO_4$ and $n = \infty$ $LaAlO_3$. Analogous

Ruddlesden-Popper phases in the $Sr_{n+1}Ti_nO_{3n+1}$ and $Sr_{n+1}Ru_nO_{3n+1}$ systems with n values up to five have recently been synthesized in single-phase form by molecular beam epitaxy (MBE).[175-177] Thus, 75% of the top 20 multicomponent oxide candidates are perovskites or perovskite-related materials.

The ability of perovskites to yield high values in Eqn. 12 is not unexpected. In the limit that the A ion is of the same size as an O^{2-} ion, the ABO_3 perovskite structure becomes a cubic closed-packed arrangement of AO_3 ions with the B ions occupying one fourth of the octahedral interstices in this cubic closed-packed solid. If we treat the ions as hard spheres and choose the ionic radii (r) of the A and B ions so as to maximize the packing density, i.e., $r_A = r_{O^{2-}}$ and $r_B = \left(\sqrt{2}-1\right)r_{O^{2-}}$, then the spherical atoms would fill $\dfrac{\pi}{24}\left(10 - 3\sqrt{2}\right)$ or 75.4% of all space. This dense packing makes for a low V_m in Eqn. 12 and if properly coupled with high α_D of the A and B ions, the denominator of Eqn. 12 is driven toward zero, leading to high predicted K for such perovskites.

From a simple hard-sphere analysis of the K expected from Eqn. 12 for perovskite structures, it is possible to see which A ions are expected to yield the highest K in ABO_3 perovskite structures as well as why a cubic closed-packed AO_3 arrangement of ions (with B ions in the interstices) can have a higher K than a cubic closed-packed arrangement of oxygen anions (with cations filling the interstices). This analysis is presented in Table 9 for the ~silicon-compatible ions from Table 2 and utilizes Shannon's dielectric polarizabilities[15] as well as his radii.[59] The latter depend on how many nearest neighbors (coordination number) an ion has. In perovskites the larger A cations are 12-coordinated, denoted by ^{XII}r, and the smaller B cations and oxygen ions are 6-coordinated, denoted by ^{VI}r. From Pauling's first rule,[178] only cations as large or larger than O^{2-} are expected to occupy the A site ($r_A \geq r_{O^{2-}}$) and only cations as large or larger than $\left(\sqrt{2}-1\right)r_{O^{2-}}$ are expected to occupy the B site ($r_B \geq \left(\sqrt{2}-1\right)r_{O^{2-}}$). These expectations correspond well to the ions that are known to occupy the A and B sites in perovskites.[179] The radii of the cations that can occupy the A and B site, respectively, are presented in the third and fourth columns of Table 9.

To achieve high K from Eqn. 12 it is desired to make V_m small and α_D^T large. The fifth and sixth columns of Table 9 compare how much dielectric polarizability per volume ($\dfrac{\alpha_D}{\frac{4}{3}\pi r^3}$) the various ~silicon-compatible ions have. Those cations that can reside on the A site are listed in column five and those that can reside on the B site are listed in column six. Some cations, e.g., most of the rare-earth ions, can reside on either site. Note that the dielectric polarizability per volume of O^{2-}, 0.24, is nearly the *lowest* of all the ions in Table 9 and that most of V_m in oxides is from oxygen. What perovskites offer is a means of substituting large A ions with substantially more dielectric polarizability per volume for one fourth of the O^{2-} ions in a cubic closed-packed array of oxygen ions. In choosing which of the A ions to substitute for O^{2-}, there is a tradeoff between its size and dielectric polarizability. As r_A increases in size compared to $r_{O^{2-}}$, V_m will increase which is undesired. What matters is whether the accompanying increase in α_D^T is sufficient to offset the increase in V_m and to result in a net increase in K (from Eqn. 12).

The final column of Table 9 indicates the net effect on K of replacing a cubic closed-packed array of oxygen ions (which would have a calculated K of 9.7) with a cubic AO_3 array, where A is one of the ions in column five of Table 9. In the hard-sphere approximation, V_m of a cubic AO_3 array is given by

$$V_m = \left[\sqrt{2} \left({}^{VI}r_{O^{2-}} + {}^{XII}r_A \right) \right]^3. \qquad (25)$$

The resulting calculated K of the AO_3 array does not include the B ion that occupies the octahedral interstices of the cubic AO_3 array to make an ABO_3 perovskite. Nonetheless, AO_3 contains the majority of the V_m and α_D^T terms, and which A ions are expected to result in the highest K can be seen. This simple analysis makes it clear why La^{3+}, Ce^{3+}, and other Re^{3+} rare-earth ions are the dominant A ions in the materials in Table 7. If the calculation were completed by including the optimal B-ions for each AO_3 structure, in accordance with attaining a charge neutral compound (e.g., $A^{2+}B^{4+}O_3$ or $A^{3+}B^{3+}O_3$) with the perovskite structure, it would be seen that structures with larger A ions can accommodate larger B ions in their octahedral interstices which contribute more dielectric polarizability than those that can be accommodated in structures with smaller A ions. This results in a relative enhancement in the calculated K of the perovskites with larger A ions, e.g., $A = Ba^{2+}$ and La^{3+}, over that shown by considering only AO_3 in column seven of Table 9.

If the calculated K values in Table 7 were quantitatively accurate, the K that could be achieved with ~silicon-stable multicomponent oxides, e.g., $CeAlO_3$ and $LaAlO_3$, would greatly exceed all binary oxides and nitrides. It is clear, however, that significant deviations exist in Table 7 between calculated and measured values. The general accuracy given by Shannon for estimating K using his dielectric polarizabilities is 2 to 3%.[15] As our aim is identifying high-K compounds, disagreement beyond the 2-3% level

62

Table 9: Hard-sphere analysis of which ~silicon-compatible ions should have the highest K in perovskite structures.

Ion	α_D (Å³)[a] from Ref. 15	^{XII}r (Å)[a] from Ref. 59	^{VI}r (Å)[a] from Ref. 59	$\dfrac{\alpha_D}{\frac{4}{3}\pi\left(^{XII}r\right)^3}$	$\dfrac{\alpha_D}{\frac{4}{3}\pi\left(^{VI}r\right)^3}$	K of AO_3[b]
Li⁺	1.20		0.90		0.39	
Be²⁺	0.19		0.59		0.22	
Mg²⁺	1.32		0.86		0.50	
Ca²⁺	3.16	1.48	1.14	0.23	0.51	6.9
Sr²⁺	4.24	1.58		0.26		6.9
Ba²⁺	6.40	1.75		0.29		7.2
Sc³⁺	2.81		0.885		0.97	
Y³⁺	3.81	1.33[c]	1.040	0.39	0.81	17
La³⁺	6.07	1.50	1.172	0.43	0.90	18
Ce³⁺	6.15	1.48	1.15	0.45	0.97	22
Pr³⁺	5.32	1.44[c]	1.13	0.43	0.88	19
Nd³⁺	5.01	1.41	1.123	0.43	0.84	19
Sm³⁺	4.74	1.38	1.098	0.43	0.85	21
Eu²⁺	4.83	1.57[d]		0.30		8.3
Gd³⁺	4.37	1.37[c]	1.078	0.41	0.83	18
Tb³⁺	4.25	1.35[c]	1.063	0.41	0.84	19
Dy³⁺	4.07	1.34[c]	1.052	0.40	0.83	18
Ho³⁺	3.97	1.33[c]	1.041	0.40	0.84	18
Er³⁺	3.81	1.32[c]	1.030	0.40	0.83	18
Tm³⁺	3.82	1.31[c]	1.020	0.41	0.86	19
Yb³⁺	3.58	1.30[c]	1.008	0.40	0.83	18
Lu³⁺	3.64	1.29[c]	1.001	0.40	0.87	20
Zr⁴⁺	3.25		0.86		1.22	
Hf⁴⁺	3.25[b]		0.85		1.26	
Al³⁺	0.79		0.675		0.61	
Si⁴⁺	0.87		0.540		1.32	
O²⁻	2.01		1.26		0.24	9.7

[a] Shannon's "crystal radii" (CR), which are believed to correspond most closely to the physical size of ions in solids [59].
[b] Calculated using Eqn. 12, Eqn. 14, and Eqn. 25.
[c] From Ref. 56 (estimated from a linear extrapolation of r vs. coordination number).
[d] Estimated from a linear extrapolation of r vs. coordination number (as done in Ref. 56 and Ref. 59 for other ions).

is expected because we are selecting compounds in which the denominator of Eqn. 12 is driven toward zero. This is why we calculated ΔK, but it is seen that the discrepancies frequently extend well beyond ΔK.

The disparity could be due to several factors: erroneous and/or incomplete K measurements, the presence of "compressed" ions in those structures that show lower measured K and "rattling" ions in those that have higher measured K, the questionable assumption that the dielectric polarizabilities of the constituent ions are additive (Eqn. 14),[180] the inaccuracy of applying the Clausius-Mossotti equation to structures that it does not rigorously apply to, the need to employ higher-order correction terms to the Clausius-Mossotti equation [181-183] when calculating high-K compounds, or the need for a revised set of dielectric polarizabilities, e.g., where the dielectric polarizability depends on the coordination of the ion in the structure. The disparity between the calculated K and measured K in Table 7 is largest in magnitude for $LaAlO_3$, with a discrepancy exceeding 1300%. The uncertainty in the calculated K for $LaAlO_3$ is, however, also large and the disagreement corresponds to a disparity of about three standard deviations in $\Delta\alpha_D^T$. Such large discrepancies have previously been noted for $LaAlO_3$ and other non-ferroelectric and non-piezoelectric perovskites and attributed to a compression of the cations in these structures.[56] If the low K were a ramification of cations being compressed to fit the perovskite structure, then amorphous $LaAlO_3$ in which the ions are freed from a confining crystalline structure and can occupy their typical ionic volumes, might be expected to have a larger K than crystalline $LaAlO_3$. We have investigated this possibility by depositing amorphous films with $LaAlO_3$ composition by molecular beam deposition (MBD) at room temperature on platinum electrodes, but find the opposite to be true. The K of amorphous $LaAlO_3$ films was recently reported to be 24.[184] Although the calculated K is an average K, the anisotropy in K of $LaAlO_3$ is insufficient to explain the large discrepancy between its calculated and measured K. Owing to its symmetry, the K tensor (a second-rank tensor with up to three independent coefficients[50]) of $LaAlO_3$ has two independent coefficients. Both of these tensor coefficients, K_{11} and K_{33}, have been measured at high frequencies on $LaAlO_3$ single crystals [185,186] and minimal anisotropy was observed. At 96 GHz $K_{11} = 24.0$ and $K_{33} = 25.2$.[185] At 145 GHz the anisotropy in K was even less, $K_{11} \approx K_{33} = 24.1 \pm 0.2$.[186] For most of the K measurements on single crystals presented in Table 7, including our own,[46] the anisotropy in the dielectric constant has not been determined. Recent first-principles calculations of the K of $LaAlO_3$ are in excellent agreement with the above values reported for single crystals.[187,188]

The material with the largest number of standard deviations in $\Delta\alpha_D^T$ between its calculated and measured K in Table 7 is Ce_2O_3. Its measured K

differs by about eight standard deviations in $\Delta\alpha_D^T$ from its calculated K. If Eqn. 16 had not been used to calculate $\alpha_D(O^{2-})$ of Ce_2O_3, the resulting calculated $K = 93 \, {}^{+ \, \infty}_{- \, 50}$ would have shown even larger disagreement. Indeed, Eqn. 16 improved the agreement between the measured and calculated K for all those materials with V_{ox} greater than 25 Å in Tables 7 and 8, with the sole exception of La_2O_3. As K has only been reported for polycrystalline Ce_2O_3,[21] this K may not be the intrinsic value.

Another possibility is that the calculation method itself is flawed and more so for perovskite and hexagonal Re_2O_3 structures than for other high-K compounds. The Clausius-Mossotti equation only rigorously applies to "diagonal cubic" crystals,[58,189] i.e., those with the rock-salt, CsCl, fluorite, or zinc-blende structure. Nonetheless, we know of no better method that can be applied to distill high-K candidates from the list of thousands of ~silicon compatible multicomponent oxides. Nearly all of the materials identified in Table 7 have sufficient measured K, compared to those currently being investigated, to warrant their further consideration. We believe that although the method that we have employed does not show as good a quantitative agreement as we hoped, that it has served to identify a tractable subset of promising materials for experimental investigation.

Although not a part of our selection criteria, it is straightforward to check whether any of the candidate materials identified in Tables 3-5 and VII-VIII might be piezoelectric or pyroelectric (and thus potentially ferroelectric) from their crystallographic symmetry.[50] Such ferroic properties are not desired for the gate dielectric of a MOSFET. None of the multicomponent oxides in Tables 5, 7, and 8 (which contain the top candidate gate oxide materials) are piezoelectric, pyroelectric, or ferroelectric. The same is also true for all of the binary oxides in Table 3 with the exception of BeO and SiO_2 (in its crystalline form as α-quartz), which are both piezoelectric. BeO is also pyroelectric. Of course the SiO_2 gate oxide used in MOSFETs is not crystalline, so it is not piezoelectric. Several of the nitrides in Table 4 are piezoelectric and of these some are also pyroelectric, i.e., AlN, Si_3N_4 (in its crystalline state), and Ge_3N_4. The absence of these undesired ferroic properties is an additional reason that the multicomponent oxides listed in Tables 5, 7, and 8 appear to be more suitable for use as the gate dielectric in MOSFETs than the nitrides listed in Table 4.

2.3.5 Epitaxial implementation of high-K candidates

Whether considering those multicomponent oxides that are known to be stable in contact with silicon (Table 5) and have the highest demonstrated K or those ~silicon-compatible candidates from Table 7 that have the highest demonstrated K, the same conclusion is evident: perovskites such as $LaAlO_3$ are promising materials for an epitaxial approach to alternative gate dielectrics. Of all the oxides in Tables 5 and 7, the one with the highest demonstrated K, as confirmed by measurements on single crystals is $La_2Be_2O_5$. The maximum K as a function of direction in this anisotropic monoclinic compound is K_{33} (tilted by $1.55°$ from the c-axis of $La_2Be_2O_5$), which has a measured K of 38.7.[15] But this compound contains beryllium and is therefore undesirable for environmental, health, and safety reasons.[190] The multicomponent oxide with the next highest K is yttria-stabilized cubic zirconia (Y_2O_3—ZrO_2) with an isotropic K of about 27.[15,191,192] This is a well known ionic conductor, however, and its high oxygen permeability will likely lead to the oxidation of the underlying silicon by the reaction described in Eqn. 24 during growth. Such oxidation commonly occurs in cubic zirconia grown on silicon [143] as well as silicon grown on cubic zirconia [153] when there is an excess of oxygen present. Further, the oxygen diffusion coefficient in cubic zirconia is more than six orders of magnitude higher than it is in SiO_2 at $900\ °C$.[142] The multicomponent oxide with the next highest demonstrated K is $LaAlO_3$. Its K (including anisotropy) is 24-25,[191,185,186,192-196] has a d-state bandgap of $5.6\ eV$ [140], and a lattice mismatch, $\left(\dfrac{a_{sub} - a_{film}}{a_{film}}\right)$ [197], of $+1.3\%$ with silicon. In addition, $LaAlO_3$ is stable in water, has a high melting point ($2110\ °C$ [198]), is stable against reduction when heated to high temperatures in vacuum or hydrogen,[199] e.g., the ~900 °C anneal needed for MOSFET fabrication,[2] and is predicted to have a band offset of $\Delta E_c = 2.1\ eV$ with the conduction band of silicon and a band offset of $\Delta E_v = 1.9\ eV$ with the valence band of silicon.[180] These properties make $LaAlO_3$ one of the most promising candidates for the dielectric in future MOSFETs. In amorphous form, high bandgap (5.7 to 6.2 eV [200,201]), high band offsets (1.8 to 2 eV for electrons and 2.6 to 3.2 eV for holes),[200,201] excellent thermal stability,[202,203] and a deposition process for depositing amorphous $LaAlO_3$ directly on silicon with $<0.2\ Å$ of SiO_2 at the $LaAlO_3/Si$ interface [204] have been achieved. The SiO_2-free interface is maintained even after annealing to temperatures in excess of 1000 °C.[202]

Tables 7 and 8 reveal that the next best materials are other perovskites closely related to $LaAlO_3$: $PrAlO_3$, $GdScO_3$, $NdAlO_3$, and $LaLuO_3$. All these perovskites have measured K in single crystals ≥ 22. The band alignment between amorphous $LaScO_3$, $GdScO_3$, $DyScO_3$, and silicon has recently been established; the observed bandgaps and band offsets are very similar to the values of amorphous $LaAlO_3$.[201] The K of these amorphous scandate films is about 22.[205]

Except for its large lattice mismatch with silicon, -8.0%, $LaLuO_3$ has several attractive properties for use in MOSFETs. It has an optical bandgap of 5.6 eV,[206] a high melting point (2120 °C [206]), and like LaAlO$_3$ it has been shown to be stable against reduction when heated to high temperatures in vacuum or hydrogen[206]. In addition, its remarkably small thermal expansion coefficient for a perovskite, averaging 4.1×10^{-6} K^{-1} between room temperature and 1000 °C,[45,207] less than half that of LaAlO$_3$ (10.8×10^{-6} K^{-1}),[208] makes it well suited for growth on silicon, which over the same temperature range has an average thermal expansion coefficient of 4.4×10^{-6} K^{-1}.[209] This excellent thermal expansion match with silicon minimizes the problem of cracking that often occurs in the growth of thick perovskite films on silicon substrates when they are cooled, i.e., due to the biaxial tensile stress in the film that results from its thermal expansion mismatch with the underlying silicon substrate.

An attractive aspect of the perovskite LaAlO$_3$ is that it is reasonably well lattice matched to silicon. In 1965 researchers at Westinghouse demonstrated the epitaxial growth of (111) Si on (111) LaAlO$_3$ single crystals.[74] More recently the epitaxial growth of (100) Si on (100) LaAlO$_3$ was achieved and cross-sectional TEM revealed an abrupt, reaction-free interface for films grown at a substrate temperature of 800°C.[210,211] Growing the inverted structure, i.e., epitaxial LaAlO$_3$ on silicon, is tricky because the excess oxidant present for the growth of the LaAlO$_3$ could oxidize the underlying silicon, resulting in a loss of the epitaxial template as well as an undesired low-K layer in series with the desired high-K dielectric. Over the past decade, McKee and Walker have developed a method of growing perovskites epitaxially on silicon by MBE [28,110,111] and have shown that atomic-level engineering of the transition from silicon to the perovskite is crucial for achieving an interface that is free of SiO$_2$ and has useful electrical properties. Although they have applied it only to titanium-based perovskites, we believe that their process or a closely-related MBE method[30,212-214] may also work for the ~silicon-compatible $A^{3+}B^{3+}O_3$ perovskites that we have identified in this paper.[32,33] Their method involves an epitaxial transition from the silicon substrate to a perovskite via an extremely thin (Ba,Sr)O / SrSi$_2$ / Si epitaxial buffer layer.[28,110-112] The dielectric thus grown can be represented by the general formula $(AO)_n(A'BO_3)_m$, where n and m are positive integers, AO is the (Ba,Sr)O transition layer, and $A'BO_3$ is the perovskite to be grown [111]. In contrast to $A^{2+}Ti^{4+}O_3$ titanium-based perovskites, the $A^{3+}B^{3+}O_3$ perovskites that we have identified are compatible in direct contact with silicon, so the AO layer in the McKee-Walker process can be made as thin as possible (i.e., $n = 1$) to avoid reaction during the ~900 °C MOSFET processing step. As BaO is known to react with silicon at elevated temperatures (see Sec. 3.1), using pure SrO in the transition layer would improve the high temperature stability of such layers. Avoiding the (Ba,Sr)O buffer layer entirely would be best, so applying the Motorola process[30,212-214] to $LaAlO_3$ / Si might work even better. When employed with $LaAlO_3$, this epitaxial approach can be represented by the chemical formula $(SrO)_1(LaAlO_3)_n$ or more simply by $SrLa_nAl_nO_{3n+1}$. The $n = 1$, $n = 2$, and $n = \infty$ members of this homologous series of compounds (Ruddlesden-Popper phases, see Fig. 5) were all identified in Table 7 due to their anticipated high K.

Although small, the +1.3% lattice mismatch between (001) $LaAlO_3$ and (001) Si, with an in-plane orientation relationship of [100] $LaAlO_3 \parallel$ [110] Si, could be reduced by alloying $LaAlO_3$ with another high-K, ~silicon-compatible perovskite to make the lattice constant of the resulting solid solution identical to that of silicon. This could decrease D_{it} as well as decrease the oxygen permeability of the epitaxial dielectric layer by removing the driving force for the formation of misfit dislocations. Dislocations, in single crystals of the perovskite $SrTiO_3$, for example, have been shown to significantly increase the oxygen diffusion coefficient.[215] Several of the perovskites in Tables 7 and 8 not only have larger lattice constants, but share the same $A^{3+}B^{3+}O_3$ perovskite chemistry as $LaAlO_3$, which lessens potential point-defect problems, and have melting points close to that of $LaAlO_3$,[45,46,216] which reduces the driving force for phase separation during growth. Employing Vegard's law,[217] lattice-matched perovskite compositions are approximately: $LaSc_{0.19}Al_{0.81}O_3$, $LaLu_{0.13}Al_{0.87}O_3$, $LaMg_{0.095}Zr_{0.095}Al_{0.81}O_3$, and $(LaAlO_3)_{0.73}$—$(GdScO_3)_{0.27}$ [218,219]. The extent of the solid solution between $LaAlO_3$ and $LaScO_3$ has been investigated and $La(Sc,Al)O_3$ single crystals have been grown.[220] Fortuitously, the solid solution extends to a $La(Sc,Al)O_3$ composition that is almost perfectly lattice matched with silicon. A general formula for ~silicon-compatible, lattice-matched, high-K epitaxial gate dielectrics that include a SrO monolayer to utilize the McKee-Walker process for epitaxial integration with silicon, is thus $SrLa_n(Sc,Al)_nO_{3n+1}$. An extra SrO layer could even be interspersed into the $La(Sc,Al)O_3$ part of the structure, e.g., as occurs in an ordered way in the Ruddlesden-Popper structures in Fig. 5. Such $SrLa_nSc_nO_{3n+1}$ structures have been shown to have K up to 18 in polycrystalline samples[221] and are attractive for an epitaxial approach where the computer-controlled shutters of MBE can be used to precisely build up the desired $SrLa_nSc_nO_{3n+1}$ structure with monolayer-level layering control.[175,176]

2.4 SUMMARY AND CONCLUSIONS

A thermodynamic approach was used to comprehensively assess the thermodynamic stability of binary oxides and binary nitrides in contact with silicon. The binary oxides were found to have significantly higher K than those few insulating nitrides that are stable or potentially stable in

68

contact with silicon. The K of nearly 2000 multicomponent oxides composed of ~silicon-compatible binary oxides was estimated using a method developed by Shannon to identify the most promising high-K silicon-compatible oxides. Although quantitative discrepancies exist, this method is a useful coarse filter to identify a limited group of multicomponent oxide candidates for experimental study as alternative gate dielectrics. One of the most promising materials for an alternative gate dielectric is the perovskite $LaAlO_3$. In addition to proven high K (24-25 [185,186]) and silicon compatibility, it has a high d-state bandgap (5.6 eV)[140], excellent high temperature stability, favorable predicted band offsets with silicon,[180] and a +1.3% lattice match to (001) Si. By combining it with the best of the other perovskites that we have identified and that have proven high K, it appears that a precisely lattice-matched solid solution could be made[222] for epitaxial integration with silicon using the McKee-Walker[28,110,111] or Motorola[30,212,214] processes. Of course high K and silicon-compatibility are just two of the many requirements that a successful alternative gate dielectric material for use in silicon MOSFETs must possess. Other factors including low D_{it}, low permeability to oxygen, low leakage, high bandgap, favorable band alignment with silicon, ease of integration into a complementary metal-oxide-semiconductor (CMOS) process, reliability, etc., must also be considered. The candidate materials identified here form a starting point for these additional considerations.

ACKNOWLEDGEMENTS

We gratefully acknowledge Eric Dowty of Shape Software (Kingsport, TN) and Paul Shlichta of Crystal Research (Olympia, WA), who modified their proprietary software so that we could perform the NIST-ICDD Crystal Data database search, Anatoly Balbashov, Georg Bednorz, and Ashot Petrosyan for supplying single crystals for K measurements, Susan Trolier-McKinstry for helpful advice with the K measurements, useful input from Robert Shannon and David Gundlach, discussions with Jon-Paul Maria, Gang Bai, Robby Beyers, Hans Christen, Brian Doyle, Ravi Droopad, Urs Dürig, Lisa Edge, Dave Fork, John Freeouf, David Gilmer, Supratik Guha, Maciej Gutowski, Arvind Halliyal, Tom Jackson, Angus Kingon, Seung-Gu Lim, Gerry Lucovsky, Jochen Mannhart, Rodney McKee, Karin Rabe, Brian Roberds, John Robertson, Jürgen Schubert, Karl Spear, Jim Speck, Stephen Streiffer, Susanne Stemmer, Baylor Triplett, and Fred Walker, and the financial support of the Semiconductor Research Corporation (SRC) and SEMATECH through the SRC/SEMATECH FEP Center. DGS gratefully acknowledges a research fellowship from the Alexander von Humboldt Foundation during his sabbatical at Augsburg University. JHH gratefully acknowledges a

Motorola/SRC Fellowship and the support of the BMBF (project 13N6918/1) during his stay at Augsburg University.

REFERENCES

1. *International Technology Roadmap for Semiconductors, 2003 Edition, Front End Processes* (Semiconductor Industry Association, San Jose, 2003), pp. 2, 23–33. Available on-line at http://public.itrs.net/Files/2003ITRS/Home2003.htm

2. *The National Technology Roadmap for Semiconductors* (Semiconductor Industry Association, San Jose, CA, 1997), p. 72.

3. A.I. Kingon, J-P. Maria and S.K. Streiffer, *Nature* **406**, 1032 (2000).

4. G.D. Wilk, R.M. Wallace and J.M. Anthony, *J. Appl. Phys.* **89**, 5243 (2001).

5. S. Zaima, T. Furuta, Y. Yasuda and M. Iida, *J. Electrochem. Soc.* **137**, 1297 (1990).

6. S.-O. Kim and H.J. Kim, *J. Vac. Sci. Technol. B* **12**, 3006 (1994).

7. J.-L. Autran, R. Devine, C. Chaneliere and B. Balland, *IEEE Electron Device Lett.* **18**, 447 (1997).

8. G.B. Alers, D.J. Werder, Y. Chabal, H.C. Lu, E.P. Gusev, E. Garfunkel, T. Gustafsson and R.S. Urdahl, *Appl. Phys. Lett.* **73**, 1517 (1998).

9. A.Y. Mao, K.A. Son, J.M. White, D.L. Kwong, D.A. Roberts and R.N. Vrtis, *J. Vac. Sci. Technol. A* **17**, 954 (1999).

10. D.C. Gilmer, D.G. Colombo, C.J. Taylor, J. Roberts, G. Haugstad, S.A. Campbell, H.-S. Kim, G.D. Wilk, M.A. Gribelyuk and W.L. Gladfelter, *Chem. Vap. Deposition* **4**, 9 (1998).

11. S.A. Campbell, H.-S. Kim, D.C. Gilmer, B. He, T. Ma, and W.L. Gladfelter, *IBM J. Res. Develop.* **43**, 383 (1999).

12. Y. Jeon, B.H. Lee, K. Zawadzki, W.-J. Qi, and J.C. Lee, in *Rapid Thermal and Integrated Processing VII*, edited by M.C. Ozturk, F. Roozeboom, P.J. Timans and S.H. Pas, (Mater. Res. Soc. Proc. **525**, Warrendale, 1998), pp. 193-198.

13. I. Barin, *Thermochemical Data of Pure Substances*, 3rd Ed., Vol. I and Vol. II (VCH, Weinheim, 1995).

14. S. Hayashi, A. Ueno, K. Okada and N. $\overline{O}$tsuka, *J. Ceram. Soc. Jpn. Int. Ed.* **99**, 787 (1991).

15. R.D. Shannon, *J. Appl. Phys.* **73**, 348 (1993).

16. W.B. Pennebaker, *IBM J. Res. Develop.* **13**, 686 (Nov. 1969).

17. J.K.G. Panitz and C.C. Hu, *J. Vac. Sci. Technol.* **16**, 315 (1979).

18. V.S. Dharmadhikari and W.W. Grannemann, *J. Vac. Sci. Technol. A* **1**, 483 (1983).

19. S. Matsubara, T. Sakuma, S. Yamamichi, H. Yamaguchi and Y. Miyasaka in *Ferroelectric Thin Films*, edited by E. R. Myers and A.I. Kingon (Mater. Res. Soc. Proc. **200**, Pittsburgh, 1990), p. 243; T. Sakuma, S. Yamamichi, S. Matsubara, H. Yamaguchi and Y. Miyasaka, *Appl. Phys. Lett.* **57**, 2431 (1990); H. Yamaguchi, S. Matsubara and Y. Miyasaka, *Jpn. J. Appl. Phys.* **30**, 2197 (1991).

20. H. Nagata, T. Tsukahara, S. Gonda, M. Yoshimoto, and H. Koinuma, *Jpn. J. Appl. Phys.* **30**, L1136 (1991).

21. *Landolt-Börnstein: Zahlenwerte und Funktionen aus Physik, Chemie, Astronomie, Geophysik, und Technik*, 6th Ed., edited by J. Bartels, P. Ten Bruggencate, H. Hausen, K.H. Hellwege, K.L. Schäfer and E. Schmidt (Springer, Berlin, 1959), Vol. II, Part 6, p. 482.

22. K.J. Hubbard and D.G. Schlom, *J. Mater. Res.* **11**, 2757 (1996).

23. M. Yoshimoto, K. Shimozono, T. Maeda, T. Ohnishi, M. Kumagai, T. Chikyow, O. Ishiyama, M. Shinohara and H. Koinuma, *Jpn. J. Appl. Phys.* **34**, L688 (1995).

70

24. T. Chikyow, S.M. Bedair, L. Tye and N.A. El-Masry, *Appl. Phys. Lett.* **65**, 1030 (1994); L. Tye, T. Chikyow, N.A. El-Masry and S.M. Bedair in *Epitaxial Oxide Thin Films and Heterostructures*, edited by D.K. Fork, J.M. Phillips, R. Ramesh, and R.M. Wolf (Mater. Res. Soc. Proc. **341**, Pittsburgh, 1994), p. 107; L. Tye, N.A. El-Masry, T. Chikyow, P. McLarty and S.M. Bedair, *Appl. Phys. Lett.* **65**, 3081 (1994).

25. T. Inoue, T. Ohsuna, Y. Obara, Y. Yamamoto, M. Satoh and Y. Sakurai, *Jpn. J. Appl. Phys.* **32**, 1765 (1993).

26. H. Mori and H. Ishiwara, *Jpn. J. Appl. Phys.* **30**, L1415 (1991); H. Ishiwara, H. Mori, K. Jyokyu, and S. Ueno in *Silicon Molecular Beam Epitaxy*, edited by J.C. Bean, S.S. Iyer and K.L. Wang (Mater. Res. Soc. Proc. **220**, Pittsburgh, 1991), pp. 595-600; B.K. Moon and H. Ishiwara, *Jpn. J. Appl. Phys.* **33**, 1472 (1994).

27. O. Nakagawara, M. Kobayashi, Y. Yoshino, Y. Katayama, H. Tabata, and T. Kawai, *J. Appl. Phys.* **78**, 7226 (1995).

28. R.A. McKee, F.J. Walker and M.F. Chisholm, *Phys. Rev. Lett.* **81**, 3014 (1998); R.A. McKee, F.J. Walker and M.F. Chisholm, *Science* **293**, 468 (2001).

29. T. Tambo, T. Nakamura, K. Maeda, H. Ueba and C. Tatsuyama, *Jpn. J. Appl. Phys.* **37**, 4454 (1998); T. Tambo, K. Maeda, A. Shimizu, and C. Tatsuyama, *J. Appl. Phys.* **86**, 3213 (1999).

30. K. Eisenbeiser, J.M. Finder, Z. Yu, J. Ramdani, J.A. Curless, J.A. Hallmark, R. Droopad, W.J. Ooms, L. Salem, S. Bradshaw and C.D. Overgaard, *Appl. Phys. Lett.* **76**, 1324 (2000).

31. B. Ho, T. Ma, S.A. Campbell and W.L. Gladfelter, in *International Electron Devices Meeting 1998* (IEEE, Piscataway, 1998), pp. 1038-40.

32. C.A. Billman, P.H. Tan, K.J. Hubbard and D.G. Schlom, "Alternate Gate Oxides for Silicon MOSFETs using High-K Dielectrics," in: *Ultrathin SiO_2 and High-K Materials for ULSI Gate Dielectrics*, edited by H.R. Huff, C.A. Richter, M.L. Green, G. Lucovsky and T. Hattori (Mater. Res. Soc. Symp. Proc. **567**, Pittsburgh, 1999), pp. 409-414.

33. D.G. Schlom and J.H. Haeni, *MRS Bull.* **27**, 198 (2002).

34. P.H. Tan and D.G. Schlom, *submitted to J. Mater. Res.*

35. R. Beyers, *J. Appl. Phys.* **56**, 147 (1984); R. Beyers, R. Sinclair and M.E. Thomas, *J. Vac. Sci. Technol. B* **2**, 781 (1984); R. Beyers, K.B. Kim and R. Sinclair, *J. Appl. Phys.* **61**, 2195 (1987); R. Beyers, Ph.D. Thesis, Stanford University, 1989, pp. 38-76.

36. J.-P. Maria, D. Wicaksana, A.I. Kingon, B. Busch, H. Schulte, E. Garfunkel and T. Gustafsson, *J. Appl. Phys.* **90**, 3476 (2001).

37. J.-P. Maria, W-H. Schulte, D.Wicaksana, B. Busch, A.I. Kingon and E. Garfunkel, *submitted to Appl. Phys. Lett.*

38. T.S. Jeon, J.M. White and D.L. Kwong, *Appl. Phys. Lett.* **78**, 368 (2001).

39. *CRC Handbook of Chemistry and Physics: A Ready-Reference Book of Chemical and Physical Data*, 77th Ed., edited by D.R. Lide (CRC Press, Boca Raton, 1996).

40. *Thermodynamic Properties of Elements and Oxides* (United States Bureau of Mines Bulletin 672, U.S. Government Printing Office, Washington D.C., 1982).

41. G.M. Lukashenko, R.I. Polotskaya and V.R. Sidorko, *J. Alloys Compd.* **179**, 299 (1992).

42. S.V. Meschel and O.J. Kleppa, *J. Alloys Compd.* **274**, 193 (1998).

43. M. Gutowski, J.E. Jaffe, C.-L. Liu, M. Stoker, R.I. Hegde, R.S. Rai and P.J. Tobin, *Appl. Phys. Lett.* **80**, 1897 (2002).

44. The $(Y_2O_3)_{0.15}$—$(HfO_2)_{0.85}$ yttria-stabilized cubic HfO_2 single crystal samples were provided by Anatoly Balbashov of the Moscow Power Engineering Institute, Moscow, Russia.

45. J.H. Haeni, A.G. Petrosyan, and D.G. Schlom (unpublished). The LaLuO$_3$ single crystal sample was provided by Ashot Petrosyan of the Institute for Physical Research, Armenian National Academy of Sciences, Ashtarak, Armenia. Details on the preparation of this sample by the Czochralski method are described in: K.L. Ovanesyan, A.G. Petrosyan, G.O. Shirinyan, C. Pedrini and L. Zhang, *J. Cryst. Growth* **198/199**, 497 (1999).

46. J.H. Haeni, J. Lettieri, M. Biegalski, S. Trolier-McKinstry, D.G. Schlom, S-G. Lim, T.N. Jackson, M.M. Rosario, J.L. Freeouf, R. Uecker, P. Reiche, A. Ven Graitis, C.D. Brandle and G. Lucovsky, *submitted to J. Appl. Phys.*

47. The SrZrO$_3$ and SrHfO$_3$ single crystals were grown by float-zone by Dima Suptel and Anatoly Balbashov of the Moscow Power Engineering Institute, Moscow, Russia.

48. The CaTiO$_3$ sample was grown by float-zone by J. Georg Bednorz of IBM Research, Rüschlikon, Switzerland.

49. M.A. Subramanian, R.D. Shannon, B.H.T. Chai, M.M. Abraham and M.C. Wintersgill, *Phys. Chem. Minerals* **16**, 741 (1989).

50. J.F. Nye, *Physical Properties of Crystals: Their Representation by Tensors and Matrices* (Oxford University Press, 1957, 1985).

51. *NIST Crystal Data 1997*, a CD-ROM database (International Centre for Diffraction Data, Newton Square, PA, 1997).

52. O.F. Mossotti, Memorie di Matematica e di Fisica della Società Italiana delle Scienze Residente in Modena **24** (Part II), 49 (1850).

53. R. Clausius, *Die Mechanische Wärmetheorie*, 2nd Ed.,Vol. 2, *Die Mechanische Behandlung der Electricität* (Vieweg, Braunschweig, 1879) pp. 62-97.

54. S. Roberts, *Phys. Rev.* **76**, 1215 (1949).

55. A.R. von Hippel, *Dielectrics and Waves* (John Wiley & Sons, New York, 1954).

56. V.J. Fratello and C.D. Brandle, *J. Mater. Res.* **9**, 2554 (1994).

57. C. Barrett and T.B. Massalski, *Structure of Metals: Crystallographic Methods, Principles, and Data*, 3rd (revised) Ed. (Permagon Press, Oxford, 1980) p. 614.

58. S. Roberts, *Phys. Rev.* **81**, 865 (1951).

59. R.D. Shannon, *Acta Cryst. A* **32**, 751 (1976).

60. R.D. Shannon (private communication). Note that the uncertainties in Table II are about twenty times larger than those given in Table III of Ref. 15. Some are as much as sixty times larger. The reason for this is the low ratio of observables to variables used in the best fit analysis used to obtain the ionic polarizabilities in Ref. 15.

61. P.R. Bevington, *Data Reduction and Error Analysis for the Physical Sciences* (McGraw-Hill, New York, 1969) pp. 56-65.

62. J.D. Filby and S. Nielsen, Br. *J. Appl. Phys.* **18**, 1357 (1967).

63. H.M. Manasevit, D.H. Forbes and I.B. Cadoff, *Trans. Metall. Soc. AIME*, **236**, 275 (1966).

64. D.K. Fork, F.A. Ponce, J.C. Tramontana and T.H. Geballe, *Appl. Phys. Lett.* **58**, 2294 (1991).

65. P. Tiwari, S. Sharan and J. Narayan, *J. Appl. Phys.* **69**, 8358 (1991).

66. G. Soerensen and S. Gygax, *Physica B* **169**, 673 (1991).

67. Y. Kado and Y. Arita, *J. Appl. Phys.* **61**, 2398 (1987); Y. Kado and Y. Arita in *Extended Abstracts of the 18th (1986 International) Conference on Solid State Devices and Materials* (Tokyo, 1986), pp. 45-48.

68. Y. Kado and Y. Arita in *Extended Abstracts of the 20th (1988 International) Conference on Solid State Devices and Materials* (Tokyo, 1988), pp. 181-184.

69. H. Behner, J. Wecker, Th. Matthée, and K. Samwer, Surf. Interface Anal. **18**, 685 (1992); Th. Matthée, J. Wecker, H. Behner, G. Friedl, O. Eibl and K. Samwer, *Appl. Phys. Lett.* **61**, 1240 (1992).

72

70. A. Bardal, O. Eibl, Th. Matthée, G. Friedl and J. Wecker, *J. Mater. Res.* **8**, 2112 (1993).

71. H. Fukumoto, T. Imura and Y. Osaka, Appl. Phys. Lett. **55**, 360 (1989); H. Fukumoto, M. Yamamoto, and Y Osaka, *Proc. Electrochem. Soc.* **90**, 239 (1990); M. Yamamoto, H. Fukumoto and Y. Osaka in *Heteroepitaxy of Dissimilar Materials*, edited by R.F.C. Farrow, J.P. Harbison, P.S. Peercy and A. Zangwill (Mater. Res. Soc. Proc. **221**, Pittsburgh, 1991), p. 35.

72. K. Harada, H. Nakanishi, H. Itozaki and S. Yazu, *Jpn. J. Appl. Phys.* **30**, 934 (1991).

73. E.J. Tarsa, J.S. Speck and McD. Robinson, *Appl. Phys. Lett.* **63**, 539 (1993).

74. T.L. Chu, M.H. Francombe, G.A. Gruber, J.J. Oberly and R.L. Tallman, *Deposition of Silicon on Insulating Substrates*, Report No. AFCRL-65-574 (Westinghouse Research Laboratories, Pittsburgh, 1965). See especially pp. 31-34 and pp. 41-44. (NTIS ID No. AD-619 992).

75. M. Morita, H. Fukumoto, T. Imura, Y. Osaka and M. Ichihara, *J. Appl. Phys.* **58**, 2407 (1985); Y. Osaka. T. Imura, Y. Nishibayashi and F. Nishiyama, *J. Appl. Phys.* **63**, 581 (1988).

76. H. Myoren, Y. Nishiyama, H. Fukumoto, H. Nasu and Y. Osaka, *Jpn. J. Appl. Phys.* **28**, 351 (1989).

77. A. Lubig, Ch. Buchal, J. Schubert, C. Copetti, D. Guggi, C.L. Jia and B. Stritzker, *J. Appl. Phys.* **71**, 5560 (1992); A. Lubig, Ch. Buchal, D. Guggi, C.L. Jia and B. Stritzker, *Thin Solid Films* **217**, 125 (1992).

78. H.M. Manasevit and W.I. Simpson, *J. Appl. Phys.* **35**, 1349 (1964); H.M. Manasevit, A. Miller, F.L. Morritz and R. Nolder, *Trans. Metall. Soc. AIME* **233**, 540 (1965).

79. H.M Manasevit, *J. Cryst. Growth* **22**, 125 (1974).

80. F.A. Ponce, *Appl. Phys. Lett.* **41**, 371 (1982).

81. M. Ishida, I. Katakabe, T. Nakamura and N. Ohtake, *Appl. Phys. Lett.* **52**, 1326 (1988); K. Sawada, M. Ishida, T. Nakamura and N. Ohtake, *Appl. Phys. Lett.* **52**, 1672 (1988); K. Sawada, M. Ishida, T. Nakamura and T. Suzaki, *J. Cryst. Growth* **95**, 494 (1989); M. Ishida, K. Sawada, S. Yamaguchi, T. Nakamura and T. Suzaki, *Appl. Phys. Lett.* **55**, 556 (1989); M. Ishida, S. Yamaguchi, Y. Masa, T. Nakamura and Y. Hikita, *J. Appl. Phys.* **69**, 8408 (1991).

82. H. Iizuka, K. Yokoo and S. Ono, *Appl. Phys. Lett.* **61**, 2978 (1992).

83. H. Nagata, M. Yoshimoto, T. Tsukahara, S. Gonda and H. Koinuma in *Evolution of Thin-Film and Surface Microstructure*, edited by C.V. Thompson, J.Y. Tsao, and D.J. Srolovitz (Mater. Res. Soc. Proc. **202**, Pittsburgh, 1991), p. 445; H. Koinuma, H. Nagata, T. Tsukahara, S. Gonda and M. Yoshimoto, *Appl. Phys. Lett.* **58**, 2027 (1991).

84. Y. Shichi, S. Tanimoto, T. Goto, K. Kuroiwa and Y. Tarui, *Jpn. J. Appl. Phys.* **33**, 5172 (1994).

85. C.W. Nieh, E. Kolawa, F.C.T. So and M.-A. Nicolet, *Mater. Lett.* **6**, 177 (1988).

86. A. Charai, S.E. Hörnström, O. Thomas, P.M. Fryer and J.M.E. Harper, *J. Vac. Sci. Technol. A* **7**, 784 (1989).

87. E. Kolawa, C.W. Nieh, J.M. Molarius, L. Tran, C. Garland, W. Flick, M-A. Nicolet, F.C.T. So and J.C.S. Wei, *Thin Solid Films* **166**, 15 (1988).

88. W.J. Meng, J. Heremans and Y.T. Cheng, *Appl. Phys. Lett.* **59**, 2097 (1991); W.J. Meng and J. Heremans, *J. Vac. Sci. Technol. A* **10**, 1610 (1992); W.J. Meng and J. Heremans, in *Interface Dynamics and Growth*, edited by K.S. Liang, M.P. Anderson, R.F. Bruinsma, and G. Scoles (Mater. Res. Soc. Symp. Proc. **237**, Pittsburgh, 1992), p. 529.

89. R.D. Vispute, J. Narayan, H. Wu and K. Jagannadham, *J. Appl. Phys.* **77**, 4724 (1995); R.D. Vispute, H. Wu, K. Jagannadham and J. Narayan, in *Gallium Nitride and Related Materials*, edited by R.D. Dupuis, J.A. Edmond, F.A. Ponce and S.J. Nakamura (Mater. Res. Soc. Symp. Proc. **395**, Pittsburgh, 1996), p. 325.

90. P. Rogl and J. C. Schuster, *Diagrams of Ternary Boron Nitride and Silicon Nitride Systems* (ASM International, Materials Park, 1992).

91. D.J. Kester, K.S. Ailey, D.J. Lichtenwalner and R.F. Davis, *J. Vac. Sci. Technol. A* **12**, 3074 (1994).

92. P.B. Mirkarimi, D.L. Medlin, K.F. McCarty, D.C. Dibbles, W.M. Clift, J.A. Knapp and J.C. Barbour, *J. Appl. Phys.* **82**, 1617 (1997).

93. K.S. Park, D.Y. Lee, K.J. Kim and D.W. Moon, *J. Vac. Sci. Technol. A* **15**, 1041 (1997).

94. I. Suni, M. Mäenpää, M.-A. Nicolet and M. Luomajarvi, *J. Electrochem. Soc.* **130**, 1215 (1983).

95. R. De Reus, F.W. Saris, G.J. van der Kolk, C. Witmer, B. Dam, D.H.A. Blank, D.J. Adelerhof and J. Flokstra, *Mater. Sci. Eng. B* **7**, 135 (1990).

96. R. Nowak and C.L. Li, *Thin Solid Films* **305**, 297 (1997).

97. F. Weitzer, J. C. Schuster, J. Bauer and B. Jounel, *J. Mater. Sci.* **26**, 2076 (1991).

98. X. Li, B.-Y. Kim and S.-W. Rhee, *Appl. Phys. Lett.* **67**, 3426 (1995).

99. K.C. Park and K.B. Kim, *J. Electrochem. Soc.* **142**, 3109 (1995).

100. C. S. Sorrell, in *Silicon-Based Structural Ceramics*, edited by B. W. Sheldon and S. C. Danforth (The American Ceramic Society, Columbus, 1993), p 115.

101. C.W. Lee, Y.T. Kim, S.-K. Min, C. Lee, J.Y. Lee and Y.W. Park, in *Interface Control of Electrical, Chemical, and Mechanical Properties*, edited by S.P. Murarka, K. Rose, T. Ohmi, and T. Seidel (Mater. Res. Soc. Symp. Proc. **318**, Pittsburgh, 1994), p. 335.

102. C.W. Lee, Y.T. Kim, and J.Y. Kee, Appl. Phys. Lett. **64**, 619 (1994).

103. C.S. Kwon, D.J. Kim, C.W. Lee, Y.T. Kim and I.-H. Choi, in *Evolution of Thin-Film, Surface Structure, and Morphology*, edited by B.G. Demczyk, E.D. Williams, E. Garfunkel, B.M. Clemens, and J.E. Cuomo (Mater. Res. Soc. Symp. Proc. **355**, Pittsburgh, 1995), p. 441.

104. M. Takeyama and A. Noya, *Jpn. J. Appl. Phys.* **36**, 2261 (1997).

105. S. A. Barnett, L. Hultman, J.-E. Sundgren, F. Ronin and S. Rhode, *Appl. Phys. Lett.* **53**, 400 (1988).

106. S. Horita, T. Tujikawa, H. Akahori, M. Kobayashi and T. Hata, *J. Vac. Sci. Technol. A* **11**, 2452 (1993); S. Horita, H. Akahori and M. Kobayahsi, *J. Vac. Sci. Technol. A* **14**, 203 (1996); K. Yoshimoto, H. Yanagisawa, and K. Sasaki, *Jpn. J. Appl. Phys.* **36**, 7302 (1997).

107. K. Holloway, P.M. Fryer, C. Cabral, Jr., J.M.E. Harper, P.J. Bailey and K.H. Kelleher, *J. Appl. Phys.* **71**, 5433 (1992).

108. K.T. Ho, C.D. Lien, M.-A. Nicolet and D.M. Scott, in *Thin Films and Interfaces II*, edited by J.E.E. Baglin, D.R. Campbell and W.K. Chu (Mater. Res. Soc. Symp. Proc. **25**, Elsevier, New York, 1984), pp. 123-129.

109. R.J. Pulham, P. Hubberstey, A.E. Thunder, A. Harper and A.T. Dadd, Int. Conf. Liq. Met. Technol. Energy Prod. **2**, 18.1 (1980).

110. R.A. McKee, F.J. Walker, J.R. Conner, E.D. Specht and D.E. Zelmon, *Appl. Phys. Lett.* **59**, 782 (1991).

111. R.A. McKee and F.J. Walker, *US Patent No. 5,225,031* (July 6, 1993); R.A. McKee and F.J. Walker, *US Patent No. 5,830,270* (November 3, 1998).

112. J. Lettieri, J.H. Haeni and D.G. Schlom, *J. Vac. Sci. Technol. A* **20**, 1332 (2002).

113. V.V. Il'chenko, G.V. Kuznetsov, V.I. Strikha and A.I. Tsyganova, *Mikroelektron.* **27**, 340 (1998) [Russ. Microelectron. **27**, 291 (1998)]; V.V. Il'chenko and G.V. Kuznetsov, *Pis'ma Zh. Tekh. Fiz.* **27**, 58 (2001) [*Sov. Tech. Phys. Lett.* **27**, 333 (2001)].

114. S.J. Wang, C.K. Ong, S.Y. Xu, P. Chen, W.C. Tjiu, J.W. Chai, A.C.H. Huan, W.J. Yoo, J.S. Lim, W. Feng and W.K. Choi, *Appl. Phys. Lett.* **78**, 1604 (2001).

74

115. C.M. Perkins, B.B. Triplett, P.C. McIntyre, K.C. Saraswat and E. Shero, *Appl. Phys. Lett.* **81**, 1417 (2002).

116. T. Ohnishi, M. Yoshimoto, G.H. Lee, T. Maeda and H. Koinuma, *J. Vac. Sci. Technol. A* **15**, 2469 (1997).

117. T. Maeda, M. Yoshimoto, T. Ohnishi, G.H. Lee and H. Koinuma, *J. Cryst. Growth* **177**, 95 (1997).

118. S. Yadavalli, M.H. Yang and C.P. Flynn, *Phys. Rev. B* **41**, 7961 (1990).

119. *Landolt-Börnstein: Numerical Data and Functional Relationships in Science and Technology*, edited by O. Madelung, M. Schulz, and H. Weiss (Springer, Berlin, 1982), New Series, Group III, Vol. 17b, pp. 31-34.

120. M. Galtier, A. Montaner and G. Vidal, *J. Phys. Chem. Solids* **33**, 2295 (1972).

121. *The Oxide Handbook*, 2nd Ed., edited by G.V. Samsonov (IFI/Plenum, New York, 1982), p. 213.

122. *Landolt-Börnstein: Numerical Data and Functional Relationships in Science and Technology*, edited by O. Madelung, M. Schulz and H. Weiss (Springer, Berlin, 1982), New Series, Group III, Vol. 17b, p. 29.

123. *Landolt-Börnstein: Numerical Data and Functional Relationships in Science and Technology*, edited by O. Madelung, M. Schulz and H. Weiss (Springer, Berlin, 1982), New Series, Group III, Vol. 17b, p. 25.

124. *Landolt-Börnstein: Numerical Data and Functional Relationships in Science and Technology*, edited by O. Madelung, M. Schulz and H. Weiss (Springer, Berlin, 1982), New Series, Group III, Vol. 17b, p. 18.

125. *Landolt-Börnstein: Zahlenwerte und Funktionen aus Physik, Chemie, Astronomie, Geophysik, und Technik*, 6th Ed., edited by J. Bartels, P. Ten Bruggencate, H. Hausen, K.H. Hellwege, K.L. Schäfer and E. Schmidt (Springer, Berlin, 1959), Vol. II, Part 6, p. 455.

126. *Landolt-Börnstein: Numerical Data and Functional Relationships in Science and Technology*, edited by K.-H. Hellwege and A.M. Hellwege (Springer, Berlin, 1984), New Series, Group III, Vol. 18, p. 201.

127. *Landolt-Börnstein: Numerical Data and Functional Relationships in Science and Technology*, edited by K.-H. Hellwege and A.M. Hellwege (Springer, Berlin, 1984), New Series, Group III, Vol. 18, p. 211.

128. *Landolt-Börnstein: Numerical Data and Functional Relationships in Science and Technology*, edited by O. Madelung, M. Scholz and H. Weiss (Springer, Berlin, 1992), New Series, Group III, Vol. 17a, pp. 149-159.

129. O.J. Gregory and E.E. Crisman, in *Integrated Circuits: Chemical and Physical Processing*, edited by P. Stroeve (American Chemical Society, Washington D.C., 1985), p. 178.

130. A. Reyes-Serrato, G. Soto, A. Gamietea and M.H. Farias, *J. Phys. Chem. Solids* **59**, 743 (1998).

131. T.D. Moustakas, R.J. Molnar and J.P. Dismukes, in *Proceedings of the First Symposium on III-V Nitride Materials and Processes*, edited by T.D. Moustakas, J.P. Dismukes, and S.J. Pearton (Electrochem. Soc, Pennington, 1996), pp. 197-204.

132. C.C. Wang, S.A. Akbar, W. Chen and V.D. Patton, *J. Mater. Sci.* **30**, 1627 (1995).

133. D. Gregusova, T. Lalinsky, Z. Mozolova, J. Breza and P. Vogrincic, *J. Mater. Sci.* **4**, 197 (1993).

134. J.M. Honig and L.L. Van Zandt, in *1975 Annual Review of Materials Science*, vol. 5, edited by R.A. Huggins (Annual Reviews, Palo Alto, 1975), pp.225-278.

135. *Landolt-Börnstein: Numerical Data and Functional Relationships in Science and Technology*, edited by O. Madelung (Springer, Berlin, 1984), New Series, Group III, Vol. 17g, pp. 321-322.

136. P.G. Steeneken, L.H. Tjeng, I. Elfimov, G.A. Sawatzky, G. Ghiringhelli, N.B. Brookes and D.-J. Huang, *Phys. Rev. Lett.* **88**, 047201 (2002).

137. J. Lettieri, V. Vaithyanathan, S.K. Eah, J. Stephens, V. Sih, D.D. Awschalom, J. Levy and D.G. Schlom, *Appl. Phys. Lett.* **83**, 975 (2003).

138. Y.G. Sukharev, I.L. Akulyushin, V.S. Mironov, A.V. Andriyanov and V.V. Zherevchuk, *Izv. Akad. Nauk SSSR, Neorg. Mater.* **30**, 556 (1994) [Inorg. Mater. (USSR) **30**, 519 (1994)].

139. A.J. Raffalovich, *Capacitor Characteristics of Anodized Thin-Film Hafnium*, Report No. ECOM-2758 (U.S. Army Electronics Command, Fort Monmouth, 1966). (NTIS ID No. AD-641 338). The thickness of the HfO_2 layer was not directly measured in this work. The anodically-formed HfO_2 layer was assumed to be 50% dense (without a confirming measurement) in the analysis leading to the reported value of K. The accuracy of the reported K is thus poor. With the true film density unknown, these measurements only bound K in the range $34 \leq K \leq 92$ (as the unknown film density ranges from 0% to 100% dense, respectively). Another work on anodically-formed HfO_2 films (in which $K = 40$ is reported) is F. Huber, W. Witt, and W.Y. Pan, in *Transactions of the Third International Vacuum Congress*, edited by H. Adam (Pergamon, Oxford, 1967) Vol. 2, Part II, pp. 359-361. These HfO_2 films made by anodic oxidation could contain water, which would function to increase the measured K over the intrinsic value for HfO_2.

140. S-G. Lim, S. Kriventsov, T.N. Jackson, J.H. Haeni, D.G. Schlom, A.M. Balbashov, R. Uecker, P. Reiche, J.L. Freeouf and G. Lucovsky, *J. Appl. Phys.* **91**, 4500 (2002).

141. *Landolt-Börnstein: Numerical Data and Functional Relationships in Science and Technology*, edited by K.-H. Hellwege and A.M. Hellwege (Springer, Berlin, 1975), New Series, Group III, Vol. 7b, pp. 67-144.

142. W.D. Kingery, H.K. Bowen and D.R. Uhlmann, *Introduction to Ceramics*, 2nd Ed. (John Wiley & Sons, New York, 1976), p. 240.

143. D.B. Fenner, A.M. Viano, D.K. Fork, G.A.N. Connell, J.B. Boyce, F.A. Ponce and J.C. Tramontana, *J. Appl. Phys.* **69**, 2176 (1991).

144. S.H. Rou, T. M. Graettinger, A.F. Chow, C.N. Soble, II, D.J. Lichtenwalner, O. Auciello and A.I. Kingon in *Ferroelectric Thin Films II*, edited by A.I. Kingon, E.R. Myers and B. Tuttle (Mater. Res. Soc. Proc. **243**, Pittsburgh, 1992), p. 81.

145. S. Stemmer, J.-P. Maria and A.I. Kingon, *Appl. Phys. Lett.* **79**, 102 (2001).

146. S. Stemmer, Z. Chen, R. Keding, J.-P. Maria, D. Wicaksana and A.I. Kingon, *J. Appl. Phys.* **92**, 82 (2002).

147. R.T. DeHoff, *Thermodynamics in Materials Science* (New York, McGraw-Hill, 1993), pp. 332-336.

148. J.J. Lander and J. Morrison, *J. Appl. Phys.* **33**, 2089 (1962).

149. F.W. Smith and G. Ghidini, *J. Electrochem. Soc.* **129**, 1300 (1982).

150. D. Starodub, E.P. Gusev, E. Garfunkel and T. Gustafsson, *Surf. Rev. Lett.* **6**, 45 (1999).

151. H. Behner, J. Wecker and B. Heines in *High T_c Superconductor Thin Films*, edited by L. Correra (North-Holland, Amsterdam, 1992), p. 623.

152. E.V. Pechen, R. Schoenberger, B. Brunner, S. Ritzinger, K.F. Renk, M.V. Sidorov and S.R. Oktyabrsky, *J. Appl. Phys.* **74**, 3614 (1993).

153. I. Golecki, H.M. Manasevit, L.A. Moudy, J.J. Yang, and J.E. Mee, Appl. Phys. Lett. **42**, 501 (1983); H.M. Manasevit, I. Golecki, L.A. Moudy, J.J. Yang and J.E. Mee, *J. Electrochem. Soc.* **130**, 1752 (1983); A.L. Lin and I. Golecki, *J. Electrochem. Soc.* **132**, 239 (1985).

154. P. Legagneux, G. Garry, D. Dieumegard, C. Schwebel, C. Pellet, G. Gautherin and J. Siejka, *Appl. Phys. Lett.* **53**, 1506 (1988).

76

155. H. Fukumoto, T. Imura and Y. Osaka, *Jpn. J. Appl. Phys.* **27**, L1404 (1988); H. Fukumoto, M. Yamamoto, Y. Osaka and F. Nishiyama, *J. Appl. Phys.* **67**, 2447 (1990); H. Fukumoto, M. Yamamoto and Y. Osaka, *J. Appl. Phys.* **69**, 8130 (1991).

156. D.K. Fork, D.B. Fenner, G.A.N. Connell, J.M. Phillips and T.H. Geballe, *Appl. Phys. Lett.* **57**, 1137 (1990); D.K. Fork, D.B. Fenner, R.W. Barton, J.M. Phillips, G.A.N. Connell, J.B. Boyce, and T.H. Geballe, *Appl. Phys. Lett.* **57**, 1161 (1990); D.K. Fork, F.A. Ponce, J.C. Tramontana, N. Newman, J.M. Phillips and T.H. Geballe, *Appl. Phys. Lett.* **58**, 2432 (1991).

157. W. Prusseit, S. Corsépius, M. Zwerger, P. Berberich, H. Kinder, O. Eibl, C. Jaekel, U. Breuer and H. Kurz, *Physica C* **201**, 249 (1992).

158. Y. Kado and Y. Arita in *Extended Abstracts of the 21st Conference on Solid State Devices and Materials* (Tokyo, 1989), p. 45.

159. H.M. Manasevit and D.H. Forbes, *J. Appl. Phys.* **37**, 734 (1966).

160. H. Seiter and Ch. Zaminer, *Z. Angew. Phys.* **20**, 158 (1965).

161. M. Ihara, Y. Arimoto, M. Jifuku, T. Kimura, S. Kodama, H. Yamawaki and T. Yamaoka, *J. Electrochem. Soc.* **129**, 2569 (1982); M. Ihara, *Microelectron. Eng.* **1**, 161 (1983).

162. E.S. Machlin and P. Chaudhari, in *Synthesis and Properties of Metastable Phases*, edited by E.S. Machlin and T.J. Rowland (The Metallurgical Society of AIME, Warrendale, 1980), pp. 11-29.

163. C.P. Flynn, *Phys. Rev. Lett.* **57**, 599 (1986).

164. R. Bruinsma and A. Zangwill, *J. Physique* **47**, 2055 (1986).

165. *Landolt-Börnstein: Numerical Data and Functional Relationships in Science and Technology*, edited by K.-H. Hellwege and A.M. Hellwege (Springer, Berlin, 1975-1985), New Series, Group III, Vol. 7.

166. For SrLaAlO$_4$, the lattice constants in Ref. 165 are significantly different than several recent measurements on single crystals. Therefore, the more accurate lattice constants from the following reference were used: A. Gloubokov, R. Jablonski, W. Ryba-Romanowski, J. Sass, A. Pajaczkowska, R. Uecker, and P. Reiche, *J. Cryst. Growth* **147**, 123 (1995).

167. G.D. Wilk and R.M. Wallace, *Appl. Phys. Lett.* **74**, 2854 (1999); G.D. Wilk and R.M. Wallace, *Appl. Phys. Lett.* **76**, 112 (2000).

168. K.F. Young and H.P.R. Frederikse, *J. Phys. Chem. Ref. Data* **2**, 313 (1973).

169. I.-S. Kim, T. Nakamura and M. Itoh, *J. Ceram. Soc. Jpn Int. Ed.* **101**, 779 (1993).

170. D. Balz and K. Plieth, *Z. Elektrochem.* **59**, 545 (1955).

171. S.N. Ruddlesden and P. Popper, *Acta Cryst.* **10**, 538 (1957); S.N. Ruddlesden and P. Popper, Acta Cryst. **11**, 54 (1958).

172. K. Lukaszewicz, *Anal. Chem.* **70**, 320 (1958); K. Lukaszewicz, *Rocz. Chem.* **33**, 239 (1959).

173. J.M. Longo and P.M. Raccah, *J. Solid State Chem.* **6**, 526 (1973).

174. J. Fava and G. Le Flem, *Mater. Res. Bull.* **10**, 75 (1975).

175. J.H. Haeni, C.D. Theis, D.G. Schlom, W. Tian, X.Q. Pan, H. Chang, I. Takeuchi and X.-D. Xiang, *Appl. Phys. Lett.* **78**, 3292 (2001).

176. W. Tian, X.Q. Pan, J.H. Haeni and D.G. Schlom, *J. Mater. Res.* **16**, 2013 (2001).

177. W. Tian, J.H. Haeni, E. Hutchinson, B.L. Sheu, M.A. Zurbuchen, M.M. Rosario, X.Q. Pan, P. Schiffer, Y. Liu and D.G. Schlom (*unpublished*).

178. F.D. Bloss, *Crystallography and Crystal Chemistry: An Introduction* (Mineralogical Society of America, Washington, D.C., 1994), pp. 230-235.

179. *Landolt-Börnstein: Numerical Data and Functional Relationships in Science and Technology*, edited by K.-H. Hellwege and A.M. Hellwege (Springer, Berlin, 1978), New Series, Group III, Vol. 12a, pp. 126-206.

180. J. Robertson, *MRS Bull.* **27**, 217 (2002)

181. T. Kaneyoshi, *Prog. Theor. Phys.* **41**, 577 (1969).

182. F. Claro, *Phys. Rev. B* **25**, 2483 (1982).

183. P. Sheng and Z. Chen, *Phys. Rev. Lett.* **60**, 227 (1988); Z. Chen and P. Sheng, *Phys. Rev. B* **43**, 5735 (1991).

184. B.-E. Park and H. Ishiwara, *Appl. Phys. Lett.* **79**, 806 (2001).

185. T.E. Harrington, J. Wosik, and S.A. Long, *IEEE Trans. Appl. Supercond.* **7**, 1861 (1997).

186. R. Schwab, R. Spörl, P. Severloh, R. Heidinger, and J. Halbritter in *Applied Superconductivity 1997: Proceedings of EUCAS 1997 Third European Conference on Applied Superconductivity*, edited by H. Rogalla and D.H.A. Blank, IOP Conf. Ser. No. 158 (Institute of Physics, Bristol, 1997), pp. 61-64.

187. C.J. Fennie and K.M. Rabe (unpublished).

188. P. Delugas, V. Fiorentini and A. Filippetti (unpublished).

189. S. Roberts, *Phys. Rev.* **77**, 258 (1950).

190. *The International Technology Roadmap for Semiconductors: 1999* (Semiconductor Industry Association, San Jose, CA, 1999), pp. 241-268.

191. G.A. Samara, *J. Appl. Phys.* **68**, 4214 (1990).

192. E.A. Giess, R.L. Sandstrom, W.J. Gallagher, A. Gupta, S.L. Shinde, R.F. Cook, E.I. Cooper, E.J.M. O'Sullivan, J.M. Roldan, A.P. Segmüller and J. Angilello, *IBM J. Res. Develop.* **34**, 916 (1990).

193. T. Konaka, M. Sato, H. Asano and S. Kubo, *J. Supercond.* **4**, 283 (1991).

194. J. Konopka and I. Wolff, *IEEE Trans. Microwave Theory Tech.* **40**, 2418 (1992).

195. J. Krupka, R.G. Geyer, M. Kuhn and J.H. Hinken, *IEEE Trans. Microwave Theory Tech.* **42**, 1886 (1994).

196. C. Zuccaro, I. Ghosh, K. Urban, N. Klein, S. Penn and N. McN. Alford, *IEEE Trans. Appl. Supercond.* **7**, 3715 (1997).

197. J.W. Matthews in *Epitaxial Growth*, Part B, edited by J.W. Matthews (Academic Press, New York, 1975), pp. 559-609.

198. *Phase Diagrams for Ceramists*, Vol. 6, edited by R.S. Roth, J.R. Dennis and H.F. McMurdie (American Ceramic Society, Westerville, 1987) p. 144 (Fig. 6438).

199. H. Fay and C.D. Brandle, *J. Phys. Chem. Solids* **51** (supplement 1), 51 (1967).

200. L.F. Edge, D.G. Schlom, S.A. Chambers, E. Cicerrella, J.L. Freeouf, B. Holländer and J. Schubert, *Appl. Phys. Lett.* **84**, 726 (2004).

201. V.V. Afanas'ev, A. Stesmans, C. Zhao, M. Caymax, T. Heeg, J. Schubert, Y. Jia, D.G. Schlom and G. Lucovsky, *Appl. Phys. Lett.* **85**, 5917 (2004).

202. L.F. Edge, V. Vaithyanathan, D.G. Schlom, R.T. Brewer, S. Rivillon, Y.J. Chabal, M.P. Agustin, Y. Yang, S. Stemmer, H.S. Craft, J-P. Maria, M.E. Hawley, B. Holländer, J. Schubert and K. Eisenbeiser (*unpublished*).

203. P. Sivasubramani, M.J. Kim, B.E. Gnade, R.M. Wallace, L.F. Edge, D.G. Schlom, H.S. Craft and J-P. Maria (*unpublished*).

204. L.F. Edge, D.G. Schlom, R.T. Brewer, Y.J. Chabal, J.R. Williams, S.A. Chambers, C. Hinkle, G. Lucovsky, Y. Yang, S. Stemmer, M. Copel, B. Holländer and J. Schubert, *Appl. Phys. Lett.* **84**, 4629 (2004).

205. C. Zhao, T. Witters, B. Brijs, H. Bender, O. Richard, M. Caymax, T. Heeg, J. Schubert, V. V. Afanas'ev, A. Stesmans and D.G. Schlom (*to be published in Appl. Phys. Lett.*).

206. K.L. Ovanesyan, A.G. Petrosyan, G.O. Shirinyan, C. Pedrini and L. Zhang, *Opt. Mater.* **10**, 291 (1998).

207. From Ref. 45 the average thermal expansion coefficients of LaLuO$_3$ between room temperature and 1000 °C along the orthorhombic axis of the unit cell are $\alpha_a \approx 6.3 \times 10^{-6}$ K^{-1}, $\alpha_b \approx 2.8 \times 10^{-6}$ K^{-1}, and $\alpha_c \approx 3.7 \times 10^{-6}$ K^{-1}.

78

[208.] B.C. Chakoumakos, D.G. Schlom, M. Urbanik and J. Luine, *J. Appl. Phys.* **83**, 1979 (1998).

[209.] *Landolt-Börnstein: Numerical Data and Functional Relationships in Science and Technology*, edited by O. Madelung (Springer, Berlin, 1987), New Series, Group III, Vol. 22a, p. 18.

[210.] S. Stemmer and D.G. Schlom, "Experimental Investigations of the Stability of Candidate Materials for High-*K* Gate Dielectrics in Silicon-Based MOSFETs," in: *Nano and Giga Challenges in Microelectronics*, edited by J. Greer, A. Korkin, and J. Labanowski (Elsevier, Amsterdam, 2003), pp. 129-150.

[211.] D.O. Klenov, D.G. Schlom, H. Li and S. Stemmer (*unpublished*).

[212.] R. Droopad, Z. Yu, J. Ramdani, L. Hilt, J. Curless, C. Overgaard, J.L. Edwards, J. Finder, K. Eisenbeiser, J. Wang, V. Kaushik, B-Y. Ngyuen and B. Ooms, *J. Cryst. Growth* **227-228**, 936 (2001).

[213.] Z.J. Yu, C.D. Overgaard, R. Droopad, J.K. Abrokwah and J.A. Hallmark, *US Patent No. 6,110,840* (August 29, 2000); Z.J. Yu, J.A. Hallmark, J.K. Abrokwah, C.D. Overgaard and R. Droopad, *US Patent No. 6,113,690* (September 5, 2000); Z. Yu, R. Droopad, C.D. Overgaard, J. Ramdani, J.A. Curless, J.A. Hallmark, W.J. Ooms and J. Wang, *US Patent No. 6,224,669* (May 1, 2001).

[214.] H. Li, X. Hu, Y. Wei, Z. Yu, X. Zhang, R. Droopad, A.A. Demkov, J. Edwards, Jr., K. Moore, W. Ooms, J. Kulik and P. Fejes, *submitted to J. Appl. Phys.*

[215.] *Landolt-Börnstein: Numerical Data and Functional Relationships in Science and Technology*, edited by D.L. Beke (Springer, Berlin, 1999), New Series, Group III, Vol. 33B1, pp. 4-15 and 4-29.

[216.] J-M. Badie, *High Temp. - High Press.* **2**, 309 (1970).

[217.] L. Vegard, *Z. Phys.* **5**, 17 (1921); L. Vegard, *Z. Krist.* **67**, 239 (1928).

[218.] *Landolt-Börnstein: Numerical Data and Functional Relationships in Science and Technology*, edited by K.-H. Hellwege and A.M. Hellwege (Springer, Berlin, 1980), New Series, Group III, Vol. 7d2, p. 179.

[219.] *Landolt-Börnstein: Numerical Data and Functional Relationships in Science and Technology*, edited by K.-H. Hellwege and A.M. Hellwege (Springer, Berlin, 1976), New Series, Group III, Vol. 7e, pp. 11, 12, 40, and 273.

[220.] R.F. Belt and R. Uhrin, *J. Cryst. Growth* **70**, 471 (1984).

[221.] I.-S. Kim, H. Kawaji, M. Itoh and T. Nakamura, *Mater. Res. Bull.* **27**, 1193 (1992).

[222.] H.M. Christen, I. Ohkubo, G.E. Jellison, Jr., C.M. Rouleau, D.H. Lowndes, S. Huang, M.E. Reeves, W. Tian, D.G. Schlom, Y. Chen and X.Q. Pan (*unpublished*).

CHAPTER 3

SCIENCE AND TECHNOLOGY OF HIGH-DIELECTRIC CONSTANT (K) THIN FILMS FOR NEXT GENERATION CMOS

Robert M. Wallace[1] and Orlando Auciello[2]

[1]University of Texas at Dallas, Department of Electronic Engineering
[2]Argonne National Laboratory, Materials Science Division

3.1 INTRODUCTION

The microelectronic revolution of the 20th Century has been based on silicon because of its exceptional electrical properties and an innate superior insulator, SiO_2, which enabled the integrated circuit (IC) technology for the microchip. Complementary metal-oxide semiconductor (CMOS) devices, the cornerstone of the Si-IC technology, have been continuously scaled down in the march towards nanoscale logic devices, to achieve faster, cheaper, energy efficient microelectronic systems, and to form the bases for FLASH, DRAM, and EEPROM memories that sustain the current multibillion-dollar memory market. However, the new generation of Si-based nanoscale CMOS devices will clearly need novel paradigms in materials, materials integration, processing, and device architecture beyond the pure Si-based current technology.

The technology roadmap[1] for the next generation of complementary metal-oxide semiconductor (CMOS) devices indicates that an equivalent oxide thickness (EOT) of less than 1.0 nm is needed to satisfy the requirement of maintaining a suitable capacitance when the gate length, and thus the area (A) of the CMOS gate, is reduced below 65 nm, as shown by the simple parallel plate capacitor equation $C = \varepsilon A/t$, where t=thickness of the gate dielectric layer and ε the relative permittivity of the dielectric. As the gate length (area) is scaled, the leakage current though the CMOS gate dielectric will be too high when the physical thickness of the SiO_2 or

SiON layer reaches ~1.0nm for some important planar CMOS applications –particularly low power applications. The alternative is to replace SiO_2 or SiON by a dielectric material with higher dielectric constant (K) that permits retention of gate capacitance (drive current) with a physical thickness that will yield a functional CMOS device. For this reason is critical to develop a high-K material to replace SiO_2 in the next generation of CMOS devices as described in the recent literature.[2,3] The development of the next generation of nanoscale CMOS will impact not only logic devices but also the next generation of memories, including non-volatile high-density memories that will drive not only the stand along, but also the embedded memory markets.

Research on high-K dielectric layers requires consideration of a number of critical material properties that affect the electrical performance of the CMOS device.[4] In addition, not only the high-K dielectric layer needs to be considered but the whole gate structure, including the gate electrode and the high-K/Si interface. Critical issues that need to be addressed include: (1) dielectric constant, (2) band structure and associated band offset that affect the carrier transport, (3) thermodynamic stability of the high-K/Si interface up to about 1000 °C, (4) interface quality, (5) film morphology and microstructure, (6) gate electrode compatibility, (7) process compatibility, (8) reliability, and obviously (9) fabrication cost.

There are efforts underway across the globe on these issues, with new reports appearing *weekly*. A key consideration for the timely technological insertion of dielectrics (or any other material) for CMOS devices must include the consideration of the *integration phase space* – the combination of the required processing variables required to produce a useful device.[5] These include consideration of thermal budgets, etching processes, constituent interdiffusion, interfaces and many others. All of which require understanding of the critical issues described above for the next generation of CMOS gates. The need for dielectric films employed in planar CMOS device fabrication to endure the 1000 °C anneals associated with source/drain dopant activation presents a particularly formidable challenge for virtually all alternative dielectric materials. Future process developments in other portions of the transistor structure would be expected to impact the integration phase space for such films dramatically.[*] Based on the discussion presented above, this chapter is divided in the following sections:

2. Materials Property Requirements (for Planar CMOS structures); Permittivity and Barrier Height; Mobility, Interfaces and thermal stability; Dielectric film morphology and microstructure; Process Integration (Compatibility with future CMOS and Metal Gates); Reliability issues.

---

[*] We note that this chapter incorporates material from previous reviews published by one of the co-authors (RMW). See refs. [2- 5].

3. Thin Film Synthesis and Characterization: Integrated film synthesis / *in situ* characterization; *Ex situ* film synthesis and characterization; Composition,
4.. Materials Candidates; State of the art Hf-based dielectrics; New Materials
Ti and/or Al-based dielectrics;
5..Conclusions

3.2 MATERIALS PROPERTY REQUIREMENTS (for planar CMOS structures)

3.2.1 Permittivity and barrier height

For non-metallic (insulating) solids, there are two main contributions of interest to the dielectric constant which give rise to the polarizability in CMOS device applications: electronic and ionic dipoles. Figure 1 illustrates the frequency ranges where each contribution dominates.[6]

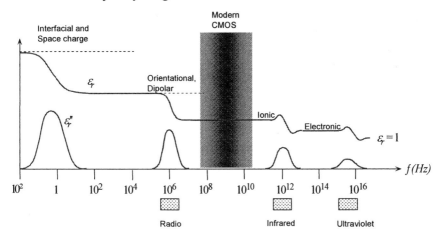

Figure 1: Schematic of the frequency dependence of the real and imaginary parts of the dielectric permittivit . Polarization mechanisms and the region of interest for modern CMOS devices is also noted. [Ref. 6]

In general, atoms with a large ionic radius (e.g. high atomic number) exhibit more electron dipole response to an external electric field, because there are more electrons to respond to the field (electron screening effects also play a role in this response). This electronic contribution tends to increase the permittivity for higher atomic number atoms.

The ionic contribution to the permittivity can be much larger than the electronic portion in cases such as in perovskite crystals of $BaSr_xTi_{1-x}O_3$ (BST), $PbZr_xTi_{1-x}O_3$ (PZT), and $SrBi_2Ta_2O_9$ (SBT), which exhibit ferroelectric behavior below the Curie point, the first of them when in the

Sr-poor ferroelectric phase. In these cases, Ti ions in unit cells throughout the crystal are uniformly displaced in response to an applied electric field (for the case of ferroelectric materials, the Ti ions reside in one of two stable, non-isosymmetric positions about the center of the Ti-O octahedra). This displacement of Ti ions causes an enormous polarization in the material, and thus can give rise to very large dielectric constants in bulk films of 2000 – 3000. Because of the high-dielectric constant, BST films in the paraelectric phase have been considered for DRAM capacitor applications.[7,8,] On the other hand, capacitors based on PZT or SBT ferroelectric films, which exhibit high remnant polarization at electric fields (thus voltage) zero have already been incorporated in non-volatile ferroelectric random access (FeRAMS) memories[9,10]. Since ions respond more slowly than electrons to an applied field, the ionic contribution begins to decrease at very high frequencies, in the infrared range of $\sim 10^{12}$ Hz, as shown in Fig. 1.

For many simple oxides, permittivities have been measured on bulk samples and (in some cases) thin films, but for the more complex materials (i.e. those with more elemental constituents), the dielectric constants may not be as well known. Shannon used a method involving the Classius-Mossoti equation for calculating ionic polarizabilities as a means to make predictions of permittivities for many dielectrics. Good agreement has been found for some materials, but there are also many cases of poor agreement between calculated and measured values. This discrepancy between calculation and measurement can be attributed to many factors, including film thickness, method of film deposition, and local electronic structure within the dielectrics. It is clear that much more experimental data is needed for measurements of dielectric constant for these high-K dielectrics, particularly below the 100 Å thickness regime.

In contrast to the general trend of increasing permittivity with increasing atomic number for a given cation in a metal oxide, the band gap E_G of the metal oxides tends to *decrease* with increasing atomic number, particularly within a particular group in the periodic table. Fig. 2 provides a comparison of this dependence for a number of oxides under investigation for the near next generation of nanoscale CMOS gates[11]. An intuitive explanation for this phenomenon is that the corresponding bonding and anti-bonding orbitals of the metal and oxygen atoms form a valence band and a conduction band, respectively. For the case of SiO_2, the σ bonds formed by the sp hybrid orbitals (which arise from Si s,p and O p orbitals) have a σ bonding orbital energy level and a higher σ^* anti-bonding orbital energy level. The energy separation between these levels defines a band gap. For the simple case of SiO_2, where there are only s and p electron orbitals that are all filled during bonding, the oxygen electron lone pair energy level actually defines the valence band maximum (rather than the σ bonding energy level).

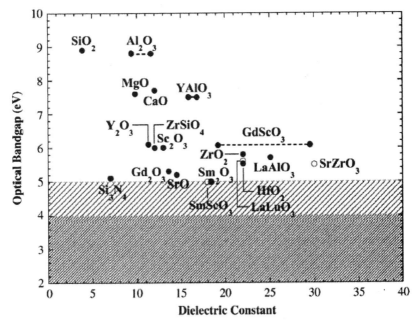

Figure 2: Plot of optical bandgap and dielectric constant for a number of oxide materials showing an inverse relationship between these properties. [Ref. 11]

The result is the observed band gap $E_G \sim 9$ eV for SiO_2. If the σ and σ* bonds defined the valence band maximum and conduction band minimum, respectively, then the band gap of SiO_2 would be larger.

Thus for the transition metal oxides, which all have five *d* electron orbitals and other non-bonding *p* orbitals, the band gaps of these oxides can be significantly decreased by the presence of partially-filled *d* orbitals, which have available states for electron occupancy. These orbital energy levels tend to lie within the gap defined by the σ and σ* orbitals. The *d* orbital levels which lie within the gap defined by σ and σ* levels are all therefore available for electron conduction at significantly lower energy levels than would be expected from σ and σ* alone. It is important to note that these partially filled and non-bonding levels are *not* the result of defects within the material, but rather are *intrinsic* to such atomic constituents where many orbitals are available for electron conduction.

This general band gap reduction for higher-K materials is a limitation that needs to be considered when selecting a suitable high-K gate dielectric. In the cases of candidate capacitor dielectrics such as Ta_2O_5 and TiO_2, both materials have small E_G values and correspondingly small ΔE_C values. These small ΔE_C values directly correlate with high leakage currents for both materials, making pure Ta_2O_5 and TiO_2 unlikely choices for gate dielectric applications.

Table 1 presents a compilation of recent data on a number of oxide materials including oxides,[11-16] silicates,[17-19] and aluminates.[20-22] Among the materials shown in Table 1, ZrO_2, HfO_2 and La_2O_3 offer relatively high values for both K and E_G. It is important to note that most of the high-K metal oxides listed in Table I, which includes those studied for gate dielectric applications, crystallize at relatively low temperature (T~500-600°C). An exception is Al_2O_3 where crystallization in thin films is observed at T~800°C.[23]

Table 1: Comparison of relevant properties for selected high-κ candidates. (mono. = monoclinic; tetrag. = tetragonal) Data from: (a) ref. 16, (b) refs. 12,13,14 (c) ref. 22 , (d) ref. 20 (e) ref. 21 (f) ref. 20 (g) ref. 22, (h) ref. 12, (i) ref. 19, (j) ref. 18 (k) ref. 16 (l) Crystallization of thin films (γ-Al_2O_3 phase) has been recently reported in ref. 24.

Material	Dielectric Constant (κ)	Band Gap E_G (eV)	ΔE_C (eV) to Si	Crystal Structure(s) (400-1050°C)
SiO_2	3.9	8.9-9.0	3.2-3.5[b]	amorphous
Si_3N_4	7	4.8[a]-5.3	2.4[b]	amorphous
Al_2O_3	9	6.7[h]-8.7	2.1[a]-2.8[b]	amorphous[l]
Y_2O_3	11[d]-15	5.6-6.1[d]	2.3[b]	cubic
Sc_2O_3	13[d]	6.0[d]		cubic
ZrO_2	22[d]	5.5[a]-5.8[d]	1.2[a]-1.4[b]	mono., tetrag., cubic
HfO_2	22[d]	5.5[d]-6.0	1.5[b]-1.9[c]	mono., tetrag., cubic
La_2O_3	30	6.0	2.3[b]	hexagonal, cubic
Ta_2O_5	26	4.6[a]	0.3[a,b]	orthorhombic
TiO_2	80	3.05-3.3	~0.05[b]	tetrag. (rutile, anatase)
$ZrSiO_4$	12[d]	6[d]-6.5	1.5[b]	tetrag.
$HfSiO_4$	12	6.5	1.5[b]	tetrag.
$YAlO_3$	16-17[d]	7.5[d]		*
$HfAlO_3$	10[e]-18[g]	5.5-6.4[f]	2-2.3[f]	*
$LaAlO_3$	25[d]	5.7[d]		*
$SrZrO_3$	30[d]	5.5[d]		*
$HfSiON$	12-17[i,j]	6.9[k]	2.9[k]	amorphous

For gate dielectric applications, however, the required permittivity must be balanced against the barrier height for the tunneling process, since electron tunneling directly relates to leakage current. Consider the band diagram shown in Fig. 3 where the electron affinity (χ) and gate electrode work-function (Φ_M) are defined. For electrons traveling from the Si substrate to the gate, this barrier is the conduction band offset, $\Delta E_C \cong q(\chi - (\Phi_M - \Phi_B))$; for electrons traveling from the gate to the Si substrate, this barrier is Φ_B. For highly defective films that have electron trap energy levels in the insulator band gap, electron transport will instead be governed by a trap-assisted mechanism such as Frenkel-Poole emission or hopping conduction.

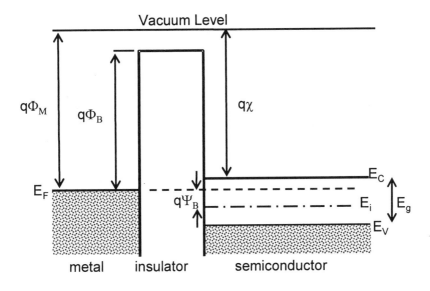

Figure 3: Schematic band diagram for MIS device structure.

In order to obtain low leakage currents, it is desirable to find a gate dielectric that has a large ΔE_C value to Si and to other gate metals that may be required. Reported values of ΔE_C for most dielectric-Si systems are scarce in the literature, but calculations indicate that some of the high-κ capacitor metal oxide and complex oxide materials, such as Ta_2O_5, TiO_2, and $SrTiO_3$, will have ΔE_C ⬜ 0.5 eV on Si, limiting their utility for gate dielectric applications with operational temperature ranges up to ~120°C.[12] For several high-K dielectrics that are currently under consideration, Robertson found that ΔE_C ~ 2.8 eV for Al_2O_3, ΔE_C ~ 2.3 eV for La_2O_3 and Y_2O_3, and ΔE_C ~ 1.5 eV for ZrO_2 and $ZrSiO_4$.[13,14] Recently, photoelectron spectroscopy measurements have provided valuable data for the Ta_2O_5, Al_2O_3 and ZrO_2 materials systems.[16] Good agreement with calculations was observed for thin films of Ta_2O_5 (2.8 nm: 0.45 eV), Al_2O_3 (5.3 nm: 2.08 eV) and ZrO_2 (3.2 nm: 1.23 eV), although values lower than 1.5eV raises concern of an insufficient energy barrier height to gate leakage (tunneling). It is also noted by Miyazaki that consideration of the film thickness dependence of the associated photoelectron energy loss features is required in the course of these measurements, particularly when comparing thin film and bulk results. Other similar photoelectron measurements for the ZrO_2 (2.6nm)/Zr-silicate (0.9 nm)/Si MIS structure (produced by pulsed laser-ablation and sputter deposition methods) indicate somewhat lower barriers (~0.8 − 1.0 eV).[25]Very recent internal photoemission work indicates that the conduction band offset for HfO_2 is 2.0 eV (regardless of the interfacial layer present),[26] in contrast with X-ray photoemission results (1.2 eV).[27] It is noted in these reports, however, that the uncertainty in the bandgap of HfO$_2$ plays an important role in the extraction of these barrier measurements.

These recent calculations and experimental measurements provide important insight into relevant barrier height (such as ΔE_C) values for many candidate dielectrics. If the experimental ΔE_C values for these oxides are even much less than 1.0 eV, it will likely preclude using these oxides in gate dielectric applications, since electron transport (from enhanced Schottky emission, thermal emission or tunneling) would lead to unacceptably high leakage currents. A large band gap E_G generally corresponds to a large ΔE_C (see Table 1), but depends upon the band offset symmetry between the valence band offset ΔE_V as well.

As noted above, it is extremely difficult to achieve the juxtaposition of these high-κ dielectrics on Si without an interfacial layer, as an SiO$_2$-like interface usually forms. This interface layer will of course alter the ΔE_C value of the system, and must be taken into consideration when comparing measured and calculated results.

Epitaxial control of the interfacial layer has been demonstrated for high-κ epitaxial SrTiO$_3$ (using molecular beam epitaxy ultrahigh vacuum deposition methods) for capacitors[28,29] and transistors.[30,31] In recent work, Jeon, et al.[29] demonstrated a Pt/SrTiO$_3$/BaSrO/Si capacitor structure with $t_{ox} = 0.54$ nm with low leakage current (7.4×10^{-4} A/cm^2 at V_{FB}-1 V) at room temperature. Reductions in the interface state density (D_{it}) for the as-deposited films ($D_{it} \sim 1 \times 10^{12}$/eVcm^2) with forming gas anneals at 450°C were reported ($D_{it} \sim 1.3 \times 10^{11}$/eVcm^2). A low conduction band offset of 0.23 eV was estimated for this structure using a conduction model and no transistor mobility results were reported.

Again, Table I shows a list of several metal oxide (and nitride) systems, some of which have been investigated as gate dielectrics, with comparison of values for K and E_G, along with ΔE_c values on Si. For these high-K materials, Table I indicates that E_G will be somewhat limited, since it can be seen that the dielectric constant and band gap of a given material generally exhibit an inverse relationship (although some materials have significant departures from this trend).

Although many researchers originally assumed that selecting a dielectric with K > 25 would be necessary to replace SiO$_2$, the more relevant consideration is whether the desired device performance (i.e. drive current) can be obtained without producing unacceptable off-state (leakage) currents, mobility degradation and reliability characteristics –a typical integration problem where compromises in the various materials properties must be carefully considered. It is therefore also important to consider dielectrics which provide even a moderate increase in K over SiO$_2$

(K ~ 12 – 20), but which also produces a sufficiently high tunneling barrier and high-quality interface to Si.

3.2.2 Mobility, interfaces and thermal stability

In the case of many of the alternate gate dielectric materials (mainly metal oxides) currently under consideration, the polarizability of the metal-oxygen ("ionic") bond is responsible for the observed low-frequency permittivity enhancement. Such highly polarizable bonds are described to be "soft" relative to the less polarizable, "stiff" Si-O bonds associated with SiO_2. Unlike the stiff Si-O bond, the polarization frequency dependence of the M-O bond is predicted to result in an enhanced scattering coupling strength for electrons with the associated low-energy and surface optical phonons.[32] This scattering mechanism can therefore degrade the electron mobility in the inversion layer associated with MOSFET devices.

A theoretical examination of this effect is provided by Fischetti, *et al*[32]. where calculations of the magnitude of the effect indicate that pure metal-oxide systems, such as ZrO_2 and HfO_2, suffer the worst degradation, whereas materials that incorporate Si-O bonds, such as silicates, fare better. In that work[32], it is also noted that the presence of a thin SiO_2 interfacial layer between the Si substrate and the high-K dielectric can help boost the resultant mobility by screening this effect, although the maximum attainable effective mobility is still below that for the ideal SiO_2/Si system. These researchers further suggest that the effect is also minimized by the incorporation of Si-O in the dielectric, as in the case of pseudo-binary systems such as silicates.[32] Comparison of HfO_2 and Hf-silicate mobility studies indicate that nMOS and pMOS poly-Si gated devices with Hf-silicate dielectrics exhibit better mobilities (approaching those of SiO_2) over those obtained using HfO_2 due to a diminished soft-phonon scattering mechanism.[33]

We note that comparison of mobility data from various sources is problematic at this time. Technologies from even "simple" planar CMOS device structures have significant variations from company to company, let alone university to university, which results in substantial changes in electric fields in the channel region. This is compounded with the various mobility measurement methodologies as well and how the data is reported in the literature (electric field versus inversion charge density, for example). Recent work has shown a *very rough* indication for the mobility performance of planar CMOS transistors with poly-Si gates produced on Si(100) substrates at electric fields ranging from ~0.8-1.0 MV/cm with nitrided Hf-silicate gate dielectrics.[34,35] This work has shown electron (hole) mobilities ranging from ~70-80% (95%) of the universal SiO_2 mobility curve for nMOS (pMOS) devices compared to ~50-80% (90-

100%) for transistors incorporating HfO_2.[36] Additional treatments in forming gas[36] and the introduction of interfacial Si-oxynitride layers at the gate/dielectric[37] and dielectric/channel[36,38,39] region has also been investigated and appear to generally result in a similar range of mobility values. Improvements in the mobility have resulted from the growth of the high-κ layer on a thin SiO_2 layer where electron mobilities close to that of SiO_2 have been reported[40,41] and thus appear to be consistent with the model proposed by Fischetti, et al. discussed above.[32] Work by Datta, et al[42], focused on examining the effects of combining strain in the channel and metal gates for HfO_2 dielectrics (reported to be "engineered" to eliminate any fixed charge), suggest that the metal gate can be utilized to screen the phonon scattering mechanism, greatly improving device performance.[42] Theory has now been developed to study the effect for thin oxides as well[43] and the effects of screening the remote charge scattering for the case of poly-Si gates have also been studied.[44]

It has also been recently reported that mobility can be improved for transistors with HfO_2 gate dielectrics by careful orientation of the device channel and crystalline Si substrate.[45] Results suggest, for example, that nMOS devices should be fabricated on Si(100) surfaces and pMOS devices on Si(110) surfaces. For planar CMOS in the near term, the prospect of changing substrate orientation is likely to meet significant resistance in the industry. However, the possibility of future 3-dimensional transistor designs, where orientations of the Si single crystal regions play an important role, may well suggest interesting opportunities for the improvement of high-κ transistor mobility performance. Improvements in mobility through combinations of channel engineering (such as strained-Si, SiGe and Ge channels) and high-k materials have also been recently reported.[42,46,47,48,49,50]

There are other possible explanations for mobility degradation that are also currently under investigation. For example, it has also been suggested by Torii, et al.,[51] that fixed-charge-induced Coulomb scattering (at the poly-Si/Al_2O_3 interface), rather than soft-phonon coupling (with the Si substrate), is the cause for mobility degradation. The recent work by Ragnersson, et al. noted above also indicates that the mobility degradation can be attributed largely to Coulombic scattering from fixed charge originated at the poly-Si/HfO_2 interface.[52] Temperature dependent studies indicate that phonon scattering could not be completely ruled out, but the dominant scattering mechanism is proposed to be remote charge scattering. It should be noted, of course, that the field dependence of the mobility is important to comprehend in regard to relevant scattering mechanisms. For example, high electric field regimes, corresponding to high inversion charge densities, result in a higher sensitivity to surface (interface) roughness scattering mechanisms.

It was also previously suggested that interdiffusion of dielectric constituents with the Si channel could lead to mobility degradation through ionized impurity scattering.[2] Interdiffusion of Zr and Hf from silicate dielectrics has recently been examined where Hf silicates (and by extension, Hf-based dielectric materials) appears to offer superior stability over Zr-based dielectric materials in this regard.[53,54,55] These results will be discussed further below. Recent work by Guha, *et al.* appears to support a correlation of mobility degradation and the interdiffusion of Al with Si from Al_2O_3.[24]

Clearly, interdiffusion of the metallic constituents between the high-κ dielectric and the Si substrate (channel) is undesirable for transistors, and may well represent the *primary* cause for mobility degradation often observed for some high-κ dielectric materials. The apparent enhanced stability of Hf-based dielectrics, discussed below, over many other high-κ dielectrics may also account for the relatively better performance that is often observed in this regard.

A comparison of recent studies on the mobility degradation mechanism indicates that further work is required to sort out what is the dominant among charge scattering and/or phonon scattering mechanisms, for high-κ materials. An attempt at a "unified mobility model" for high-k dielectric gate stacks, where the relative contributions of mechanisms such as remote phonon scattering, fixed charge, interfacial dipoles, interface roughness and crystallization has been very recently described by Saito, *et al.* as a function of the effective electric field.[56]

For all gate dielectrics, the interface with the Si substrate plays a key role, and in most cases is the dominant factor in determining the overall electrical properties. Most of the high-κ metal oxide systems investigated thus far have unstable interfaces with Si: that is, they react with Si under equilibrium conditions to form an undesirable interfacial layer. These materials therefore require an interfacial reaction barrier, as mentioned previously. Any ultrathin interfacial reaction barrier with $t_{eq} < 20$ Å will have the same quality, uniformity and reliability concerns as SiO_2 does in this thickness regime. This is especially true when the interface plays a determining role in the resulting electrical properties. It is important to understand the thermodynamics of these systems, and thereby attempt to control the interface with Si.

As also noted above, similar concerns for the control of the dielectric interfacial layers with capacitor electrodes must also be addressed primarily to maintain the required memory cell capacitance with scaling.

The relative stability of a particular three-component system for device applications can be examined through consideration of the ternary phase diagrams shown schematically in Fig. 4.[57-59] An analysis of the Gibbs free energies governing the relevant chemical reactions for the Ta-O-Si and Ti-

O-Si ternary systems indicates that Ta_2O_5 and TiO_2 (or mixtures with Si), respectively, are not stable with respect to SiO_2 formation when placed next to Si. Rather, the tie lines show that Ta_2O_5 and TiO_2 on Si will tend to phase separate into SiO_2 and metal oxide (M_xO_y, M = metal), and possibly silicide (M_xSi_y) phases. This instability resulting in SiO_2 formation has been observed experimentally for both of these metal oxides, which leads to the necessity for an additional thin interfacial barrier layer to minimize the undesirable reactions with the silicon substrate.

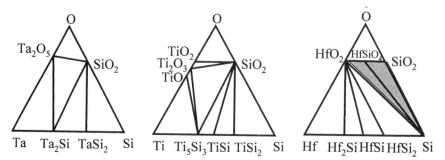

Figure 4: Schematic ternary phase diagrams for Ta-O-Si, Ti-O-Si and Hf-O-Si systems. [Refs. 2, 57,59]

In contrast to the Ta and Ti systems, the tie lines expected for the phase diagram for the Hf-O-Si system, based upon analogy with the Zr-O-Si materials system, are shown in Fig. 4.[59-62] These tie lines indicate that, *under equilibrium conditions*, the metal oxide HfO_2 and the compound silicate $HfSiO_4$ will both be stable in direct contact with Si up to 950°C. The gray shaded area denotes a large phase field of $(HfO_2)_x(SiO_2)_{1-x}$ compositions which are expected to exhibit enhanced stability in direct contact with Si up to the high temperatures (~1000 °C) utilized in CMOS processing.[2] Other $(HfO_2)_x(SiO_2)_{1-x}$ compositions outside of the gray area are also stable on Si, but since it is desirable to prevent any Hf-Si (silicide) bonding, film compositions within the gray area may be preferable.

Care must be taken in selecting the appropriate composition, however, because phase separation into ZrO_2, HfO_2 and SiO_2 has been reported for these mixtures for T < 900°C and x > 0.25.[63] This suggests that compositions in the right upper region of the gray shaded area may be preferable for obtaining a stable film which maintains a high-quality interface to Si for transistor applications. This fundamental difference from the Ta and Ti systems is extremely important for gate dielectric applications, because it suggests that there is potential to control the dielectric-Si interface.

A recent review of thermodynamic considerations in regard to gate dielectric selection has been provided by Schlom and Haeni.[11] Recent work has shown, however, that deposited films of metal oxides such as

Al_2O_3,[64,65]Y_2O_3,[66,67] La_2O_3,[68] and ZrO_2[64,69] , predicted to be thermodynamically stable, do result in the formation of an interfacial layer, most likely a silicate or SiO_2, stressing the need to consider reaction kinetics during the deposition process which may be far from equilibrium. Recent work has demonstrated that even for the case of Zr-silicate deposited directly on Si, an extremely thin 3.5 Å SiO_2 layer still forms at both Si interfaces.[70]

For the cases of Zr- and Hf-silicates, a range of dielectric constants from 5 to 15 has been reported for a range of compositions (metal content from 3 to 27 at.%) using sputtering and CVD techniques.[25,71-75] Clearly, the surface preparation and deposition conditions have a strong impact on the resulting quality and dielectric constant of the films. Post deposition exposure of uncapped films to oxygen can also result in the formation of an interfacial layer, as has been reported for Y_2O_3,[66] La_2O_3,[68] ZrO_2,[76] and HfO.[22] Recently, the role of OH species in the context of interfacial reactions with poly-Si electrodes has also been examined.[77]

3.2.3 Dielectric film morphology and microstructure

Many of the alternate gate dielectrics studied to date exhibit either poly-crystalline or single crystal microstructure. As shown in Table I, nearly all bulk metal oxides of interest, with the exception of Al_2O_3, will form a poly-crystalline film during deposition or upon modest thermal treatments: HfO_2 and ZrO_2 are no exceptions. It is important to note, however, that the phases listed in Table I are bulk properties, and some suppression of crystallization for very thin films such as gate dielectrics is anticipated at temperatures where bulk crystallization would otherwise be expected to occur. The extent of crystallization suppression for a given oxide will depend on composition (metal content), the annealing ambient (capped and uncapped films) as well as the thermal budget (i.e. temperature and time) associated with processing (i.e. kinetic factors). For example, *uncapped* Al_2O_3 films appear to exhibit crystallization upon annealing above T=650°C, whereas *capped* films apparently do not.[78] Previous work by Wilk and Wallace indicate that, for low concentrations of Hf and Zr, Hf and Zr-silicates appear to remain amorphous upon post-deposition annealing to 1050°C for 20s.[17,71,72] Films with somewhat higher concentrations of Hf are expected to exhibit some crystallization and this effect has been reported.[19]

Poly-crystalline gate dielectrics could be problematic because grain boundaries may serve as high-leakage or diffusion paths, and this may lead to the need for an amorphous interfacial layer to reduce leakage current. In addition, grain size and orientation changes throughout a poly-crystalline film can cause significant variations in K, leading to irreproducible properties, especially if gate dimensions approach that of the dielectric grain size.

In regard to enhanced leakage currents from nano-crystallization, several recent studies appear to offer counter-examples from electrical properties for ZrO_2 films deposited by atomic layer deposition (ALD)[79] and HfO_2 layers deposited by ALD and metallo-organic chemical vapor deposition (MOCVD) are reported.[22,80-83] It should be noted, however, that both studies also used a gate dielectric stack, with the dielectric film on top of an amorphous SiO_2 layer. It is unclear at this point to what extent the amorphous SiO_2 layer affords the encouraging electrical properties, but this issue will become important, as the SiO_2 layer presents a limit to the minimum achievable t_{eq} value for these structures.

Recent work on the crystallization kinetics of ALD HfO_2 has determined that grain growth is thermally activated, as the average grain size is primarily determined by anneal temperature, while anneal time has little effect.[84] Interestingly, it was also found in this study that HfO_2 films grown on a 10 Å SiO_2 layer grown by dry thermal oxidation exhibit a small-grain poly-crystalline structure, containing monoclinic and either tetragonal or orthorhombic phases, whereas HfO_2 films deposited under equivalent conditions on a 10 Å SiO_2 layer grown by wet chemical oxidation appear completely amorphous by bright-field, high-resolution transmission electron microscopy (TEM). Further investigation in Z-contrast scanning TEM using fluctuation electron microscopy, however, revealed strong signatures of order at a sub-nanometer level. The morphologies for these two cases become indistinguishable after a 700°C anneal, as only a poly-crystalline monoclinic phase is observed for both cases.[84]

Perhaps a more significant issue is the impurity (dopant) diffusion through the dielectric, typically enhanced by the presence of grain boundaries, resulting in significant threshold and flatband voltage shifts. Recent work. has demonstrated B penetration from poly-Si capped amorphous Al_2O_3 [78,81] and polycrystalline ZrO_2 thin films into the Si substrate.[86] Phosphorous penetration was also reported for Al_2O_3 [86] and ZrO_2.[87] Similar results have been reported for HfO_2 films as well.[88,89] For pseudo-binary systems, such as Hf-silicate, recent studies indicate that films with sufficient Hf content resulting in polycrystalline films exhibit enhanced dopant diffusion upon annealing.[90-92] For example, as shown in Fig. 5, aggressively annealing thin films of Hf-silicate ("HfSiO") results in structural changes as these films have sufficient Hf content (~10-12 at. %). In contrast, nitrided Hf-silicate ("HfSiON") exhibits an amorphous structure.[93]

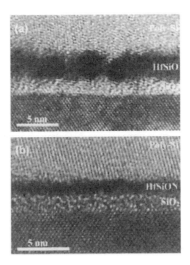

Figure 5: HRTEM results for films after a 60-s RTA at 1050 °C (a) HfSiO, and (b) HfSiON. Note the crystalline regions in the HfSiO films. No crystallization is observed in HfSiON films. [Ref. 92]

In the case of B diffusion through these film stacks, one notes a significant enhancement of B diffusion into the underlying substrate for the HfSiO films compared to the HfSiON films from SIMS measurements (Fig. 6). This is postulated to be due to enhanced grain boundary diffusion of B.[90-92] It seems clear from these recent reports that the gate dielectric film morphology will play an important role to inhibit dopant diffusion (and possible metal gate constituents) as well.

The use of N incorporation to inhibit dopant penetration, particularly for amorphous SiO_2, has been extensively reported and recently reviewed.[23,94] Suppression of dopant diffusion through the incorporation of N (either within the film or as an interfacial barrier layer) has been considered for alternate gate dielectrics.[93] Recent work indicates that N incorporation in high-κ materials such as Hf-silicate,[19,34,91,92,95] ZrO_2,[85,96] Al_2O_3,[78,97] and HfO_2[88,91,98-100] is a promising approach to control properties. N incorporation also appears to provide sufficient modification of the bonding network in Hf-silicate so as to preserve an amorphous structure even with elevated Hf content[19,95] As in the case of Si-oxynitride dielectric films, the concentration depth profile of N throughout the film has been shown to impact transistor performance, with control of the N concentration near the channel/dielectric interface crucial to avoid mobility degradation.[99] Recent work on plasma nitridation of Hf-silicate demonstrates that the N profile can be controlled to a greater degree than that provided by thermal nitridation methods.[101]

Evaluations of mobility degradation with increased crystallization for Hf-silicate films has been recently reported by Yamaguchi, *et al.*[102] Films

94

with 5-7% Hf content resulted in nanocrystalline regions within the poly-Si/HfSiO/Si gate stack upon annealing at 900-1000°C for 60 s in N_2. Peak

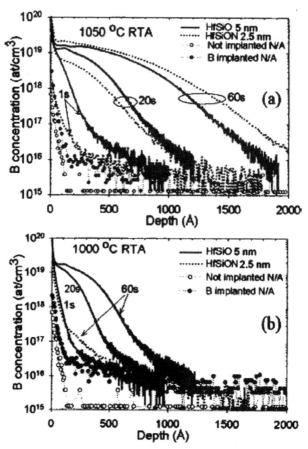

Figure 6:. B profiles in the Si substrate after poly-Si and dielectric film (5-nm HfSiO and 2.5-nm HfSiON) removal (a) after 1050 °C RTA and (b) after 1000 °C RTA. For comparison, the profiles for nonimplanted /not annealed and B-implanted/not-annealed samples are also shown (N/A: not annealed). [Ref. 92]

electron mobilities from these annealed nMISFET structures appear to degrade by a factor of 1.3-1.6 with annealing temperature. These authors propose that the formation of nanocrystals in the films results in enhanced Coulomb scattering in the channel.

Very recent studies [103] focused on examining the effect of residual polarization of amorphous HfO_2 and polycrystalline ZrO_2 and HfO_2 after the *removal* of an applied voltage – called dielectric relaxation (DR).[104] The effect occurs in all amorphous dielectric (glassy) materials, but is several orders of magnitude smaller for SiO_2 compared to high-κ metal-oxide insulators, apparently due to the constrained sp^3 hybridization associated with the SiO_2 bonding network (and therefore the limited distribution of geometric atomic configurations). A "perfect" (non-ferroelectric) metal-oxide dielectric crystal would not exhibit the effect as the available atomic configurations since the constituent atoms are constrained. As noted above, many high-κ metal oxides incorporate transition metals with d orbitals, thus allowing a significantly larger distribution of atomic configurations. This DR effect can produce undesirable time-dependent shifts in the threshold voltage for transistors, which has been most often identified with charging due to trap states in the dielectric or at the interface.

In contrast, Jameson and coworkers provide a theory, based on the earlier work by Phillips [105] and Anderson,[106] where the changes in the atomic configuration of the dielectric constituent atoms are due to quantum mechanical tunneling between a manifold of two stable configuration states (potential wells), resulting in the residual polarization effect.[103] Although it is not currently possible to distinguish the source of the observed V_T shifts between charge-trapping or DR, the theory does appear to provide an explanation for the current decay time constant over several orders of magnitude, as well as other properties commonly associated with glassy materials. It is noted that disorder at grain boundaries for even nanocrystalline metal-oxide dielectrics may result in the observed DR effects. The residual DR currents observed for amorphous HfO_2 appears to be somewhat lower than that for crystalline HfO_2 and ZrO_2.

Given the concerns regarding poly-crystalline and single crystal films, it appears that an amorphous film structure remains the ideal one for the gate dielectric – particularly in the near term. The prospects for longer term single crystal dielectrics remain dependent upon (1) adequate manufacturing tools that can address large volume throughput associated with mainstream Si technology and (2) a modified CMOS process flow that incorporates metal gates in a replacement-gate scheme where thermal budgets are lower than conventional CMOS process flows.

3.2.4 Process integration (compatibility with future CMOS and metal gates)

As noted in the Introduction, the "integration phase space" is a key concern in the evaluation of potential alternative dielectric materials. For example, extensive studies examining the transport of species from post-deposition vacuum, N_2 and O_2 annealing of Al_2O_3,[107-110] ZrO_2,[111-113] HfO_2,[114] and their associated silicates[53,54,55,115,116] and aluminates[117,118] have been recently reported. An excellent review of transport studies within these materials (as well as Yttrium, Lanthanum and Gadolinium oxides/silicates/aluminates) and a description of modeling such transport has been recently provided by de Almeida and Baumvol.[23] Using isotopic labeling methods, a significant distinction between the oxidation kinetics observed for SiO_2 and various high-κ dielectrics have been noted.[23,119] For example, isotopic oxygen exchange reactions throughout the high-κ films is apparent, in contrast to SiO_2 films where exchange tends to occur at the interfacial region through the associated bulk Si oxidation kinetics.[7] This emphasizes that the deposition and post—deposition processes have kinetic components that must be controlled for an optimized gate stack. Such work also demonstrates the need to focus very early in the research of materials on the *integration requirements* (as-processed device structure, thermal cycling, annealing ambient, O_2 reaction sensitivity, etc.) associated with conventional CMOS process flows rather than only acquiring a low t_{eq} value.

Moreover, very recent work by Quevedo-Lopez, *et al.* demonstrates that fundamental stability issues are observed for Zr-based dielectric materials where Zr interdiffusion with the Si substrate is observed in annealed Zr-silicate films (Fig. 7(a)).[53] In this study,[55] films were aggressively annealed and then removed by chemical etching methods.[120] The remaining Si substrate was then examined using a special Time-of-flight Secondary Ion Mass Spectrometry (ToFSIMS) "multi-crater" technique. The multi-crater approach permits the measurement of accurate depth profiles in the region very near the Si surface (where channel mobility is most sensitive) by avoiding ion-beam mixing artifacts.[55] As may be seen in Fig. 7, Zr interdiffusion was observed to occur well into the Si substrate upon aggressive thermal annealing.

Such profiles permit simple diffusion models to be developed that further permit extrapolations of near surface impurity concentrations upon reasonable thermal budgets (e.g. a 1050°C, 1 sec activation anneal). In the case of Zr, concentrations near the surface in excess of 10^{16}/cm^3 are anticipated from such extrapolations and raise concern in regard to mobility degradation effects. In contrast, Hf-silicate instability due to Hf interdiffusion with the Si substrate (Fig. 7(b)) under similar annealing conditions was *not* observed within the depth resolution of the ToFSIMS "multi-crater" technique employed (i.e. <1nm). As noted in that work, remnant Hf interdiffused with the substrate, if present, was below the detectible limits of the technique.[54] Recent surface science (scanning tunneling microscopy) investigations of uncapped Hf-silicate films (produced from Hf/SiO_2 layers annealed in ultra-high vacuum) have identified the formation of Hf-silicides on the Si surface.[122] The reported reaction of Hf with the Si substrate appears to be limited to depths of <1 nm and is consistent with the ToFSIMS profiles reported by Quevedo-Lopez, *et al.*[54] Recent work examining ALD HfO_2 and ZrO_2 films also appear to be in agreement with these earlier studies.[122]

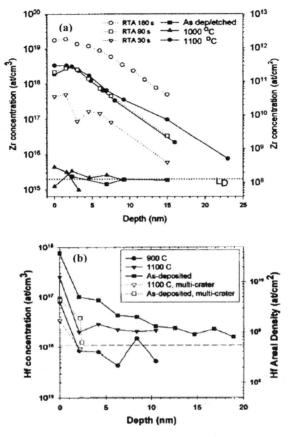

Figure 7: ToFSIMS depth profiles for (a) Zr and (b) Hf in Si from thermal annealing of their respective silicates. [Refs.53, 54, 55]

As noted in Section 2.2, the impact of the presence of dielectric metal constituents in regard to electrical properties has been reported[56,123,124] Guha, *et al.* reported on the effects of Al interdiffusion with the Si substrate from thermal activation and clearly observed penetration from SIMS measurements.[56] The introduction of a Si-oxynitride interfacial layer results in an apparent reduction of Al/Si interdiffusion as determined by SIMS, presumably due to the presence of N in the layer. Interestingly, a concomitant degradation in mobility by ~2× was also noted for transistors incorporating an Al_2O_3 dielectric upon thermal annealing at 1000°C for 5 sec in N_2, where Al interdiffusion was observed in the SIMS experiments.

The impact of the presence of Zr and Hf on minority carrier lifetime was recently reported by Hegde, *et al.*[123] The effects of remnant Zr and Hf originating from 3-5nm thick films of ZrO_2, HfO_2 (deposited by chemical vapor deposition) and $ZrSi_xO_y$ (deposited by sputtering methods) were studied by electrolytic metal tracer techniques after annealing at 1025°C for 20 s. It was found that techniques such as total-reflection x-ray fluorescence results in only the detection of remnant Zr for the CVD ZrO_2 films, and appears to be consistent with previous reports on the relative stability of Hf vs. Zr-based dielectrics.[53-55] However, apparent contamination (Fe and Ni) were observed for the $ZrSi_xO_y$ films, presumably from the sputter target source. The average minority carrier lifetime (MCL) measurements from these experiments were reported to be 449 μs for HfO_2 films, 278 μs for ZrO_2 films, and less than 10 μs for $ZrSi_xO_y$ films. Compared to a SiO_2 control sample (546 μs), these authors report that the values for the HfO_2 and ZrO_2 films are comparable to the control, and that the difference of the MCL between these dielectrics (a factor of ~1.6) was attributed to the variations in the starting wafer material, and *not* due to the presence of detected Zr. The large reduction in MCL for the Zr-silicate appears to be associated largely with contamination of the sputter source, however.

Kang *et al.*, recently reported on the effects of implanted Hf in the silicon channel through a SiO_2 "pad" layer and subsequent annealing at 950°C for 60 s, followed by pad oxide removal and regrowth of a ~5nm SiO_2 dielectric.[125] Implantation parameters were reported to provide the projected range of Hf in the vicinity of the substrate/dielectric interface, although no independent SIMS measurements were reported of the resultant Hf profiles to verify the presence of Hf in the channel region. Nevertheless, both capacitance-voltage and mobility measurements as a function of Hf dose indicated that there was no degradation in the associated capacitor or transistor performance. These results appear to be in agreement with the enhanced stability properties associated with Hf-based dielectrics previously reported.[55]

Even in the case of significant process flow modifications, such as a "replacement gate" flow that allows the high-temperature processing to be done *before* deposition of the final gate dielectric, the continued thermal cycling afterward is sufficient to result in poor electrical properties for materials such as Ta_2O_5.[125] It is therefore important to select a materials system in which the desired *final state* is stable, from a process integration point of view.

As many of the material candidates examined to date result in mixed interfacial alloys upon deposition or further processing, the use of materials such as silicates may allow for control of the Si interface composition, which may solve a key problem for the high-κ gate dielectric materials approaches. The κ values of materials such as $(HfO_2)_x(SiO_2)_{1-x}$ are substantially lower than those of their pure metal oxide counterparts (in this case HfO_2), but this tradeoff for interfacial control and relatively higher mobility will be acceptable as long as the resulting leakage currents are low enough.

There also are applications of high-κ dielectric materials on processes where process temperatures are more rigorously constrained to lower values than that obtained in Si CMOS processing. Examples include capacitor structures incorporated in the "back-end-of-line" for integrated circuits and even circuits on plastic, flexible substrates. Low temperature (≤20°C) oxidation methods, such as UV/O_3 exposures, have been recently explored to produce high-κ dielectric films.[126-130]

Si-based gate electrodes have dominated CMOS technology because precise control of dopant implant energy and flux can accurately tune the desired threshold voltage, V_t, for both nMOS and pMOS. Additionally, the process integration components (annealing, etching, contact silicidation etc.) are well established in industry. It is therefore still desirable to employ a gate dielectric which will be compatible in direct contact with Si-based gates. We also note that other efforts are focused on using poly-Si$_{1-x}$Ge$_x$ gates for achieving higher boron activation levels and therefore better performance in pMOS devices, and potentially better performance in nMOS devices as well.[131-133] Control of dopant penetration through any gate dielectric will remain an important issue since doped poly-Si is the incumbent gate electrode material.

For CMOS scaling in the longer term, however, current roadmap predictions indicate that poly-Si gate technology will likely be phased out beyond the 70 nm technology node, after which a metal gate electrode substitute appears to be required.[1] It is therefore also desirable to focus efforts on dielectric materials systems that are compatible with potential metal gate materials. Specifically, metal gates such as TiN and Pt have been previously used with most of the high-κ gate dielectrics mentioned above for material evaluation purposes due to their anticipated stability toward adverse reactions with various dielectrics. As noted above, metal gates are very desirable for eliminating dopant depletion effects and sheet resistance/performance constraints for scaled CMOS. In addition, use of metal gates in a replacement gate process can lower the required thermal budget by eliminating the need for a gate dopant activation anneal as for the case of a poly-Si electrode.[134-136]

A key issue for gate electrode materials research will be the control of the gate electrode work function (Fermi level) at the dielectric interface after CMOS processing. Moreover, many of the potential advanced gate dielectrics investigated require metal gates, due to the instability when in direct contact with Si. There are two basic approaches toward achieving successful insertion of metal electrodes: a single "midgap" metal or two separate metals (which may include two compositions of the same alloy system). The energy diagrams that illustrate these two approaches are shown in Fig. 8.

The first approach is to use a metal (such as TiN) that has a work function that places its Fermi level at the midgap of the Si substrate, as shown in Fig. 8(a). These are generally referred to as midgap metals. The main advantage of employing a midgap metal arises from a symmetrical V_T value for both nMOS and pMOS, because by definition the same energy difference exists between the metal Fermi level and the conduction and valence bands of Si. This affords a simpler CMOS processing scheme, since only one mask and one metal and no ion implantation would be required for the gate electrode.

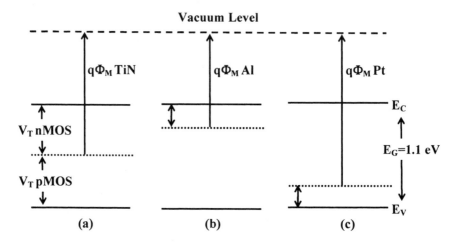

Figure 8: Energy diagrams of threshold voltages for nMOS and pMOS devices using (a) midgap metal gates and (b) dual metal gates. [Ref. 3]

For the case of bulk CMOS devices, however, a major drawback is that the band gap of Si is fixed at 1.1 eV, thus the threshold voltage for any midgap metal on Si will be $\sim$ 0.5 V for both nMOS and pMOS. Since voltage supplies are expected to be $\sim$ 1.0 V for sub-0.13 μm CMOS technology, $V_T \sim 0.5$ V is much too large, as it would be difficult to turn on the device. Typical threshold voltages for these devices are expected to be 0.2 – 0.3 V. Compensation implants can be made in the channel to lower the V_T, but other concerns then arise regarding increased impurity ion scattering, which would degrade the electron channel mobility. Furthermore, midgap work function metal gate systems have been predicted not to provide a performance improvement worthy of the added process complexity to replace Si-based gates.[137,138] Recent work focused on examining TiN (and TiSiN) metal gates in conjunction with HfO$_2$ gate dielectrics on n- and p-MOSFETs [136,139] as well as TaN gates.[140] Recently, TaN gates have been examined with Hf-silicate dielectrics as well.[73]

We also briefly note here that alternative device technology research and development, such as fully-depleted silicon-on-insulator (FDSOI) CMOS technology, focused on midgap metal gate electrode materials such as TiN,[141] Ta,[142] and most recently TaSiN in conjunction with HfO_2.[143,144] Promising performance were reported for these devices from midgap metal electrodes as the transistor is isolated from the bulk substrate, allowing the adjustment of the transistor threshold voltage through electrostatic means.[145]

The second approach toward metal electrodes involves two separate metals, one for pMOS and one for nMOS devices, and is often referred to as a "dual metal gate" approach. As shown schematically in Figs. 8(b) and (c), two metals could be chosen by their work functions, Φ_M, such that their Fermi levels line up favorably with (i.e., closer to) the conduction and valence bands of Si as appropriate for nMOS and pMOS devices. In the ideal case depicted in Fig. 8(b), the (vacuum) Φ_M value of Al could achieve $V_T \sim 0.2$ V for nMOS, while the higher (vacuum) Φ_M value of Pt could achieve $V_T \sim 0.2$ V for pMOS. In practice, Al is not a feasible electrode metal because it will reduce nearly any oxide gate dielectric to form an Al_2O_3-containing interface layer. Other metals with relatively low work functions, such as Ta, TaN, and TaSiN however, appear to be feasible gate metals for nMOS.[146]

Similarly for pMOS, Pt is not a practical choice for the gate metal, since it is not easily processed, does not adhere well to most dielectrics, and is high cost. Other elemental metals with high Φ_M values such as Au are also not practical, for the same reasons as for Pt. It has been suggested that the work-functions for each metal should be within 0.2 eV of the conduction and valance band edges of Si.[138] However, metals such as Al, Ta, Ti, Hf and Zr have been demonstrated to reduce dielectrics when in direct contact.[147]

Recently, the difference between vacuum-based work-function ($\phi_{m,vac}$) values and the effective work-function ($\phi_{m,eff}$) obtained from stack structures has been examined by employing dipole theory for the associated intrinsic interface states at the metal/dielectric interface.[148-151]

Comparisons to a set of metal oxide dielectric materials were performed. From that work, it is suggested that the metal gate materials should exhibit a vacuum work-function $\phi_{vac} < 4.05$ eV for nMOS and $\phi_{vac} > 5.17$ eV for pMOS. As noted by Yeo, et al., this requirement leads to the consideration of relatively inert metal gate materials for p-MOS devices, while metal gate materials that are reactive are suitable for n-MOS devices.[148] The use of reactive metals in direct contact with insulating oxides would be expected to result in interfacial reactions, and thus the formation of extrinsic interface states as well. Yeo, et al. also suggested that a thin layer (monolayer) of SiO_2 between the inert or reactive metal gate and the dielectric could relax the above vacuum work-function requirement, albeit with a penalty of a reduced stack capacitance. Of course, reactions between such thin SiO_2 layers and reactive metals would also be expected, and so one would anticipate work on the utilization of N incorporation at such interfaces to help mitigate such reactions.

As an alternative to elemental metals, conducting metal oxides such as IrO_2 and RuO_2, which have been studied and used for years in DRAM applications, [1,7,8] can provide high Φ_M values in addition to affording the use of standard etching and processing techniques. Alloys of these and similar conducting oxides can also potentially be fabricated to achieve a desired work function. Regarding potential gate electrodes for pMOS devices, Zhong et al. made initial measurements of the important properties of RuO_2, including thermal stability up to 800°C, a low resistivity of 65 $\mu\Omega$-cm, and a measured RuO_2 work function of $\Phi_M = 5.1$ eV. [152,153] Recently, metal alloys of TaN and TaSiN, which have been employed as diffusion barriers for Cu in "backend" portions of the CMOS process flow, have also been examined for nMOS applications. [154,155]

The concept of a "tunable" work-function metal gate electrode would, of course, be appealing as this is one of the key successes associated with poly-Si gate technology. Misra et al. have recently examined the feasibility of the deposited Ta-Ru alloy system to accomplish this with Ru as a pMOS gate and $Ru_{0.6}Ta_{0.4}$ as an nMOS gate. [153,156,157] Compatibility of TaN, TiN, $Ru_{0.9}Ta_{0.1}$ and $Ru_{0.5}Ta_{0.5}$ alloys with HfO_2 have also recently been reported by this group, where similar work-functions were obtained from previous studies on SiO_2, after correcting for charging in the HfO_2 dielectric. [158] Alteration of a metal through implantation, as in the case of poly-Si gates, is obviously appealing as well. Recently, the effects of N implantation in Mo on the Mo work-function have been examined [148,159] Metal interdiffusion (for example Ni and Ti) has also been suggested as a process efficient method of controlling metal work functions. [160,161]

Recently, extensive work has been conducted on low-temperature (~300 – 400°C) processing to complete silicidation of poly-Si gates, called "fully silicided" (FUSI), with metals such as Ni and Co to control the gate work-function. Such processes are appealing as they may extend currently adopted self-aligned silicide (SALICIDE) gate contact technologies. Material phases such as NiSi provide advantages over current silicide technologies, such as $TiSi_2$, where a phase-dependent line-width increase in sheet resistance is observed, or $CoSi_2$, where a high rate of consumption of Si and increased junction leakage is observed. [159] This approach has been recently examined for conventional CMOS processes, [163,164] as well as alternative 3-d device technologies such as thin channel, non-planar channel field effect transistors (called "FinFETs" as the channel region resembles the shape of a vertical "fin") or Fully-depleted Silicon on

Insulator (FDSOI) structures, for example.[165] The FUSI NiSi approach has very recently been integrated with high-κ La_2O_3 dielectrics, where NiSi was observed to exhibit superior leakage properties compared to similar gate stacks fabricated with $CoSi_2$.[166] The physical mechanism that results in control of the resultant gate alloy work function remains under investigation. Both dopant type in the poly-Si bulk (which subsequently undergoes silicidation with a metal such as Ni or Co) and metal substitutional impurities at the dielectric/silicide interface appear to play a role in the resultant work-function.[165]

The interdiffusion of gate electrode constituents through dielectrics with crystalline morphologies remains to be explored for metal gate dielectric materials. This issue has been extensively studied, however, for capacitor applications where conductive diffusion barriers have been introduced albeit with additional process complexity.[8] Materials such as TaN and TaSiN are also employed for such diffusion barrier applications in interconnect technologies.

3.2.5 Reliability issues

The electrical reliability of a new gate dielectric must also be considered critical for application in CMOS technology and represents a rigorous challenge for any dielectric candidate that must be addressed. The determination of whether or not a high-κ dielectric satisfies the strict reliability criteria requires a well-characterized materials system – a prospect not yet available for the alternate dielectric materials considered here. The nuances of the dependence of voltage acceleration extrapolation on dielectric thickness and the improvement of reliability projection arising from improved oxide thickness uniformity, have only recently become understood, despite decades of research on SiO_2.[167] This further emphasizes the importance and urgency to investigate the reliability characteristics of alternative dielectrics, as these materials are sure to exhibit subtleties in reliability that differ from those of SiO_2. A brief summary of the reliability issues for high-κ dielectrics is recently provided by Oates.[168]

However, some preliminary projections for reliability, as determined by stress induced leakage current (SILC), time-dependent dielectric breakdown (TDDB) and mean time to failure (MTTF) measurements, appear to be encouraging for many candidate materials.[4] Recently, the role of Si interdiffusion with the gate dielectric and the Si substrate.[90,91] has been examined in view of reliability constraints.[169] The susceptibility to hot-carrier charge trapping has also been recently investigated.-[170] Thickness studies of the breakdown dependence of sputter deposited HfO_2 have been recently reported, and the resultant Weibull slope for thin (<2 nm) films appears to be greater than that obtained for thin SiO_2, suggesting adequate reliability performance for thin HfO_2 films.[171,172]

Recent work focused on the development of pulsed charge electrical stressing measurement methods, where the detrapping rate can be adequately controlled to permit the accurate measurement of threshold voltage shifts in high-κ dielectrics.[173,174] For the ALD HfO_2/SiO_2 bilayer system, it is proposed in this work that a defect band (originating from defects within the HfO_2 layer) occurs at energies above the Si conduction band edge and shifts rapidly with gate bias. This rapid shift is thought to be responsible for the efficient charging and discharging of defects located near the interface of the HfO_2/SiO_2 bilayer. This rapid charging and discharging effect can lead to errors in the electrical evaluations of such dielectrics that are performed without a rapid transient measurement approach.[174] Examinations of charge trapping and break down measurements for nitrided Hf-silicates have also been reported.[175,176] It was noted that HfSiON appears to have better electrical stability compared to HfO_2.[176] Studies of the initial HfO_2 bulk trap density have indicated a correlation to yield and reliability, and the application of percolation models, such as those used in the SiO_2 dielectric, appears to be useful in regard to breakdown.[177]

Treatments involving deuterium in post deposition anneals or during the film growth process have also been recently reported.[180,181] The use of D_2O during ALD growth of HfO_2 appears to have useful effects in regard to electrical stability and reliability.[181] Improvements have also been reported for anneals incorporating D_2 at temperatures up to 600°C.[178] The process presumably permits a preferential placement of D at the high-κ/Si interface, resulting in better stability against interfacial bond depassivation as has been reported in the literature for the SiO_2/Si interface.[180-183] It has been noted that the transistor gate stack materials, including the side wall material, plays an important role in the efficient delivery of D (and H) to the interfacial region.[184-186] High temperature forming anneals (400°C to 600°C) have also been examined as a means to improve device reliability for HfO_2 and HfSiON gate stacks.[187]

Recent work by Zafar, et al. examining HfO_2 and Al_2O_3 dielectrics (also produced by ALD methods) have suggested that the defects responsible for charge trapping appear to be intrinsic to these materials, and that pulsed electrical stressing results in no detectible additional bulk or interfacial traps produced from such stressing, as indicated by the behavior of the gate current density and the sub-threshold voltage shift, respectively.[188] Additionally, charge trapping models in very good agreement with experimental measurements were developed in that work [188] to enable predictions of the shift in threshold voltage shift with stressing time, enabling predictions of 10 year reliability lifetimes.

3.3 THIN FILM SYNTHESIS / CHARACTERIZATION

3.3.1 Integrated film synthesis / *in situ* characterization systems and techniques

3.3.1.1 Integrated sputter deposition/time-of-flight ion scattering/ mass recoil spectroscopy/x-ray photoelectron spectroscopy/ spectroscopic ellipsometry

The understanding of film growth process and gate electrode/high-k dielectric/Si interfaces is critical in order to control them and produce the heterostructure gate suitable to drive the next generation CMOS devices. This can be better achieved using integrated film synthesis/*in situ* characterization systems capable of providing information on both during film growth and processing.

One of the authors (OA) and colleagues have developed in recent years integrated sputter-deposition / time-of-flight ion scattering and recoil spectroscopy (TOF-ISARS) / X-ray photoelectron spectroscopy (XPS)[189,190] and (TOF-ISARS) / Spectroscopic ellipsometry systems[192] as unique tools for fundamental and applied science focused on studies of integration of complex oxide with semiconductor substrates.

The TOF-ISARS technique provides a remarkably wide range of information directly relevant to the growth of single and multicomponent component thin films and layered heterostructures of dissimilar materials, probably exceeding that obtainable by any other surface analytical technique. Information obtainable with the TOF-ISARS technique includes surface composition, atomic structure of the first few monolayers, lattice defect density, trace element analysis and phonon characteristics [189,190] in thin films and layered hetrostructures. The long source-sample-detector distances (0.5 m) inherent in the ToF detection scheme permit the introduction of thin film deposition equipment in the analysis chamber, and also permit the use of differential pumping of the ion source and detector regions, thereby allowing the region around the substrate to be at the elevated pressure of e.g. oxygen required for the growth of the desired oxide phase when growing complex oxide thin films. Operation at pressures approaching one Torr, i.e. 6-8 orders of magnitude higher than any other analysis technique that provides a comparable degree of surface specificity has been demonstrated. [189,190] These capabilities make the TOF-ISARS analysis technique ideal for *in-situ*, real-time studies of thin film growth and interface formation processes, particularly for the complex oxide / semiconductor integration research required for the development of the next generation of nanoscale CMOS gates. An schematic of the TOF-ISARS / XPS systems in shown in Fig. 9 detailed description of the film deposition / *in* situ characterization systems and techniques mentioned above can be found elsewhere. [189,190]

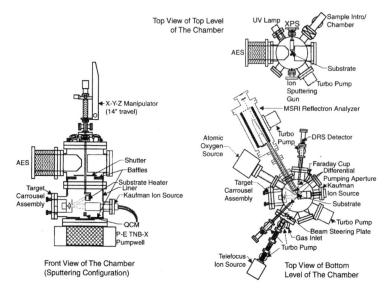

Figure 9: Schematic of the TOF-ISARS system that is operational at Argonne National Laboratory. It consists of a bottom chamber containing a broad beam ion source, a rotating target holder with three targets, the pulsed probe inert gas ion beam and the ion scattering spectroscopy (ISS) line, the direct recoil spectroscopy (DRS) line and the mass spectroscopy of recoil ions (MSRI) line. The top interconnected chamber contains a cylindrical mirror analyzer and an X-ray source to implement in situ Auger and XPS analysis.

A system similar to that shown in Fig. 1 was developed and installed at the University of North Carolina-Chapel Hill.[191] However, the UNC system features spectroscopic ellipsometry (SE) instead of XPS, since SE enables studies of the buried interface being developed during the growth of the films.

A discussion of the use of these systems to study high-K dielectric layers for the next generation of CMOS devices is presented in Section 4, when discussing the science and technology of a new amorphous $Ti_xAl_{1-x}O_3$ (TAO) and $Br_xSr_{1-x}TiO_3$ (BST) layers as alternative high-K dielectrics. However, since MSRI has been the most used method of the TOF-ISARS approach to study high-K dielectric film growth and interfaces, we present here a brief description of the fundamentals of this method. Mass Spectroscopy of Recoiled Ions (MSRI) is a variation of time-of-flight direct recoil spectroscopy (DRS)[189,190]. MSRI involves the use of a time-of-flight scheme to detect the mass of ionized atoms ejected from the surface of a solid, in a forward direction, via a direct impact from a primary ion.

Primary ions with kinetic energy in the range 5-10 keV are generally used as a probe. MSRI can be used to characterize the surface composition and structure of solid surfaces and particularly thin films with monolayer-specific resolution.

The detection of surface atoms ejected in an ionized state provides MSRI with a much higher mass resolution than the alternative DRS method, which detects ejected atoms in a neutral state. The higher MSRI resolution is due to the use of a reflectron mass analyzer with time-focusing capability.[189] The spread in kinetic energy of the recoiled ionized atoms from the surface, due to any deviation from an ideal binary collision process causes broadening of DRS peaks, thus, low resolution. The capability of adjusting the energy spread in the MSRI reflectron analyzer provides the exceptional MSRI resolution (isotopic masses can be resolved). All ions of a given mass can be forced to arrive at a detector at the same time by adjusting MSRI reflectron potentials. As a result, MSRI can provide the capability for detecting all elements, including hydrogen, in a single spectrum with unit mass resolution. A detailed description of DRS and MSRI techniques can be found elsewhere.[189,190]

MSRI involves the use of differentially pumped ion source and detectors, which provide the capability to perform in-situ, real time compositional analysis of films during or after growth, and surfaces in general, under various deposition / environmental conditions. Unlike the conventional surface analytical techniques, which are subject to gas phase scattering, the detected recoiled ions in MSRI with kinetic energy of several keV can survive transmission to a detector through ambient gas pressures up to several mTorr.[191,192] The high-pressure analytical capability of MSRI is demonstrated in Fig. 10, which shows and MSRI spectrum resulting from the analysis of a Si surface by a 10 keV Ar$^+$ ion beam. As can be seen, the peak positions that provide element identification do not

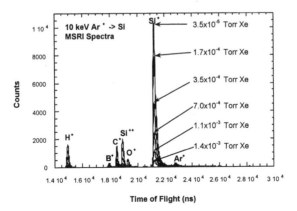

Figure 10: MSRI spectra for 10 keV Ar+ primary ions incident on a boron doped silicon surface in various ambient xenon pressures. [Ref. 190]

shift for ambient pressures up to the maximum pressure tested (1.4 mTorr). In addition, Fig. 10 shows the other great capability of MSRI, which is the detection of all masses from hydrogen to the heaviest mass that might be on the surface of the solid under analysis.

Further details of the use of the integrated film synthesis/analytical techniques described above are discussed in Section 4.

3.3.1.2 Integrated molecular beam epitaxi / *in situ* X-ray synchrotron standing wave (XSW) technique for nanoscale studies of complex oxide film growth

In a traditional XSW experiment, the XSW is generated by a strong Bragg reflection from a bulk perfect single crystal.[192] The outgoing diffracted plane wave interferes with the incident plane wave and forms a standing wave field. This standing wave field has the same periodicity as the diffraction planes. As the crystal is advanced in angle through the arc-second wide "total reflection", the antinodes of the standing wave field shift from a location halfway between the diffraction planes to a position coincident with the planes. This inward shift of the antinodes causes an angle-dependent modulation in an atom's fluorescence signal that can be used to determine that atom's lattice position(s). This traditional method is not in general applicable to thin-film overlayers, since the period of the atomic layers in the film would not in general be equal to the period of the XSW probe generated by the substrate diffraction-planes. Therefore, Bedzyk et al.[193] use a variation of XSW where the XSW is generated by the weak kinematical Bragg diffraction from the film itself. Since the thin-film reflection is weak, the modulation in the fluorescence signal is also weak. Therefore a high intensity x-ray source such as that available in an undulator line of the Advanced Photon Source (one of the two worldwide existing third generation X-ray synchrotron operating at Argonne National Laboratory) is required to observe this phenomena with the required counting statistics to extract atomic-scale structural information from the analysis of the data. (see for example ref. 193). The XSW technique is just starting to be used to study high-K dielectric film growth and interfaces and it will play a critical role in future research on high-K dielectric layers.

3.3.1.3 Integrated metalorganic chemical vapour deposition (MOCVD) and atomic layer deposition (ALD)/*in situ* characterization techniques

Metalorganic chemical vapour deposition (MOCVD) and atomic layer deposition (ALD) are two variations of deposition process that are the most compatible with fabrication of complex-oxide/silicon integrated film-based devices. In fact, MOCVD synthesis of perovskite ferroelectric thin films is currently being used by Matsushita-Panasonic for the fabrication of low-density ferroelectric random access memories (FeRAMs) that are currently in the market in "smart cards".[9,10] Both MOCVD [194] and ALD [195] processes involve the cracking of metalorganic precursors, containing the atoms required to produce the desired films, on the surface of the heated substrate used to sustain the desired oxide film. The main differences between MOCVD and ALD is that the first can yield crystalline films much easier than the second, but require generally higher temperatures (400-700 °C) [195] to produce the films, while ALD yields mainly amorphous films, but at low temperatures (150-300 °C).[195] The key feature of the ALD process is the supply of the precursors in one cycle and of water vapour on the subsequent cycle. A self-limiting growth process that yields one monolayer of film per cycle characterizes the ALD technique. Both MOCVD and ALD produce films with excellent conformality on three-dimensional micro and nanostructures, which is required for the fabrication of the next generation of high-density memories and other microelectronic devices. More recently, however, researchers demonstrated that it might be possible to growth crystalline oxide films by ALD at low temperature, using new precursors. Previously, attempts at growing crystalline oxide films we unsuccessful because researchers were using β-diketonate compounds used in MOCVD, which do not react with water or oxygen as required for the ALD process.[195] Recently, researchers demonstrated that $SrTiO_3$ and $BaTiO_3$ crystalline films can be obtained by the ADL method [196] via deposition with new precursors such as strontium bis (triisopropylcyclopentadienyl) $[Sr(C_5\text{-}i\text{-}Pr_3H_2)_2]$ and barium bis (pentamethylcyclopentadienyl) $[Ba_2(C_5Me_5)_2]$. The films were synthesized in the 250-325 °C temperature range. However, the as-deposited film exhibited relatively poor crystallinity and low permittivity (90) and acquired better crystallinity and high permittivity (170-180)[196] only after a pulsed thermal annealing step in air at about 500 °C, which is still lower than the temperature needed (550-750 °C) to produce similar films by MOCVD, which exhibit only slightly higher permittivity ($\sim K = 210$). These first studies indicate that ALD may provide a suitable method to produce crystalline high-K dielectric films at thermal budgets compatible with CMOS fabrication processes, and therefore should be pursued more vigorously.

Of the two MOCVD and ALD techniques discussed here, MOCVD has been integrated with an *in situ* characterization method. One of the authors (OA) and colleagues developed an MOCVD system integrated with one of the ports in the Advanced Photon Source at Argonne, which has been used for the last five years to perform the first comprehensive studies of perovskite ferroelectric film growth process in the relatively high pressure environment of the MOCVD method, using X-ray scattering methods.[197,198] This can be achieved because of the high energy (~ 25 keV) and high brightness of the APS beam that enables to have enough signal even after the beam traversing the walls of the quartz chamber used in the MOCVD system. The *in situ* synchrotron X-rays scattering technique recently enabled to determine the ultimate limit in thickness for the existence of polarizability in a PbTiO$_3$ (PTO) thin film. The Argonne group determined that a 3-unit cell thick PTO film is the thinnest ferroelectric film of the PbZr$_x$Ti$_{1-x}$O$_3$ family that exhibits polarizability.[199] This work opened the way for investigation of atomic scale complex oxide thin films, which is critical to studies of high-K dielectric thin films for the next generation CMOS gates. Bedzyk and Auciello are currently developing an ALD system to be integrated with a port of the APS synchrotron to study high-K dielectric thin films using the *in situ* XSW technique.

3.4 MATERIALS CANDIDATES

3.4.1 Current: Hf-based dielectrics

At this time (end of 2004), work on Hf-based dielectric materials, mainly Hf-silicates and HfO$_2$, dominates the recent literature (see Fig. 11) and appear to exhibit useful properties for integrated circuit scaling down to the 20nm node. These properties include a relative stability for interfacial reactions that prefer silicate formation thus can avoid a lower overall dielectric constant of the high-κ layer, due to the formation of an uncontrolled SiO$_2$ layer. The papers on these materials systems are too numerous to list individually here and so the reader is referred to the reviews cited on the topic.[2,3,4]

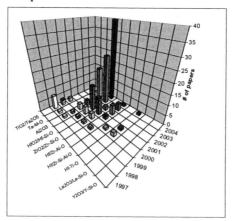

Figure 11: Distribution of papers presented at recent IEEE conferences indicate the emphasis of Hf-based materials systems examined for gate dielectric research and development.

Of particular importance for gate dielectrics have been investigations of the stability (in particular, changes in morphology and interdiffusion) of these films upon thermal processing in view of the required integration constraints for CMOS applications.[5] A significant hurdle for integration of all dielectrics currently under consideration for conventional CMOS integration is the stability of all gate stack constituents (stable film morphology, minimal interdiffusion, etc.) under a thermal budget of ~1000C for 10 s for adequate process integration margin. Future scaling may well require a modification (reduction) of such thermal budget desires to incorporate metal gate electrodes in CMOS, and thus may open the door to the consideration of other gate dielectric materials that exhibit stability at somewhat lower temperatures. However, key issues that must be addressed under such a scenario include alternative source/drain engineering (e.g. dopant activation at lower thermal budgets) and/or gate electrode formation in the device fabrication flow (e.g. gate electrode insertion after high temperature anneals). However, previous research on such CMOS process modifications has indicated that a variety of challenges exist to address such scenarios with adequate process margin and yield.

As noted above in the case of Hf-silicates, film morphology (*viz*, suppression of crystallization) and diffusion barrier properties can be controlled through the incorporation of N, as in the case of SiO_2.[2,4] Currently, HfSiON dielectrics appear to provide desirable properties in this regard.

3.4.2 Potential New Gate Dielectric Materials

3.4.2.1 New $Ti_xAl_{1-x}O_y$ Alloy Oxide

As discussed in prior sections of this chapter, TiO_2 attracted attention because of its high K (> 50), arising from soft phonons involving Ti ions. However, TiO_2 exhibits high leakage current, due to a near zero offset barrier to Si.[13] and undesirable crystallization at low temperature (~400 °C). Alternatively, Al_2O_3 exhibits the largest band gap ($E_g = 8.8$ eV) next to SiO_2 with a conduction band offset of 2.8 eV with respect to Si, 4.9 eV valence band offset, and thermodynamic stability against Si, but has relatively low K = 8-10. Based on these prior work and on the universal plot of bandgap vs. pemittivity of potential high-K dielectric materials shown in Fig. 2. Auciello proposed to investigate a $Ti_XAl_{1-x}O_y$ (TAO) alloy oxide as a potential new generation gate oxide. The idea of the TAO high-K evolved form prior work focused on developing a $Ti_{0.5}Al_{0.5}$ alloy layer as oxygen diffusion barrier for integration of complex oxide layers with Cu-based electrodes.[200] Those studies showed that the TiAl oxygen barrier resulted from formation of an amorphous TAO thin (~ 4-6 nm thick) surface layer, including TiO_2 and Al_2O_3 (the two extremes of high permittivity and high bandgap materials shown in Fig. 2, respectively). Thus, TAO (with Ti:Al = 75:25 at.%) appeared as a potential dielectric material with combined high permittivity, low leakage current, and high thermal stability. In addition to the conceptual design described above, the idea of synthesizing TAO layers needed to account for the limitation to achieve a sub-1 nm *EOT* dielectric layer in terms of the formation of an interfacial low-K SiO_2 layer that generally is produced during high temperature growth of high-K film, which lowers the MOS capacitance when in series with a high-K layer. Therefore, a low partial pressure of oxygen / growth temperature processes was considered as desirable to inhibit or minimize the formation of the SiO_2 interface layer.

The initial studies of the TAO layer discussed here were performed using the unique sputter-deposition system integrated with time-of-flight ion scattering (TOF-ISARS) and mass spectroscopy of recoil ions (MSRI) and x-ray photoelectron spectroscopy (XPS),[190,191] described in Section 3 in this chapter.

TAO layers were produced via room temperature (RT) sputter-deposition of thin TiAl alloy films in, followed by *in situ* oxidation. TiAl (75/25 at%) layers were deposited on hydrofluoric acid finished n⁻ Si (100) (ρ = 4-6 $\Omega \cdot$cm) and p⁻ Si (100) (ρ = 12-16 $\Omega \cdot$cm) substrates using a broad beam of 500 eV Xe ions. ToF-MSRI, used to monitor the TiAl growth on Si, indicated that a pinhole free 0.8 nm TiAl layer covers the Si substrate. Atomic oxygen delivered by an Oxford source (at 10^{-4} Torr) was used to convert the TiAl metallic into fully oxidized TAO via post-deposition oxidation at 25-700 °C. [130]

In situ XPS analysis of the TiAl layers after oxidation revealed complete transitions of Ti^0 to Ti^{4+} and Al^0 to Al^{3+} within the 25 - 700 °C range (Figs. 12a and 12b), indicating that stoichiometric alloy oxides with identical chemical bonding are produced by oxidation with atomic oxygen at all temperatures, unlike molecular oxygen (O_2). The XPS results were confirmed by near-edge x-ray absorption fine structures (NEXAFS) studies of the O K-edge and Ti L-edge of TAO thin layers on Si. NEXAFS is sensitive to the local bonding environment, providing information on crystal structures and forms of titanium oxides. [130]

Cross-sectional transmission electron microscopy (TEM) studies of the RT oxidized TAO thin layer, performed in a TEM with a spherical C_s aberration corrector to avoid contrast delocalization,[201] largely confirmed the XPS studies. Two amorphous layers were distinguished in the cross-section TEM sample (Fig. 13a, i.e., a darker top layer (L2) (~ 3-4 nm thick) and a brighter intermediate layer (L1) (< 2 nm thick). The line scan across the image with a line width of three nanometers shows a sharp boundary

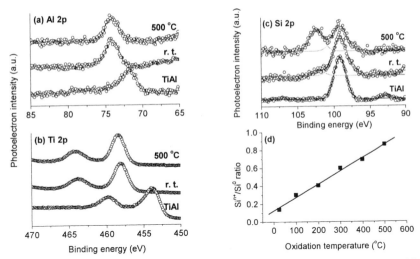

Figure 12: In situ XPS analysis of a TiAl thin layer on Si before and after oxidation with atomic oxygen. The Al 2p (**a**) and Ti 2p (**b**) spectra show complete transitions to Al^{3+} and Ti^{4+} for both oxidation processes both at 500 ˚C and RT. The Si 2p spectra (**c**) and the Si^{n+}/Si^{0} intensity ratio vs. oxidation temperature (**d**) reveal inhibition of SiO_2 interfacial layer formation for the RT oxidation, where the XPS peak (between 105 and 100 eV), corresponding to SiO2 (appearing at 500 ˚C), is practically eliminated. [Ref. 130]

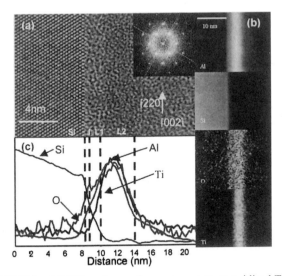

Figure 13: HRTEM and EELS analyses of room-temperature oxidized TAO layer grown on Si(100). (**a**) HRTEM image, (**b**) elemental maps and (**c**) integrated line scans for Al, Si, O and Ti, respectively. [Ref. 130]

between the crystalline Si substrate and the amorphous TAO layer. A Fourier transformation of the line-scan at the L1 and L2 layers location did not reveal any periodicities. Energy filtering TEM (EFTEM) was used to obtain a qualitative mapping of the elements Si, O, Ti and Al in a 200 kV

114

TEM (TECNAI F20ST). For energy width of 10 eV, a resolution of better than 0.6 nm can be expected from the dampling envelopes of spatial and temporal coherence.[202] Elemental maps using the Al-L, Si-L, Ti-L and O-K edges were obtained using the three-window method[203] (Fig. 13b). The TAO-layer (L2) and a layer containing mostly Al and oxygen (L1), which provides a much more stable interface than having direct contact between Ti and Si, were recognizable in the high-resolution TEM image. An intermediate layer (I) consisting mostly of Si and O (SiO_2 or SiO_x?) was evident between the substrate and L1, in agreement with the XPS analysis showing a minimal Si^{3+} intensity on the TAO layer oxidized at RT (Fig. 12).

The C-V analysis of capacitors defined with 100-500 μm diameter top Pt electrodes, with various TAO physical thicknesses, demonstrated a permittivity of 30 for the TAO dielectric layer with the particular composition used in the first studies.[130] This is among the highest values reported for binary and ternary amorphous oxides. A significant increase for the accumulation capacitance density was observed for the TAO dielectric layer produced at temperatures < 200 °C with atomic oxygen, due to a large suppression of the interfacial SiO_2 formation. Figure 14 shows the C-V characteristics for 4 nm thick TAO layers produced on n⁻ Si and n⁺ Si by RT oxidation. The dual frequency technique [204] was used to extract the C-V behaviour of TAO layers grown on n⁻ Si, to eliminate the parasitic resistance influence from the substrate. High capacitance densities of 7.7 - 8.3 μF/cm² (EOT = 0.42 – 0.48 nm) were obtained on all n⁻ and p⁻ Si substrates, consistent with those measured from the same TAO layers grown on heavily doped Si (ρ = 0.004 Ω·cm). The leakage current across the TAO layers, at 1 V, was about 5 A/cm² on n⁻ and p⁻ substrates (i.e., 4-5 decade lower than that for SiO_2 with a similar EOT).

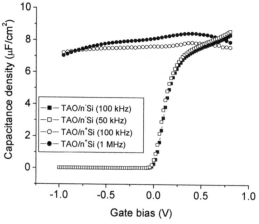

Figure 14: C-V characteristics of a TAO-based MOS capacitors on n⁻ and n⁺ Si with Pt top electrodes, measured at different frequencies.

Preliminary bandgap measurements via spectroscopic ellipsometry and EEELS revealed a bandgap of ~ 4 eV for the TAO composition reported here, Initial measurements of the trapping density of states revealed values of ~ 8×10^{11} to 10^{12}. It is encouraging that by adding a very small amount of Al_2O_3 into a TiO_2 matrix, the bandgap of the latter is increased by ~ 1 eV, and the density of trapping states is within the same order of magnitude as for frontrunners amorphous oxide candidate materials. Atomic force microscopy (AFM) analysis demonstrated that the surface morphology of the thin TAO layer remained practically the same as that of the original atomically smooth Si surface after 30 sec rapid thermal annealing up to 950 °C, indicating that no crystallization or interfacial reaction has taken place. There is substantial room for compositional improvement in the TAO layer to optimize its performance as a high-K dielectric. However, further work is required to fully explore the potential of TAO layer as a candidate material for the next generation CMOS devices.

3.4.2.2 Amorphous LaAlO₃ High-K Dielectrics

Attention has focused recently on amorphous and crystalline $LaAlO_3$ (LAO) as potential high-K dielectric materials for the next generation of CMOS gates (45nm node). The science and technology of crystalline LAO thin films is discussed by Schlom in another chapter of this book.[205] Therefore, only amorphous LAO thin films will be briefly discussed in this chapter for completion of the treatment of amorphous high-k gates.

The reasons for such attention are based on the demonstrated properties of this material, which include: (a) relatively high permittivity (20-26), [205-207] (b) suitable band-gap (~ 5 eV), and (c) thermal stability on Si up to 2100 °C.[208] However, the synthesis of LAO thin films involve processes and structure transformation that exhibit disadvantages such as relatively high deposition temperature and a cubic-to-rhombohedral phase transition a about 500 °C.[209] This introduces difficulties in preserving amorphous structural quality of the material.[210] LAO films have been synthesized by various methods; including magnetron sputter deposition,[211] electron-beam evaporation,[211] MBE,[208, 209, 212] and MOCVD. Control of film stoichiometry and resulting properties appear to depend strongly on the synthesis method and variations in parameters employed for post deposition annealing.

For example, films produced by sputter deposition, including ion-assisted growth (via substrate bias at –90 V), [211] where the films were bombarded by low energy ions during growth, resulted in amorphous $(La_xAl_{1-x})_2O_3$ films (with x > 0.5) that generally were less dense (~ 10%) than the crystalline counterpart, resulting in significantly reduced dielectric constant (~13.4) with respect to that characteristic of crystalline LAO (~20-26) as indicated above. So, although ion-assisted deposition was used to presumably increase film density, the desired effect was apparently not achieved. One question that comes to mind is whether the ion bombardment during growth resulted in the well know preferential sputtering effect with preferential loss of oxygen characteristic of ion bombarded complex oxides, thus affecting the structure and density of the films.[214] However, the work cited here on sputter-deposition of amorphous LAO showed that these films, as similar to those synthesized by the other methods, are thermodynamically stable on Si.

The question is whether that stability might depend on the presence of AlO_4 structures in the material. In this respect, it is known from magnetic resonance measurements[215] that $(La_xAl_{1-x})_2O_3$ has AlO_4, AlO_5, AlO_6 network units in numbers that depend on both relative concentrations of Al and La and the annealing temperature used to produce the material. AlO_4 units were found in $LaAlO_3$ films. [211] Therefore, an interesting question to answer is whether the stability of the amorphous LAO films on Si is due to this unusual bonding in the lattice. In interesting to notice that similar AlO_x structures are found, via calculations, in Al_2O_3. For this case, it was suggested that low density of amorphous Al_2O_3 with respect to its crystalline counterpart might result in part from the presence of under-coordinated Al atoms in AlO_4 complexes. This under-coordination might occur in ion-assisted deposition, as in ref. 211 from preferential sputtering of oxygen during growth. Unfortunately, no electrical characterization of the LAO films was reported in ref. 211, so we cannot judge how the sputter-deposition process affected the electrical properties.

Recent studies of amorphous LAO films produced by MBE revealed interesting aspects of these films produced by this method.[212] Films were grown using and e-beam evaporator for Al and a high-temperature effusion cell for La. The oxygen partial pressure during growth was about 5×10^{-6} Torr, and active oxygen was produced by an RF plasma source. Amorphous LAO films of about 11 nm physical thicknesses were produced on Si at 400 °C. Top contacts of 50 nm Mo films were produced to define capacitors for electrical measurements. Both I-V measurements at different temperatures and C-V measurements at RT were performed. The results, according to the authors, indicate that the p-type Si substrate used for the measurements enter a non-equilibrium state when positive voltage is applied on the gate electrode, making the electrical behaviour similar to that of an MOS tunnel diode, where the current saturates for reverse bias and is dominated by the generation-recombination current of the minority carriers in Si. The energy bands of Si at the oxide interface are pinned by the gate metal Fermi level and the concentration of the inversion does no depend on the bias. Consequently, the Si substrate enters into a deep depletion state and the high-frequency C-V curve does not reach its minimum in the inversion regime. This identification of non-equilibrium behaviour in a MOS structure with and amorphous LAO gate produced by

MBE points to a possible drawback of extracting structure parameters from capacitance measurements, which might apply to other systems materials.

On the other hand, the work on low pressure MOCVD synthesis of amorphous LAO films[213] revealed that films grown at 400-700 °C in N_2 ambient containing O_2 exhibit interfacial problems related to the formation of a thin compositional graded Al-poor La-Al-Si-O interfacial layer. In addition, I-V measurement of the MOCVD LAO films revealed relatively high leakage currents that were attributed to the incorporation of C atoms into the film, where the C atoms originate from the MOCVD precursors. This points to a weakness of the MOCVD process for synthesizing amorphous LAO that wil need to be addressed.

The brief discussion of the recent work on LAO films presented above indicate that this material might be a promising candidate for the next generation of CMOS gates, but definitely much more work is required to address its suitability for such an application.

In general, suppression of crystallization, required for amorphous Al-based films, has also been observed for aluminates (depending upon Al content), but there are fewer studies of interdiffusion with the Si substrate available at this time. However, in the case of Al_2O_3 thin films, Al penetration into the Si substrate has been reported upon thermal annealing budgets appropriate for dopant activation, and a concomitant (perhaps coincidental) reduction in channel mobility was noted.[24] Such penetration could be a concern for any high-κ dielectric incorporating Al as well.

3.5 CONCLUSIONS

The technology roadmap for the next generation of complementary metal-oxide semiconductor (CMOS) devices indicates that an equivalent oxide thickness (EOT) of SiO_2 of less than 1.0 nm is needed to satisfy the requirement of maintaining a suitable capacitance when the gate length, and thus the area (A) of the CMOS gate, is reduced below 65 nm. The development of the next generation of nanoscale CMOS will impact not only logic devices but also the next generation of memories, including non-volatile high-density memories that will drive not only the stand along, but also the embedded memory markets. Industry, academia and national laboratories are currently investigating high-K dielectric materials as potential replacement for the SiO_2 gate. The experimental and theoretical work discussed in this chapter indicates that progress has been made in identifying the critical issues that need to be addressed and on developing alternative materials and processing for high-K dielectrics, However, substantial work remains to be done to develop the best possible high-K layer and the processing that will provide the integration into CMOS gates that will yield the properties required by the next generation of nanoscale CMOS devices.

ACKNOWLEDGEMENTS

RMW thanks G. Wilk for his collaborations and many contributions to the content of this chapter, and further acknowledges the support of the Texas Advanced Technology Program. OA acknowledges the great contribution of W. Fan (PhD student and postdoc at Argonne) and B. Kabius to the work on the new TAO high-K dielectric thin film described in this review. Also, OA acknowledges support from the Department of Energy-Office of Science-Basic Energy Science under Contract No. W-31-109-ENG-38.

REFERENCES

[1.] International Technology Roadmap for Semiconductors, 2001 edition, Available on-line at http://public.itrs.net/Files/2003ITRS/Home2003.htm.

[2.] G.D. Wilk, R.M. Wallace, J.M. Anthony, *J. Appl. Phys.* **89**, 5243 (2001).

[3.] R.M.Wallace and G.D, Wilk, "High-k Dielectric Materials for Microelectronics," *Critical Reviews in Solid State and Materials Sciences* **28**, 231 (2003).

[4.] R.M. Wallace and G. Wilk (Eds), MRS Bulletin, articles in special issue on "Alternative Gate Dielectrics for Microelectronics", vol 27 (March 2002).

[5.] R.M. Wallace, *Appl. Surf. Sci.* **231-232**, 543 (2004).

[6.] S.O. Kasap, *"Principles of Electrical Engineering Materials and Devices,"* McGraw-Hill, New York (2002).

[7.] S. Wolf, *"Silicon Processing in the VLSI Era,"* Vol. 4, Lattice Press, Sunset beach, CA (2003).

[8.] D.-S.Yoon, J.S.Roh, H.K.Baik, and S.-M.Lee, *Critical Reviews in Solid State and Materials Sciences* **27**, 143 (2002).

[9.] O. Auciello, J.F. Scaott, R. Ramesh, *Physics Today* (July 1998) p. 22-27.

[10.] C.A. Paz de Araujo, O. Auciello, and R. Ramesh (Eds.Science and Technology of Integrated Ferroelectrics: Past Eleven Years of the International Symposium on Integrated Ferroelectrics Proceedings",), Gordon and Breach Publishers, *"Ferroelectricity and Related Phenomena,"* vol. **11** (2000).

[11.] D. Schlom and J. Haeni, Mat. Res. Soc. Bull. **27**, 198 (2002).

[12.] J. Robertson and C.W. Chen, *Appl. Phys. Lett.* **74**, 1168 (1999).

[13.] J. Robertson, *J. Vac.Sci. Technol.* B **18**, 1785 (2000).

[14.] J. Robertson, J. Non-Cryst. Sol., 303, 94 (2002).

[15.] H. Nohira, W. Tsai, W. Besling, , E. Young, J. Petry, T. Conard, W. Vandervorst, S. De Gendt, M. Heyns, J. Maes, and M. Tuominen, *Journal of Non-Crystalline Solids* **303**, 83 (2002).

[16.] S. Miyazaki, *J. Vac.Sci. Technol.* B **19**, 2212 (2001).

[17.] G.D. Wilk, R.M. Wallace, J.M. Anthony, *J. Appl. Phys.* **87**, 484 (2000).

[18.] M. S. Akbar, S. Gopalan, H.-J. Cho, K. Onishi, R. Choi, R. Nieh, C. S. Kang, Y. H. Kim, J. Han, S. Krishnan and J. C. Lee, *Appl. Phys. Lett.* **82**, 1757 (2003).

[19.] M.R. Visokay, J.J. Chambers, A.L.P. Rotondaro, A. Shanware, L. Colombo, *Appl.Phys. Lett.* **80**, 3183 (2002).

20. H. Y. Yu, M. F. Li, B. J. Cho, C. C. Yeo, M. S. Joo, D.-L. Kwong, J. S. Pan, C. H. Ang, J. Z. Zheng, and S. Ramanathan, *Appl. Phys. Lett.* **81**, 376 (2002).

21. E., Zhu, T.P Ma, T. Tamagawa, Y. Di, J. Kim, R. Carruthers, M. Gibson and T. Furukawa, *Tech. Dig. Int. Electron Devices Meet.*, **20.4.1** (2001).

22. G. Wilk, M. Green, M.Y. Ho, B. Busch, T. Sorsch, F. Klemens, B. Brijs, R. van Dover, A. Kornblit, T. Gustafsson, E. Garfunkel, S. Hillenius, D. Monroe, P. Kalavade, J. Hergenrother , *Symp. VLSI Tech. Technical Digest of Papers* **88** (2002).

23. R.M.C. de Almeida and I.J.R. Baumvol, *Surface Science Reports* **49**, 1 (2003).

24. S. Guha, E.P. Gusev, H. Okorn-Schmidt, M.C. Copel, L.A. Ragnarsson, N.A. Bojarczuk, P. Ronsheim, *Appl. Phys. Lett.* **81**, 2956 (2002).

25. T. Yamaguchi, H Satake, N. Fukushima, A Toriumi, *Technical Digest Int. Electron Dev. Meet.*, **19** (2000).

26. V.V. Afanas'ev, A. Stesmans, F. Chen, X. Shi, S.A. Campbell , *Appl. Phys. Lett.* **81**, 1053. (2002).

27. S. Sayan, E. Garfunkel, S. Suzer, *Appl. Phys. Lett.* **80**, 2135 (2002).

28. R. A. McKee, F. J. Walker, and M. F. Chisholm, *Phys. Rev. Lett.* **81**, 3014 (1998).

29. S. Jeon, F. J. Walker, C. A. Billman, R. A. McKee, and H.Hwang, *Technical Digest Int. Electron Dev. Meet.*, **26.7** (2002).

30. K. Eisenbeiser,J. M. Finder, Z. Yu, J. Ramdani, J. A. Curless, J. A. Hallmark, R. Droopad, W. J. Ooms, L. Salem, S. Bradshaw, and C. D. Overgaard, *Appl. Phys. Lett.* **76**, 1324 (2000).

31. Z. Yu, J. Ramdani, J. A. Curless, J. M. Finder, C. D. Overgaard, R. Droopad, K. W. Eisenbeiser, J. A. Hallmark, W. J. Ooms, J. R. Conner and V. S. Kaushik, *J. Vac. Sci. Technol.* B **18**, 1653 (2000).

32. M.V. Fischetti, D.A. Nuemayer, E.A. Cartier, *J. Appl. Phys.* **90**, 4587 (2001).

33. Z.Ren, M.V.Fischetti, E.P.Gusev, E.A.Cartier and M.Chudzik, *Technical Digest Int. Electron Dev. Meet.* **793** (2003).

34. A.L.P. Rotondaro, M.R. Visokay, J.J. Chambers, A. Shanware, R. Khamankar, H. Bu, R.T. Laaksonen, L. Tsung, M. Douglas, R. Kuan, M.J. Bevan, T. Grider, J. McPherson, L. Colombo, *Symp. VLSI Tech. Technical Digest of Papers*, **148** (2002).

35. S.Inumiya, K.Sekine, S. Niwa, A. Kaneko, M. Sato,T. Watanabe, H. Fukui, Y. Kamata, M. Koyama, A. Nishiyama, M. Takayanagi, K. Eguchi and Y. Tsunashima, *VLSI Tech. Technical Digest of Papers*, **T03.A1** (2003).

36. K. Onishi, C. S. Kang, R. Choi, H.-J. Cho, S. Gopalan, R.Nieh, S. Krishnan, and J. C. Lee, *VLSI Tech. Technical Digest of Papers*, **T03.P1** (2002).

37. Y. Morisaki, T. Aoyama, Y. Sugita, K. Irino, T. Sugii, and T. Nakamura, *Technical Digest Int. Electron Dev. Meet.*, **34.4** (2002).

38. C. H. Choi, S. J. Rhee, T. S. Jeon, N. Lu, J. H. Sim, R. Clark, M. Niwa and D. L. Kwong, *Technical Digest Int. Electron Dev. Meet.* **34.3** (2002).

39. Y. Kim, C. Lim, C.D. Young, K. Matthews, J. Barnett, B. Foran, A. Agarwal, G. A. Brown, G. Bersuker, P. Zeitzoff, M. Gardner, R. W. Murto, L. Larson, C. Metzner, S. Kher, and H. R. Huff, *Symp. VLSI Tech. Technical Digest of Papers*, **T12A.5** (2003).

40. A.Morioka, H.Watanabe, M. Miyamura, T. Tatsumi, M. Saitoh, T. Ogura, T. Iwamoto, T. Ikarashi, Y. Saito,Y.Okada, H. Watanabe, Y. Mochiduki, and T. Mogami, *VLSI Tech. Technical Digest of Papers*, **T12A.4** (2003).

41. B. Tavell, X. Garros, T. Skotnicki, F. Martin, C. Leroux, D. Bensahel, M.N. Séméria,Y. Morand, J.F. Damlencourt, S. Descombes, F. Leverd, Y. Le-Friec, P. Leduc, M Rivoire, S. Jullian, and R.Pantel, *Technical Digest Int. Electron Dev. Meet.*, **17.01** (2002).

42. S. Datta, G. Dewey, M. Doczy, B.S. Doyle, B. Jin, J. Kavalieros, R. Kotlyar, M. Metz, N. Zelick and R. Chau, *Technical Digest Int. Electron Dev. Meet.*, **28.1** (2003).

43. F. Gámiz, J. B. Roldán, J. E. Carceller, and P. Cartujo, *Appl. Phys. Lett.* **82**, 3251 (2003).

44. F. Gámiz and M.V.Fischetti, *Appl. Phys. Lett.* **83**, 4848 (2003).

45. M.Yang, E.P.Gusev, M.Ieong, O.Gluschenkov, D.C.Boud, K.K.Chan, P.M.Kozlowski, C.P.D'Emic, R.M.Sicina, P.J.Jamison, and A.I.Chou, *IEEE Electron Dev. Lett.* **24**, 339 (2003).

46. T. Ngai, W. J. Qi, R. Sharma, J. Fretwell, X. Chen, J. C. Lee, and S. Banerjee, *Appl. Phys. Lett.* **76**, 502 (2000), and **78**, 3085 (2001).

47. Z.Shi, D.Onsongo, K. Onishi, J. C. Lee, S. K. Banerjee, *IEEE Elect. Dev Lett.* **24**, 34 (2003).

48. C.O.Chui, S. Ramanathan, B. B. Triplett, P. C. McIntyre, K. C. Saraswat, *IEEE Elect. Dev. Lett.,* **23**, 473 (2003).

49. C.O.Chui, H.Kim, P. C. McIntyre, and K. C. Saraswat, *Technical Digest Int. Electron Dev. Meet.,* **437** (2003).

50. A.Ritenour, S.Yu, M.L.Lee, N.Lu, W.Bai, A.Pitera, E.A.Fitzgerald, D.L.Kwong, and D.A.Antionadis, *Technical Digest Int. Electron Dev. Meet.,* **433** (2003).

51. K. Torii, Y. Shimamoto, S. Saito, O. Tonomura, M. Hiratani, Y. Manabe, M. Caymax, J.W. Maes, Symp. *VLSI Tech. Technical Digest of Papers,* **188** (2002).

52. L.-Å.Ragnarsson, L.Pantisano, V.Kaushik, S.-I.Saito, Y.Shimamoto, S.Degent and M.Heyns, *Technical Digest Int. Electron Dev. Meet.,***87** (2003).

53. M. Quevedo-Lopez, M.El-Bouanani, S. Addepalli, J.L. Duggan, B.E. Gnade, R.M. Wallace, M.R. Visokay, M. Douglas, M.J. Bevan, L. Colombo, *Appl. Phys. Lett.* **79**, 2958 (2001).

54. M. Quevedo-Lopez, M.El-Bouanani, S. Addepalli, J.L. Duggan, B.E. Gnade, R.M. Wallace, M.R. Visokay, M. Douglas, L. Colombo, *Appl. Phys. Lett.* **79**, 4192 (2001).

55. M. Quevedo-Lopez, M.El-Bouanani, B.E. Gnade, R.M. Wallace, M.R. Visokay, M. Douglas, M.J. Bevan, L. Colombo, *J. Appl. Phys.* **92**, 3540 (2002).

56. S.Saito, D.Hisamoto, S.Kimura and M.Hiratani, *Technical Digest Int. Electron Dev. Meet.,* **797** (2003).

57. R. Beyers *J. Appl. Phys.* **56**, 147 (1984).

58. K.J. Hubbard and D.G. Schlom, *J. Mat. Res.* **11**, 2757 (1996).

59. S.Q. Wang and J.W. Mayer, *J. Appl. Phys.* **64**, 4711 (1988).

60. I. Barin and O. Knacke, *Thermochemical Properties of Inorganic Substances,* (Springer-Verlag, Berlin) (1973).

61. L.B. Pankratz , *Thermodynamic Properties of Elements and Oxides,* (U.S. Dept. of Interior, Bureau of Mines Bulletin 672, U.S. Govt. Printing Office, Washington, D.C., 1982) (1982).

62. S.P. Murarka, *Silicides for VLSI Applications,* (Academic Press, New York) (1983).

63. D.A. Neumeyer, E. Cartier *J. Appl. Phys.* **90**, 1801 (2001).

64. B.W Busch, O Pluchery, YJ Chabal, DA Muller, RL Opila, J Kwo, E Garfunkel, *Mat. Res. Soc. Bulletin,* **27**, 206 (2002).

65. D. Niu, R.W. Ashcraft, M. J. Kelly, J.J. Chambers, T.M. Klein, G.N. Parsons, *J. Appl. Phys.* **91**, 6173 (2002).

66. B.W Busch, J Kwo, M Hong, JP Mannaerts, BJ Sapjeta, WH Schulte, E Garfunkel, T Gustafson, *Appl. Phys. Lett.* **79**, 2447 (2001).

67. J.J Chambers, GN Parsons, *Appl. Phys. Lett.* **77**, 2385 (2000).

68. J.P. Maria, D. Wickasana, A.I. Kingon, B. Busch, H. Schulte, E. Garfunkel, T. Gustafson , *J Appl. Phys.* **90**, 3476 (2001).

69. J.P. Chang, Y.S. Lin, S. Berger, A. Kepton, R. Bloom, S. Levy, *J. Vac. Sci. and Tech.* B **19**, 2137 (2001).

70. D.A. Muller, G.D. Wilk, *Appl. Phys. Lett.* **79**, 4195 (2001).

71. G.D. Wilk, R.M. Wallace, *Appl. Phys. Lett* **74**, 2854 (1999).

72. G.D. Wilk, R.M. Wallace, *Appl. Phys. Lett.* **76**,112 (2000).
73. S. Gopalan, K. Onishi, R. Nieh, C.S. Kang, R. Choi, H.J. Cho, S. Krishnan, and J.C. Lee, *Appl. Phys. Lett.* **80**, 4416 (2002).
74. W.J. Qi, R. Nieh, E. Dharmarajan, B.H. Lee, Y. Jeon, L. Kang, K. Onishi, J.C. *Lee, Appl. Phys. Lett.* **77**, 1704-6 (2000).
75. A. Callagari, E. Cartier, M. Gribelyuk, H. Okorn-Schmidt, T. Zabel, *J. Appl. Phys.* **90**, 6466 (2001).
76. H. Watanabe, *Appl. Phys. Lett.* **78**, 3803 (2001).
77. T Gougousi, M Jason Kelly, GN Parsons, *Appl. Phys. Lett.* **80**, 4419 (2002).
78. D.G. Park, H. Cho, I.S. Yeo, J.A. Roh, J.M. Hwang, *Appl. Phys. Lett.* **77**, 2207(2000).
79. M. Houssa, V.V. Afanas'ev, A. Stesmans, M.M. Heyns , *Appl. Phys. Lett.* **77**, 1885 (2000).
80. J.M. Hergenrother, G.D. Wilk, T. Nigam, F.P. Klemens, D Monroe, P.J. Silverman, T.W. Sorsch, B Busch, M.L. Green, M.R. Baker, T. Boone, M.K. Bude, N.A. Ciampa, E.J. Ferry, A.T. Fiory, S.J. Hillenius, D.C. Jacobson, R.W. Johnson, P. Kalavade, R.C. Keller, C.A. King, A. Kornblit, H.W. Krautter, J.T.C. Lee, W.M. Mansfield, J.F. Miner, M.D. Morris, S.H. Oh, J.M. Rosamilia, B.J. Sapjeta, K. Short, K. Steiner, D.A. Muller, P.M. Voyles, J.L. Grazul, E.J, Shero, M.E. Givens, C. Pomarede, M. Mazanec, C. Werkhoven , *Technical Digest Int. Electron Dev. Meet.*, **51** (2001).
81. E.P. Gusev, D.A. Buchanan, E. Cartier, A. Kumar, D. DiMaria, S. Guha, A. Callegari, S. Zafar, P.C. Jamison, D.A. Nuemayer, M. Copel, M.A. Gribulyek, H. Okorn-Schmidt, C. D'Emic, P. Kozlowski, K. Chan, N. Bojarczuk, L.-A. Ragnarsson, P. Ronsheim, K. Rim, R.J. Fleming, A. Mocuta, A. Ajmera, *Technical Digest Int. Electron Dev. Meet.* **451** (2001).
82. C. Hobbs, H. Tseng, K. Reid, B. Taylor, L. Dip, L. Hebert, R. Garcia, R. Hegde, J. Grant, D. Gilmer, A. Franke, V. Dhandapani, M. Azrak, L. Prabhu, R. Rai, S. Bagchi, J. Conner, S. Backer, F. Dumbuya, B. Nguyen, P. Tobin, *Technical Digest Int. Electron Dev. Meet.,* **651** (2001).
83. H. Kim, P. McIntyre, K. Saraswat, *Appl. Phys. Lett.* **82**, 106 (2003).
84. M.Y. Ho, H. Gong, G.D. Wilk, B.W. Busch, M.L. Green, P.M. Voyles, D.A. Muller, M. Bude, W.H. Lin, A. See, M.E. Loomans, S.K. Lahiri, P.I. Räisänen , *J. Appl. Phys.* **93**, 1477 (2003).
85. D.G. Park, K.Y. Lim, H.J. Cho, J.J. Kim, J.M. Yang, J. Ko, I.S. Yeo, J.W. Park, H. de Waard, M. Tuominem, *J.Appl. Phys.* **91**, 65 (2002).
86. J.H. Lee, K. Koh, N.I. Lee, K.H. Cho, Y.K. Kim, J.S. Jeon, K.H. Cho, H.S. Shin, M.H. Kim, K. Fujihara, H.K. Kang, J.T. Moon, *Technical Digest Int. Electron Dev. Meet.*, **645** (2000).
87. K.Y. Lim, D.G. Park, H.J. Cho, J.J. Kim, J.M. Yang, I.S. Choi, I.S. Yeo, J.W. Park, *J. Appl. Phys.* **91**, 414 (2002).
88. K Onishi, L Kang, R Choi, E Dharmarajan, S Gopalan, Y Jeon, C Kang, B Lee, R Nieh, JC Lee, Symp. *VLSI Tech. Technical Digest of Papers*, **131** (2001).
89. K. Onishi, L. Kang, R. Choi, H.J. Cho, S. Gopalan, R. Nieh, E. Dharmarajan, J.C. Lee, *Technical Digest Int. Electron Dev. Meet.*, **30.3.1** (2001).
90. M. Quevedo-Lopez, M. El-Bouanani, M.J. Kim, B.E. Gnade, M.R. Visokay, A. LiFatou, M.J. Bevan, L. Colombo, R.M. Wallace, *Appl. Phys. Lett.* **81**, 1074 (2002).
91. M. Quevedo-Lopez, M. El-Bouanani, M.J. Kim, B.E. Gnade, M.R. Visokay, A. LiFatou, M.J. Bevan, L. Colombo, R.M. Wallace, *Appl. Phys. Lett.* **81**, 1609 (2002).
92. M. Quevedo-Lopez, M. El-Bouanani, M.J. Kim, B.E. Gnade, M.R. Visokay, A. LiFatou, J.J. Chambers, L. Colombo, R.M. Wallace, *Appl. Phys. Lett.* **82**, 4669 (2003).
93. R.M. Wallace, R.A. Stolz, G.D. Wilk, U.S. Patents 6,013,553; 6,020,243; 6,291,866; 6,291,867 (2000).

122

94. M.L.Green, E.P.Gusev, R. DeGraeve, and E.L.Garfunkel, *J. Appl. Phys.* **90**, 2057 (2001).
95. M. Koyama, A.Kaneko, T. Ino, M. Koike, Y. Kamata, R. Iijima,Y. Kamimuta, A Takashima, M. Suzuki, C.Hongo, S. Inumiya, M. Takayanagi and A. Nishiyama, *Technical Digest Int. Electron Dev. Meet.,* **34.1** (2003).
96. S. Jeon, C.J. Choi, T.Y. Seong, H. Hwang , *Appl. Phys. Lett.* **79**, 245 (2001).
97. Y Tanida, Y Tamura, S Miyagaki, M Yamaguchi, C Yoshida, Y Sugiyama, H Tanaka , Symp. *VLSI Tech. Technical Digest of Papers,* **190** (2002).
98. C.S. Kang, H.J. Cho, K. Onishi, R. Choi, R. Nieh, S. Goplan, S. Krishnan, J.C. Lee, Symp. *VLSI Tech. Technical Digest of Papers,* **146** (2002).
99. Y. Morisaki, T. Aoyama, Y. Sugita, K. Irino, T. Sugii, and T. Nakamura, *Technical Digest Int. Electron Dev. Meet.,* **34.04** (2002).
100. M.L. Green, M.Y. Ho, B. Busch, G.D. Wilk, T. Sorsch, T. Conard, B. Brijs, W. Vandervorst, P.I. Räisänen, D. Muller, M. Bude, and J. Grazul, *J. Appl. Phys.* **92**, 7168 (2002).
101. K.Sekine, S.Inumiya, M.Sato, A.Kaneko, K.Eguchi, and Y.Tsunashima, *Technical Digest Int. Electron Dev. Meet.,* **103** (2003).
102. T.Yamaguchi, R. Iijima, T. Ino, A.Nishiyama, H. Satake, and N. Fukushima, *Technical Digest Int. Electron Dev. Meet.,* **26.3** (2002).
103. J.R.Jameson, P.B.Griffin, A.Agah, J.D.Plummer, H.-S.Kim, D.V.Taylor, P.c.McIntyre and W.A.Harrison, *Technical Digest Int. Electron Dev. Meet.,* **91** (2003).
104. H.Reisenger, et al., *Technical Digest Int. Electron Dev. Meet.,* **267** (2001).
105. W.A.Phillips, *J. Low Temp. Phys.* **7**, 351 (1972).
106. P.W.Anderson, *Phil. Mag.,* **25, 1** (1972).
107. T.Klein, D.Niu, W.Li, D.M.Maher, C.C.Hobbs, R.I.Hedge, I.J.R.Baumvol, G.N.Parsons, *Appl. Phys. Lett.* **75**, 4001 (1999).
108. C. Krug, E.B.O. da Rosa, R.M.C. de Almeida, I.J.R.Baumvol, T.D.M. Salgado, F.C. Stedile, *Phys. Rev. Lett.,* **85**, 4120 (2000); **86**, 4714 (2001).
109. E. B.O. da Rosa, I.J.R.Baumvol, J.Morais, R.M.C. de Almeida, R.M.POapaleo, F.C. Stedile, *Phys. Rev. B* **85**, 121303 (2002)
110. M.Kundu, M.Ichikawa, N.Miyata, *Appl. Phys. Lett.,* **78**, 1517 (2001); **91**, 492 (2002); **92**, 1914 (2002).
111. M.Copel, M.Gribelyuk, E.Gusev, *Appl. Phys. Lett.* **76**, 436 (2000).
112. B.W.Busch, W.H.Schulte, E.Garfunkel, T.Gustafsson, W.Qi, R.Nieh, J.Lee, *Phys. Rev. B* **62**, 13290 (2000).
113. T.Gustafsson, H.C.Lu, B.W.Busch, W.H.Schulte, E.Garfunkel, Nucl. Instrum. *Meth. Phys. Res. B* **183**, 146 (2001).
114. K.P.Bastos, J.Morais, L.Miotti, R.P.Pezzi, G.V.Soares, I.J.R.Baumvol, H.-H.Tseng, R.I.Hedge, P.J.Tobin, *Appl. Phys. Lett.* **81** (2002) 1669.
115. J.Morais, E.B.O. de Rosa, L.Miotti, R.P.Pezzi, I.J.R.Baumvol, A.L.P.Rotondaro, M.J.Bevan and L. Colombo, *Appl. Phys. Lett.* **78**, 2446 (2001).
116. J.Morais, L.Miotti, G.V.Soares, R.P.Pezzi, S.R.Teixeira, K.P.Bastos, I.J.R.Baumvol, J.J.Chambers, A.L.P.Rotondaro, M.Visokay and L. Colombo, *Appl. Phys. Lett.* **81**, 2995 (2002).
117. J.Morais, E.B.O. de Rosa, R.P.Pezzi, L.Miotti, I.J.R.Baumvol, *Appl. Phys. Lett.* **79**, 1998 (2001).
118. E.B.O. da Rosa, J.Morais, R.P.Pezzi, L.Miotti, I.J.R.Baumvol, *J. Electrochem. Soc.* **148**, G44 (2001).
119. I.J.R.Baumvol, *Surface Science Reports,* **36** 1 (1999).
120. M. A. Quevedo-Lopez, M. El-Bouanani, R. M. Wallace, and B. E. Gnade, *J. Vac. Sci. Technol.,* **A20**, 1891 (2002).
121. .H. Lee and M. Ichikawa, *J.Appl.Phys.,* **92**, 1929 (2002).

122. Y.S.Lin, R.Puthenkovilaken, J.P.Chang, *Appl. Phys. Lett.* **81**, 2041 (2002).

123. R.I. Hegde, C.C. Hobbs, L.Dip, J. Schaeffer, and P.J. Tobin, *Appl. Phys. Lett.* **80**, 3850 (2002).

124. C.S. Kang, K. Onishi, L. Kang, and J.C. Lee, *Appl. Phys. Lett.* **81**, 5018 (2002).

125. A. Chatterjee, R.A. Chapman, K. Joyner, M. Otobe, S. Hattangady, M. Bevan, G.A. Brown, H. Yang, Q. He, D. Rogers, S.J. Fang, R. Kraft, A.L.P. Rotondaro, M. Terry, K. Brennan, S.W. Aur, J.C. Hu, H-L. Tsai, P. Jones, G. Wilk, M. Aoki, M. Rodder, I-C. Chen , *Technical Digest Int. Electron Dev. Meet.* 777 (1998).

126. For example, see: G.D.Wilk and R.M.Wallace, US Patent 6,245,606 (2001).

127. S. Ramanathan, G.D.Wilk, D.A.Muller, C.-M.Park, P.C.McIntyre, *Appl. Phys. Lett.* **79**, 2621 (2001).

128. S. Ramanathan, D.A.Muller, G.D.Wilk, C.-M.Park, P.C.McIntyre, *Appl. Phys. Lett.* **79**, 3311 (2001).

129. P. Punchaipetch, G. Pant, M. Quevedo-Lopez, H. Zhang, M. El-Bouanani, M.J. Kim, R.M. Wallace,B.E. Gnade, *Thin. Sol. Films* **425**, 68 (2003).

130. O. Auciello, W. Fan, B. Kabius, S. Saha, and J. A. Carlisle, R. P. H. Chang, C. Lopez and E. A. Irene, R. A. Baragiola, *Appl. Phys. Lett.* (in press, 2005).

131. T.J. King, J.P. McVittie, K.C. Saraswat, and J.R. Pfiester *IEEE Trans. Electron Dev.*, **41**, 228 (1994).

132. K. Uejima, T. Yamamoto, and T. Mogami, *Technical Digest Int. Electron Dev. Meet.*, **445** (2000).

133. Q. Lu, H. Takeuchi, X. Meng, T.J. King, C. Hu, K. Onishi, H.J. Cho, and J. Lee, *Symp. VLSI Tech. Technical Digest of Papers*, **86** (2002).

134. A. Chatterjee, M. Rodder, I.-C. Chen, *IEEE Trans. Electron Dev.* **45**, 1246 (1998).

135. Y. Ma, Y. Ono, L. Stecker, D.R. Evans, S.T. Hsu , *Technical Digest Int. Electron Dev. Meet.*, **149** (1999).

136. W.Tsai, L.-Å.Ragnersson, L.Pantisano, P.J.Chen, B.Onsia, T.Schram, E.Cartier, A.Kerber, E.Young, M. Caymax, S.DeGendt, and M. Heynes, *Technical Digest Int. Electron Dev. Meet.*, **311** (2003).

137. S. Murtaza, J. Hu, S. Unnikrishnan, M. Rodder, and I. Chen, *Proc. SPIE* **3506**, 49 (1998).

138. I. De, D. Johri, A. Srivastava, and C.M. Osburn, *Sol.-State Electron.* **44**, 1077 (2000).

139. S.B. Samavedam, H.H. Tseng, P.J. Tobin, J. Mogab, S. Dakshina-Murthy, L.B. La, J. Smith , J. Schaeffer, M. Zavala, R. Martin, B.-Y. Nguyen, L. Hebert, O. Adetutu, V. Dhandapani, T.Y. Luo, R. Garcia, P. Abramowitz, M. Moosa, C.C. Gilmer, C. Hobbs, W.J. Taylor, J. Grant, R. Hegde, S. Bagchi, E. Luckowski, V. Arunachalam, and M. Azrak , *Symp. VLSI Tech. Technical Digest of Papers*, **24** (2002).

140. C.H. Lee, J.J. Lee, W.P. Bai, S.H. Bae, J.H. Sim, X. Lei, R.D. Clark, Y. Harada, M. Niwa, and D.L. Kwong, *Symp. VLSI Tech. Technical Digest of Papers*, **82** (2002).

141. J.Chen, B.Maiti, D.Connelly, M.Mendicino, F.Huang, O.Adetutu, Y.Yu, D.Weddington, W.Wu, J.Candelaria, D.Dow, P.Tobin and J.Mogab, *Symp. VLSI Tech. Technical Digest* of Papers, **25** (1999).

142. H.Shimada, Y.Hirano, T.Ushiki, K.Ino, and T.Ohmi, *IEEE Trans. Elect. Dev.*, **44**, 1903 (1997).

143. A.Vandooren, S.Egley, M.Zavala, A.Franke, A.Barr, T.White, S.Samavedam, L.Mathew, J.Schaeffer, D.Pham, J.Conner, S.Dakshina-Murthy, B.-Y.Nguyen, B.White, M.Orlowski, and J.Mogab, *IEEE SOI Int. Conf.*, Oct. 7-20, 205 (2002).

144. A.Vandooren, A.Barr, L.Mathew, T.R.White, S.Egley, D.Pham, M.Zavala, S.Samavedam, J.Schaeffer, J.Conner, B.-Y.Nguyen, B.E.White Jr., M.K.Orlowski, and J.Mogab, *IEEE Elect. Dev. Lett.* **24**, 342 (2003).

145. A. Vandooren, A.V.Y. Thean, Y. Du, I. To, J. Hughes, T. Stephens, M. Huang, S. Egley , M. Zavala, K. Sphabmixay, A.Barr, T. White, S. Samavedam, L. Mathew, J.

124

Schaeffer, D. Triyoso, M. Rossow, D. Roan, D. Pham, R. Rai,, B.-Y.Nguyen, B. White, M. Orlowski, A. Duvallet , T. Dao, and J. Mogab, *Technical Digest Int. Electron Dev. Meet.*, **11.5** (2003).

[146.] S. B. Samavedam, L. B. La, J. Smith, S. Dakshina-Murthy, E. Luckowski, J. Schaeffer, M. Zavala, R.Martin, V. Dhandapani, D. Triyoso, H. H. Tseng, P. J. Tobin, D. C. Gilmer, C. Hobbs, W. J. Taylor, J.M. Grant, R. I. Hegde, J. Mogab, C. Thomas, P. Abramowitz, M. Moosa, J. Conner, J. Jiang, V.Arunachalam, M. Sadd, B-Y. Nguyen and B. White, *Technical Digest Int. Electron Dev. Meet.,* **17.02** (2002).

[147.] V. Misra, G.P. Heuss, H. Zhong , *Appl. Phys. Lett.* **78**, 4166 (2001).

[148.] Y.C.Yeo, T.J. King, and C. Hu, *J. Appl. Phys.* **92**, 7266 (2002).

[149.] Q. Lu, R. Lin, P. Ranade, Y.C. Yeo, X. Meng, H. Takeuchi, T.J. King, C. Hu, H. Luan, S. Lee, W. Bai, C.H. Lee, D.L. Kwong, X. Guo, X. Wang, T.P. ,*Technical Digest Int. Electron Dev. Meet,* **641** (2000).

[150.] Y.C.Yeo, P. Ranade, Q. Lu, R. Lin, T.J. King, and C. Hu , *Symp. VLSI Tech. Technical Digest of Papers,* **49** (2001).

[151.] Y.C.Yeo, P. Ranade, T.J. King, and C. Hu , *IEEE Electron Device Letters* **23**, 342 (2002).

[152.] H. Zhong, G.P. Heuss, V. Misra, H. Luan, C.H. Lee, D.L. Kwong , *Appl.Phys. Lett.* **78**, 1134 (2001).

[153.] H. Zhong, S.N. Hong, Y.S. Suh, H. Lazar, G. Heuss, V. Misra ,*Technical Digest Int. Electron Dev. Meet,* **567** (2001).

[154.] Y.S. Suh, G.P. Heuss, H. Zhong, S.N. Hong, V. Misra, Symp. *VLSI Tech. Technical Digest of Papers,* **47**(2001).

[155.] Y.S. Suh, G.P Heuss, V. Misra, *Appl. Phys. Lett.* **80**,1403 (2002).

[156.] V. Misra, H. Zhong, H. Lazar, *IEEE Elect. Dev. Lett.* **23**, 354 (2002).

[157.] J.H. Lee, H.Zhong, Y.-S.Suh, G. Heuss, J. Gurganus, B.Chen and V.Misra, *Digest Int. Electron Dev. Meet,* **14.02** (2002).

[158.] J.H. Lee, Y.-S. Suh, H. Lazar, R. Jha, J. Gurganus, Y. Lin and V. Misra, *Digest Int. Electron Dev. Meet,* **323** (2003).

[159.] Q. Lu, R. Lin, P. Ranade, T.J. King, C. Hu , Symp. *VLSI Tech. Technical Digest of Papers,* **45** (2001).

[160.] I. Polishchuk, P. Ranade, T.J. King, and C. Hu , *IEEE Electron Dev. Lett.* **22**, 444 (2001).

[161.] I. Polishchuk, P. Ranade, T.J. King, and C. Hu , *IEEE Electron Dev. Lett.* **23**, 200 (2002).

[162.] P.S. Lee , D. Mangelinck , K.L. Pey , J. Ding , D.Z. Chi , T. Osipowicz ,J.Y. Dai , and A. See, *Microelect. Eng.* **60**, 171 (2002).

[163.] J.P. Lu, D. Miles, J. Zhao, A. Gurba, Y. Xu, C. Lin, M. Hewson, J. Ruan, L. Tsung, R. Kuan, T. Grider, D. Mercer, and C. Montgomery*, Digest Int. Electron Dev. Meet,* **14.05**, 371 (2002).

[164.] J. Kedzierski, D. Boyd, P.Ronsheim, S.Zafar, J.Newbury, J.Ott, C. Cabral, M. Ieong, and W. Haensch, *Technical Digest Int. Electron Dev. Meet,* **315** (2003); data for B dopants was presented at the meeting, but is not described in the publication.

[165.] J. Kedzierski, E. Nowak, T. Kanarsky, Y. Zhang, D. Boyd, R.Carruthers, C. Cabral, R. Amos, C. Lavoie, R. Roy, J.Newbury, E. Sullivan, J. Benedict, P. Saunders, K.Wong, D. Canaperi, M. Krishnan, K.-L. Lee, B. A. Rainey, D. Fried, P. Cottrell, H.-S. P. Wong, M. Ieong, and W. Haensch, *Technical Digest Int. Electron Dev. Meet,* **10.01**, 247 (2002).

[166.] C.Y.Lin, M.W.Ma, A.Chin, y.C.Yeo, C.Zhu, M.F.Li,and D.-L.Kwong, *IEEE Electron Dev. Lett.,* **24**, 348 (2003).

[167.] For example, see: E.Wu, B.Linder, J.Stathis and W.Lai, *Technical Digest Int. Electron Dev. Meet,* **919** (2003).

168. A.S.Oates, *Technical Digest Int. Electron Dev. Meet*, **923** (2003).

169. Y. Harada, M. Niwa, S. Lee, D.L. Kwong, *Symp.VLSI Tech. Technical Digest of Papers,* **26** (2002).

170. A. Kumar, T.H. Ning, M.V. Fischetti, E. Gusev, *Symp.VLSI Tech. Technical Digest of Papers,* **152** (2002).

171. Y.H.Kim, K.Onishi, C. S. Kang, H.-J. Cho, R. Choi, S. Krishnan, M. S. Akbar, and J. C. Lee, *IEEE Elect. Dev. Lett.* **24**, 40 (2003).

172. Y. H. Kim, K. Onishi, C. S. Kang, R. Choi, H. -J. Cho, R. Nieh, J. Han, S. Krishnan, A. Shahriar, J. C. Lee, *Technical Digest Int. Electron Dev. Meet*, **26.05**, (2002).

173. A.Kerber, E. Cartier, L. Pantisano, R. Degraeve, T. Kauerauf, Y. Kim, A. Hou, G. Groeseneken, H. E. Maes, and U. Schwalke, *IEEE Elec. Dev. Lett.* **24**, 87 (2003).

174. A.Kerber, E. Cartier, L. Pantisano, R. Degraeve, Y. Kim, A. Hou, G. Groeseneken, H. E. Maes, and U. Schwalke, as presented at the *IEEE Semiconductor Interface Specialists Conference*, San Diego, CA., (2002).

175. M.Koyama, H.Satake, M.Koike, T.Ino, M.Suzuki, R.Iijima, Y.Kamimuta, A.Takashima, C. Hongo, and A.Nishiyama, Technical Digest Int. Electron Dev. Meet, **931** (2003).

176. A.Shanware, M.R.visokay, J.J.Chambers, A.L.P.Rotonndaro, J.McPherson, L. Colombo, G.A.Brown, C.H.Lee, Y.Kim, M.Gardner and R.W.Murto, *Technical Digest Int. Electron Dev. Meet*, **939** (2003).

177. R.Degraeve, A.Kerber, Ph. Roussel, E.Cartier, T.Krauerauf, L.Pantisano and G.Groeseneken, *Technical Digest Int. Electron Dev. Meet*, **935** (2003).

178. H.Sim and H. Hwang, "Effect of deuterium postmetal annealing on the reliability characteristics of an atomic-layer-deposited HfO_2/SiO_2 stack gate dielectrics," *Appl. Phys. Lett.* **81**, 4038 (2002).

179. H.-H. Tseng, M. E. Ramón, L. Hebert, P.J. Tobin, D. Triyoso, J. M. Grant, Z. X. Jiang, D. Roan, S. B. Samavedam, D. C. Gilmer, S. Kalpat, C. Hobbs, W. J. Taylor, O. Adetutu, and B. E. White, *Technical Digest Int. Electron Dev. Meet,* **4.1** (2003).

180. J.Lyding, K.Hess and I. Kizilyalli, *Appl. Phys. Lett.*, **68**, 2526 (1996).

181. I.Kizilyalli, J.Lyding and K.Hess, *IEEE Electron Dev. Lett.* **18**, 81 (1997).

182. H.C.Mogul, L.Cong, R.M.Wallace, T.A.Rost and K.Harvey, *Appl. Phys. Lett.* **72**, 1721 (1998).

183. R.M.Wallace, P.J.Chen, L.B.Archer and J.M.Anthony, *J. Vac. Sci. Technol.* B **17**, 2153 (1999).

184. P.J.Chen and R.M.Wallace, *Appl. Phys. Lett.* **73**, 3441 (1998).

185. P.J.Chen and R.M.Wallace, *Mat. Res. Soc. Symp. Proc.* **513**, 325 (1998).

186. P.J.Chen and R.M.Wallace, *J. Appl. Phys.* **86**, 2237 (1999).

187. J. C. Lee H. J. Cho, C. S. Kang, S. J. Rhee, Y. H. Kim, R. Choi, C. Y. Kang, C. H. Choi, M. Akbar, *Technical Digest Int. Electron Dev. Meet*, **4.4** (2003).

188. S. Zafar, A. Callegari, E. Gusev, and M.V. Fischetti, J. Appl. Phys. 93, 9298 (2003).

189. O. Auciello and A. R. Krauss (Eds), "Ion Beam Deposition and Surface Characterization of Thin Multicomponent Oxide Films During Growth", Book Chapter: Characterization of Thin Film Growth Processes and Surfaces via In Situ Techniques, *John Wiley and Sons,* Inc. (2001).

190. O. Auciello, A.R. Krauss, J. Im, and J.A. Schultz, *"Studies of Multicomponent Oxide Thin Films and Layered Heterostructure Growth Processes via In Situ, Time-of-Flight Ion Scattering and Direct Recoil Spectroscopy,"* in *"Annual Review of Materials Science,"* vol. 28, O. Auciello and R. Ramesh (Eds.) (1998) P.375.

191. A.R. Krauss, O. Auciello, A.M. Dhote, J. Im, E.A. Irene, Y. Gao, A.H. Mueller, S. Aggarwal, and R. Ramesh, *Integrated Ferroelectrics* vol. **32**, 121 (2000).

192. B.W. Batterman, *Phys. Rev. Lett.* **22**, 703 (1969).

126

193. D. L. Marasco, A. Kazimirov, M. J. Bedzyk, T. -L. Lee, S. K. Streiffer, O. Auciello, and G. -R. Bai, *Appl. Phys. Lett.* **79**, 515 (2001).

194. O. Auciello, C.M. Foster, and R. Ramesh, *in "Annual Review of Materials Science,"* vol. 28, O. Auciello and R. Ramesh (Eds.) (1998) p.501-531.

195. M. Leskela and M. Ritala, *J. Phys.* IV, **5**, C5 (1995).

196. M. Vehkamäki, T. Hatanpää, T. Hänninen, M. Ritala, and M. Leskelä, *Electrochem. And Solid State Lett.***2**, 504 (1999).

197. S. K. Streiffer, J. A. Eastman, D. D. Fong, C. Thompson, A. Munkholm, M. V. Ramana Murty, O. Auciello, G. R. Bai, G. B. Stephenson, *Phys. Rev. Lett.* **89** (2002) 067601.

198. M. V. Ramana Murty, S. K. Streiffer, G. B. Stephenson, J. A. Eastman, G.-R. Bai, A. Munkholm, O. Auciello, C. Thompson, *Appl. Phys. Lett.* **80**, 1809-1811 (2002).

199. D.D. Fong, G.B. Stephenson, S.K. Streiffer, J.A. Eastman, O. Auciello, P.H. Fuoss, C. Thompson, *Science*. **304**, 1650-1653 (2004).

200. W. Fan, B. Kabius, J. M. Hiller, S. Saha, J. A. Carlisle, O. Auciello, R. P. H. Chang, and R. Ramesh, *J. Appl. Phys.* **94**, 6192 (2003).

201. M. Haider, S. Uhlemann, E. Schwan, H. Rose, B. Kabius, and K. Urban, *Nature* **392**, 768 (1998).

202. K. Ishizuka, *Ultramicroscopy* **5**, 55 (1980).

203. R. F. Egerton, *Electron Energy-Loss Spectroscopy in the Electron Microscope*, (Plenum Press, New York, 1996).

204. K. J. Yang and C. Hu, *IEEE Trans. Electr. Dev.* **46**, 1500 (1999).

205. D. Schlom Chapter in this book and references there in.

206. G. Samara, *J. Appl. Phys.* **68**, 4214 (1990).

207. E.A. Giess, R.L. Sandstrom, W.J. Gallagher, A. Gupta, S.L. Shinde, R.F. Cook, E.I. Cooper, E.J.M. O'Sullivan, J.M. Roldan, A.P. Segmeller, J. Angilello, *IBM J. Res. Dev.***34**, 916 (1990).

208. B.E. Park and H. Ishiwara, *Appl. Phys. Lett.* **82**, 1197 (2003).

209. W. Xiang, H. Lu, L. Yan, H. Guo, L. Liu, Y. Zhou, G. Yang, J. Jinag, H. Cheng, Z. Chen, *J. Appl. Phys.* **93**, 533 (2003).

210. G. Sung, K. Kangand, S. Park, *J. Amer. Ceram. Soc* **74**, 437 (12991).

211. R.A.B. Devine, *J. Appl. Phys.* **93**, 9938 (2003).

212. B. Mereu, G. Sarau, A. Dimoulas, G. Apostolopoulos, I. Pintilie, T. Botila, L. Pintilie, M. Alexe, *Mater. Sci. and Engin.* B **109**, 94 (2004).

213. A-D. Li, Q-Y. Shao, H-Q. Ling, J-B. Cheng, D. Wu, *Appl. Phys. Lett.* **83**, 3540 (2003).

214. O. Auciello and R. Kelly (Eds.), "Ion Bombardment Modification of Surfaces: Fundamentals and Applications," *Elsevier Scientific Publishing Co.*, 1984.

215. D. Iuga, s. Simon, E. de Boer, A.P.M. Kentgens, *J. Phys. Chem.* B **103**, 7591 (1999).

216. G. Gutierrez and B Johansson, *Phys. Rev.* B **65**, 104202 (2002).

Magnetic Memory and Spintronics

CHAPTER 4

MATERIALS REQUIREMENTS FOR MAGNETIC RANDOM-ACCESS MEMORY (MRAM) DEVICES

Wolfgang Raberg

Infineon Technologies
ALTIS Semiconductor
224 Boulevard John Kennedy
91105 Corbeil-Essonnes, France

Arunava Gupta

Center for Materials for Information Technology (MINT)
Dept. of Chemistry, Chemical and Biological Engineering
The University of Alabama
Tuscaloosa, Al. 35487, USA

4.1 INTRODUCTION

Over the past several decades, the constant drive towards miniaturization of electronic devices has often led to new challenges and opportunities for each successive generation. As we step into the age of nanoelectronics, it is becoming increasingly apparent that the limits of conventional semiconductor charge-based electronics are being approached and revolutionary new concepts will need to be explored to help scale down to much smaller dimensions. One such concept that has been gaining increasing attention in recent years is the exploitation of the electron spin, in addition to its charge, to create a new class of spintronic (short for spin-based electronic) devices that, besides offering a path to future scaling, can provide additional functionality.[1, 2]

Our increasing ability to control the geometry and composition of materials in the nanometer scale has led to a number of significant new advances in thin film magnetism during the past two decades. The discovery of novel physical phenomena at practical temperatures in ferromagnetic metal-based systems, such as spin-dependent scattering [3] and tunneling,[4] has resulted in the rapid development of advanced storage and memory devices. Storage hard drives utilizing giant magnetoresistance (GMR) heads have already been in the market for several years and have helped revolutionize the industry. Significant progress has also been made in the development of tunneling magnetoresistance (TMR)-based memory, sensor and storage devices and in bringing them closer to commercialization. In particular, intensive development work on TMR-based magnetic random access memory (MRAM) devices is ongoing in a number of industrial laboratories world-wide, with the expectation that the first commercial products will be introduced in the market soon. [5,6] In addition to non-volatility, MRAM promises to provide radiation hardness, nondestructive readout, very fast read and write speeds (< 50 nsec), low power consumption, as well as unlimited read/write endurance. These combined attributes make it very attractive as a universal memory technology with a potentially diverse range of applications both as stand-alone and embedded-memory devices. Unlike conventional memories such as DRAM and Flash that rely on stored charge, in an MRAM element a bit of information is stored in the orientation of its bistable magnetization vector.

In this chapter we provide an overview of the material aspects related to fabrication of MRAM devices. The primary focus will be on the TMR-based magnetic stack that is at the core of the present-day memory cell being actively investigated. We address the function of the different stack layers and their key requirements. While a number of the materials issues related to the magnetic layers are similar to GMR-based stacks, the tunnel barrier is unique to TMR-based stacks and is covered in some detail.

4.2 STACK FUNCTION

The choice of materials for the memory cell is governed by the requirements for the two general mechanisms of reading from and writing the information into the cell. In an MRAM device information is stored in a multilayer stack consisting of the basic components of a ferromagnetic reference (pinned) layer, a non-magnetic spacer layer and a ferromagnetic information (free) layer as shown schematically in fig. 1. Depending on the relative orientation of the magnetization in the ferromagnetic layers the multilayer structure exhibits either a high or a low resistance to a current flowing through the device. To achieve a large magnetoresistive (MR) effect it is important that the ferromagnetic materials on either side of the spacer layer are highly spin-polarized. The spacer layer on the other hand must not cause any significant spin scattering of electrons moving from one ferromagnetic layer to the other. Two different magnetoresistive effects are utilized in MRAM devices, namely GMR and TMR-effect which are discussed below.

In order to be able to write information into an MRAM device there has to be a mechanism for changing the relative orientation of the ferromagnetic layers. This is usually achieved by generating a local magnetic field using two mutually perpendicular wires to switch the magnetization of one of the two layers. The required magnetic field depends on the intrinsic anisotropy and magnetization of the free layer, its thickness and the shape of the junction. At the same time the reference layer must be sufficiently rigid so as not to switch at the applied field strength. It is certainly desirable to use the smallest possible switching field consistent with highly reliable device operation so as to reduce the power consumption of the MRAM chip. In order to accommodate all these challenging requirements an elaborate structure for the magnetic stack has been developed in recent years. The key features of this structure will be discussed in the subsequent sections.

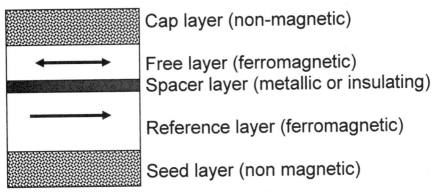

Figure 1: Schematic drawing illustrating the basic components of a typical stack utilized in MRAM devices where two ferromagnetic layers are separated by a non-magnetic spacer. The information is stored in the relative orientation of the magnetization of the two ferromagnetic layers and can be read out by the resulting resistance change of the structure.

4.3 REFERENCE LAYER

The simplest approach to realize a reference layer is to use a ferromagnetic material that has a much higher coercivity than the free layer. An alternative method, that is much more practical and is commonly utilized, is to use a softer magnetic material such as CoFe and exploit the mechanism of exchange biasing by depositing this magnetic material in direct contact with an anti-ferromagnetic material, such as NiO, FeMn, IrMn, PtMn, etc. Due to the exchange coupling at the interface between the antiferromagnetic material and the ferromagnet, the hysteresis loop related to the reversal of the ferromagnet can be shifted (biased) to one side of the field axis. Thereby one obtains a clear separation of the switching of the free layer from that of the reference layer.

Among the antiferromagnetic materials employed, NiO [7] is an insulator and is of use primarily in current-in-plane (CIP) GMR-based structures, since it reduces the amount of shunting current. In TMR-based devices, where the flow of current is perpendicular to the stack, conducting Mn-based antiferromagnets (AF's) are commonly utilized. These can roughly be divided into two groups. [8] The first consisting of materials like IrMn, FeMn or RhMn, which have an fcc-structure similar to the soft magnetic materials that they are intended to pin, and are antiferromagnetic as deposited. The Mn-content in these alloys is ordinarily quite high (around 80%), which is a potential cause for degradation at high temperatures. [9] They also tend to be more susceptible to corrosion. Nonetheless, IrMn exhibits good pinning characteristics even when used as very thin layers (below 10 nm) [10] which can be beneficial for the subsequent patterning steps and also has a positive impact on materials cost reduction. The blocking temperature (the temperature at which the exchange biasing reduces to zero) is around 230°C for IrMn. This does not pose a serious issue for MRAM devices since these are operated at much lower temperatures. For FeMn the blocking temperature is around 150 - 180°C which would still be acceptable for normal operation but might cause problems of depinning during the process integration steps needed for device fabrication. [11] The pinning properties of both FeMn and IrMn can be improved by a moderate anneal step (below 200°C) in the presence of a magnetic field.

A second group of antiferromagnetic materials include NiMn and PtMn. In the as-deposited metastable fcc-phase they are non-magnetic and require an annealing step to undergo the phase transformation into the antiferromagnetic fct-phase. The anneal has to be carried out at a minimum temperature of 230-250°C in a magnetic field in the presence of an adjacent ferromagnet to imprint the desired pinning orientation. Due to the reordering of the atoms in the unit cell during transformation the films need to have a certain minimum thickness in order to form a stable phase and they show a rather large change in the stress on forming the fct phase. [12] From a positive perspective, in comparison to IrMn, the lower Mn-content (50 at. %) makes these AF's much more stable against degradation resulting from Mn-diffusion.

The strength of the exchange bias effect is strongly dependent on the crystalline structure quality of the antiferromagnet and can be improved by providing adequate texturing underlayer(s) on which the antiferromagnets are grown. For example, it is known that a very thin underlayer of Cu can significantly improve the (111)-texture of IrMn. [13] In addition, Ta, NiFeCr,

NiCr, NiFe are also often used to provide the proper texture for the growth of IrMn, or PtMn. [14]

Even though one can design an adequate reference layer/free layer system by using the exchange bias effect, such a so-called simple-pinned structure is not readily usable in an MRAM device. The reason for this is that the two ferromagnetic electrodes will interact (couple) via magnetostatic stray fields once the stack is patterned as shown in fig. 2. This interaction will lead to a shift of the switching curve of the free layer because it will prefer to be aligned anti-parallel to the reference layer. To avoid such an effect, so-called artificial or synthetic antiferromagnetic (SAF) structures have been developed for GMR stacks that are used in sensor heads. [15,16] They can be readily applied to TMR stacks as well. In these structures the reference layer consists of an additional ferromagnetic layer which is separated from the first ferromagnetic layer by a thin non-magnetic spacer layer. The spacer layer material is chosen in such a way that it allows the mediation of interlayer exchange coupling between the two ferromagnetic layers. This exchange coupling exhibits an oscillatory behavior with spacer thickness, switching between the antiparallel and parallel states. [17] If one now selects a spacer thickness that establishes an antiparallel configuration between the two ferromagnetic layers one can tune the magnetic moments of these ferromagnetic layer in such a manner that their stray fields are cancelled out and do not introduce a shift in the hysteresis loop of the free layer. The most commonly used materials for these spacer layers are copper and ruthenium. The latter, for example, shows a strong antiferromagnetic coupling maximum around 7-8 Å. Such an AP-coupled system is in fact so rigid that it can, in principle, be used without an antiferromagnetic biasing layer as suggested, for example, by van den Berg et al. for the Co-Cu- [18] and the Co-Rh-system. [19]

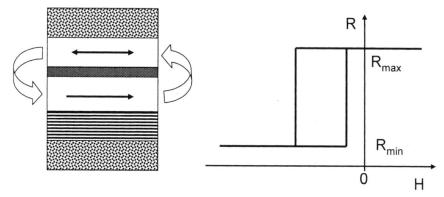

Figure 2: An uncompensated patterned magnetic layer in the reference electrode will shift the hysteresis loop off from the center due to magnetostatic stray fields that will likely preclude information storage (i.e., through the availability of two stable states at zero magnetic field).

4.4 TUNNEL BARRIER

Although MRAM devices were first realized using GMR stacks, [20] the fact that the current in this geometry flows in-the-plane (CIP) makes scaling to small feature sizes very challenging and most of the present effort on the development of high density memory technology is thus focused on TMR junctions, also commonly called magnetic tunnel junctions (MTJs). In these devices, the non-magnetic spacer layer between the free layer and the reference layer is replaced by a very thin (1-2 nm) insulating spacer, which is thin enough to allow electron tunneling between the two ferromagnetic layers. In contrast to CIP GMR, the transport is now perpendicular to the plane of the multilayer stack which theoretically enables the realization of very small devices that are competitive with today's mainstream memory devices such as DRAM, SRAM or Flash. In addition to the needed scalability, the MR-signal that can be realized with MTJ-stacks is significantly higher than that for GMR-stacks. For a fixed barrier thickness, the resistance of the cell would increase when scaling to smaller dimension requiring the use of thinner barrier layers to maintain the resistance. This is a challenging task and it is expected that with scaling to sub-100 nm dimensions GMR devices with current perpendicular to the plane (CPP) geometry will become more competitive because of the easier control of their resistance.

In magnetic tunnel junctions the MR depends on the spin polarization of the two ferromagnetic layers at the interface with the tunnel barrier, as described by the commonly used Julliere model: [21]

$$MR \propto \frac{2 P_1 P_2}{1 - P_1 P_2},$$

with P_1, P_2 being the spin polarizations at the Fermi level for the two electrodes. This is a simplified model and of course assumes that the spin of the electrons remains unchanged during the tunneling process.

4.4.1 Aluminum Oxide Barrier

The challenge from a materials perspective is to ensure that the spin-dependent tunneling is the dominant transport mechanism. Thus it is essential to avoid pinholes that would enable ordinary conductance through the barrier. Also impurities in the barrier will have a detrimental effect since they could act as spin scattering centers. [22] Much effort has been expended in recent years on the development of materials and techniques that enable the production of high quality tunnel barriers with excellent reproducibility. Based on the experience of fabricating superconducting tunnel junctions, aluminum-oxide (AlO_x) tunnel barriers have evolved as the standard in most MTJs due to the relative ease of their production and the wide range of resistance values over which they display reasonable performance in terms of the MR. This behavior can in the most part be attributed to the excellent wetting properties of Al and its ability to be oxidized with excellent control. Indeed, the major breakthrough that led to practical applications of MTJ's, with the first report of room temperature MR values in excess of 10% by Moodera *et al* in 1995, was realized using aluminum oxide as the barrier material.[23]

While there have been a few reports of the use of oxide ceramic targets,[24] reactive sputtering,[25] or ALCVD,[26] tunnel barriers based on aluminum oxide are almost exclusively fabricated by first depositing the multilayer metal stack in an UHV-PVD system up to the aluminum layer and then oxidizing the top aluminum layer. A variety of methods are now available for this process, including plasma or radical shower oxidation,[27] natural oxidation (exposure to oxygen atmosphere or air), ozone, etc. As an example, fig. 3 shows the observed *RA* (resistance x area)-product for a 1.2 nm thick Al-layer that had been exposed to ambient air as a function of the exposure time. Not surprising one finds that the resistance increases logarithmically with exposure. Nevertheless, other than for obtaining relatively low resistance junctions, the time required for oxidation using such a process becomes unacceptably long. Moreover, with the relatively thicker barriers required for achieving appropriate resistance values for MRAM devices, the repeatability of the natural oxidation process is inadequate. This is likely because of the fact that the aluminum will be readily oxidized along grain boundaries and subsequently more slowly inside the grains.[28] The resulting oxidation of the aluminum in the barrier is microscopically very non-uniform and thus causes variations in the resistance as well as in the MR.

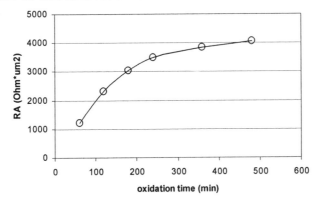

Figure 3: The resistance-area (*RA*) product of a magnetic tunnel junction with 1.2 nm thick aluminum layer as a function of the exposure time to ambient air (data obtained within the Infineon/IBM MRAM development alliance).

The oxidation homogeneity can be improved by annealing the stack, [29, 30, 31, 32] but natural oxidation performed within acceptable process times almost invariably lead to under-oxidized barriers. For this reason alternative techniques, such as plasma oxidation, exposure to oxygen radicals, or ozone treatment, are preferred for achieving RA values desirable for MRAM tunnel barriers ($\sim 500 - 10000$ Ω-μm^2).

A number of studies have shown that, unlike natural oxidation, a more complete oxidation can be achieved using plasma oxidation with relatively short process times. Indeed, a variety of different plasma oxidation techniques have been successfully utilized for obtaining barriers that exhibit reliable performance. [33,34] However, one need to be careful not to over-oxidize the aluminum using these more reactive processes, since this might lead to oxidation of the ferromagnetic layer underneath the barrier and result in the loss of the spin polarization and lead to reduced MR. Achieving the optimum barrier properties requires a well balanced combination of Al-thickness, oxidation and thermal treatment. [35] This includes any thermal budget introduced by subsequent process integration steps in the fabrication of the devices. A recent proposal to avoid over-oxidation involves utilizing a mixture of Kr/O_2 which is considered to permit a more rapid direct oxidation of the Al grains rather than the much slower oxidation path through grain boundary diffusion. [36]

4.4.2 Alternative Barrier Materials

While aluminum oxide remains by far the most commonly used barrier material for tunnel junctions, a number of other materials have also been recently investigated, particularly for producing low-resistance junctions. Aluminum nitride and oxynitride (AlN and AlON) barriers [37, 38] qualitatively exhibit a similar performance to that of AlO_x, but are usually more difficult to grow. Barriers using aluminum oxide with an inserted layer of hafnium (Hf) have shown good properties in terms of MR, bias-voltage and thermal dependence. [39] The properties of aluminum oxide barriers doped with boron have also been investigated for low resistance junctions. [40]

Ta_2O_5 is also one of the more attractive materials to produce low resistance junctions due its low barrier height (0.9 eV) and its electronegativity. Although junctions can be grown using Ta_2O_5, the observed MR of 10 % at room temperature is rather low. [41,42] Other candidates that have been studied for low resistance tunnel junctions are ZrO and ZrAlO, which have shown TMR values up to 15-20 % for very low RA values. [43] Tunnel junctions using the semiconducting ZnS barrier have been fabricated with MR values up to 5%. [44]

A novel approach for obtaining high MR in MTJs is the use of fully epitaxial structures in which coherent electron tunneling occurs across the barrier, as theoretically predicted in epitaxial structures such as Fe(001)/MgO(001)/Fe(001).[45,46] The mechanism of such a high MR ratio differs from that of the simple Julliere's model discussed earlier. Recent experimental results have confirmed that high MR can be achieved in such structures, as detailed in the last section of this chapter.

An interesting result has been observed using the perovskite oxide $SrTiO_3$ as the barrier material in MTJs between Co and the half-metallic oxide ferromagnet $La_{0.7}Sr_{0.3}MnO_3$.[47] In this case the tunnel magnetoresistance is in fact reversed, i.e., the low resistance state is found to be the antiparallel orientation of the two magnetic layers. It has been suggested that the inverse MR results from the properties of the ferromagnet-barrier interface which preferentially enables the tunneling of d-electrons in contrast to the s-electrons that dominate in the AlO_x barrier system. An inverse magnetoresistance effect has also been observed in tunnel junctions of the half-metallic oxide CrO_2 with Co counter-electrode.[48] Here a reduced layer of insulating Cr_2O_3, which naturally forms on the surface of CrO_2, operates as the barrier layer. Although the observation of inverse magnetoresistance has thus far been limited to a few systems, and that too mostly at low temperatures (77 K and below), the effect in principle opens the door for future applications with interesting device configurations.

4.5 FREE LAYER

As compared to the above discussed structures and processes needed for the growth of the reference layer and the tunnel barrier, the formation of the free layer would appear much more straightforward, since one needs "only" a soft ferromagnetic layer for the switching. However, in reality the properties of the free layer are the most important and difficult to control. In order to obtain a sizable read signal to enable differentiation of the distribution of the low-resistance '0' state from the high resistance '1' state, it is of course desirable to attain a substantial magnetoresistive effect. Furthermore, the stack design has to ensure that the easy axis of the free layer is oriented exactly parallel to the uniaxial anisotropy of the reference layer to ensure the maximum resistance change.

While a large number of materials have been investigated for use as the free layer in MTJ's, almost all of them involve the use of the transition magnetic elements Fe, Co or Ni. The principal reason being that the spin polarization of these elements and their alloys is among the highest to be readily obtainable using commonly accessible materials. Their spin polarization values range between 40-50% on the high end as compared to maximum values of less than 10% for the magnetic rare-earth elements and their alloys.[49] The most frequently utilized free layer materials are the alloys NiFe, CoFe and NiCoFe of different compositions. The measured magnetoresistance using these individual alloys as free layers is typically in the range of 30 – 40 % for junction RA values generally needed for MRAM tunnel barriers.[50] It is possible to obtain a further increase in the MR using a bi-layer structure consisting of, for example, a thin interface layer of CoFe in combination with a NiFe free layer.[51] However, this enhancement is usually achieved at the expense of a higher switching field. Figure 4 shows the cross-section TEM micrograph of a complete MTJ stack including the top and bottom contact electrodes. Note that even the thinnest layers in the structure (Ru and AlO_x) are continuous and provide smooth interfaces.

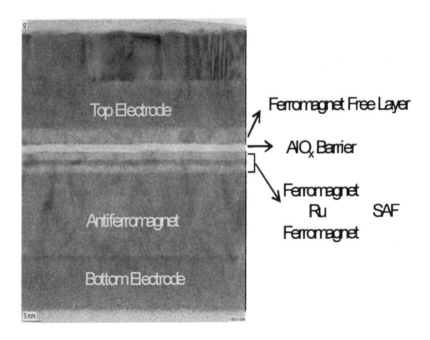

Figure 4: Cross-section TEM image of a MTJ stack clearly showing the bottom electrode, antiferromagnet layer, pinned SAF, AlO_x tunnel barrier, ferromagnetic free layer, and the top contact electrode.

There have been some recent reports detailing the use of nanocrystalline and/or amorphous magnetic layers as the free layer material in MTJs. In particular, the use of an amorphous CoFeB alloy has been shown to result in a substantial MR enhancement.[52] For example, MR values as high as 70% have been obtained in a multilayer MTJ stacks that incorporates CoFeB on both sides of the tunnel barrier, i.e., both in the free layer and also the top of the SAF pinned layer structure. [53, 54]

The usually quoted MR value, as in the examples discussed above, is measured at a very low bias across the junction and is well known to decrease with increasing bias voltage. This bias dependence results in a reduced signal being available to the read circuit under actual operating conditions and is thus not at all desirable for device applications. Indeed, most MTJ stacks loose roughly half of its MR at moderate bias voltages of 300 – 500 mV. [55] The reduction in the MR with bias has been attributed to the excitation of magnons in the magnetic electrodes that is believed to result in the randomization of the spin tunneling current. [56]

Even more challenging than reading the tunnel junction element in a MRAM array is the ability to reliably write the memory elements, without disturbing the neighboring cells. Moreover, the write field that can be generated by the metal lines in the MRAM chip places limitations on the use of certain free layer materials, excluding those with high coercivity.

As shown in fig. 5, the switching of a single ferromagnetic particle can be characterized by the Stoner-Wohlfarth (SW) astroid, [57] which provides the critical field for switching the magnetization state. The SW-model describes the rotation of a single domain element with uniaxial anisotropy in the presence of two orthogonal magnetic fields, similar to the one generated by the currents driven through the metal wires below and above the TJ-elements in an array. As can be seen, the field necessary to switch the magnetization from one resistance state to the other along the easy-axis decreases if there is an additional field applied along the hard-axis direction. Due to this fact it is possible to arrange a number of magnetic tunnel junctions in an array formed by metal lines. By selectively applying currents that locally generate magnetic fields just outside the astroid area one can in principle switch a single cell at the cross point of the two metal lines. However, in order to achieve such selective switching several requirements have to be fulfilled. First of all the free layer of the magnetic stack is not necessarily of the size of a single magnetic domain, especially if the junction size is of the order of a micron or larger.

In the presence of several domains the switching in the cell becomes more complicated and the hysteresis loops exhibit kinks at various field values. These domains will preferably form at the tips of an elongated cell (so called end-domains). The influence of these end-domains can be reduced by opting for cell shapes with a large aspect ratio (length versus width) and an optimum shape. There have been a number of studies aimed at determining the optimal shape for reliable switching. [58] At present it is fairly well established that elliptical shapes, or their asymmetric derivatives, provide the most suitable switching characteristics – although by no means ideal. Other complicated shapes like the 'Saturn-shape' have also been suggested. [59] However, the patterning of these structures is much more challenging, particularly for smaller dimensions. Fortunately, for high density applications, the single domain limit is approached when the cell size becomes smaller – typically below 100 nm.

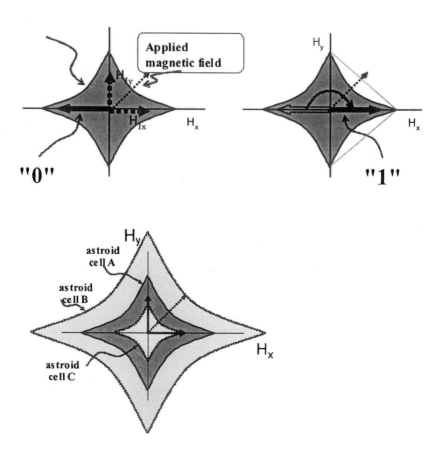

Figure 5: Stoner-Wohlfarth astroids describing the switching of the magnetization of a magnetic element in the single domain limit (top). The lower figure indicates the failure mechanism that limits the functionality of an MRAM array in the case of varying cell switching characteristics.

Secondly, the size of the switching astroid is a function of the magnetic material used for the free layer and its shape. Any variation in the local magnetic properties as well as in the shape and aspect ratio of the cell will cause variations in the astroid size and thus narrow the operational window as is shown in fig. 5. For a sufficiently large array this might lead to failing bits either due to half-select switching (i.e., the field generated by a single metal line is sufficiently large to switch a cell different from the one that is selected), or due to high coercivity cells that cannot be written even at full select. To be more quantitative, an array of magnetic cells with mean switching field H_{sf} is expected to have a distribution of width σ (H_{sf}) due to materials and process variations. The ratio H_{sf} / σ (H_{sf}) defines a quality factor, termed array quality factor (AQF), which captures this variability. A high value for the AQF (>10) is necessary in order to increase the operating margin and thereby the write yield. [60]

A third factor of concern is the centering of the astroid. A shift in one direction along the easy axis will directly reduce the operating window, since the field that has to be added in one direction might already be sufficient to switch a cell under half-select condition in the other direction. A shift along the hard-axis indicates that the two states of the free layer are not exactly antiparallel to each other and will consequently result in a lower signal. Such a shift can result from several factors: as mentioned above, magnetostatic stray fields from the reference layer can couple with the free layer and cause an offset always in direction of the antiparallel configuration in case of a simple ferromagnetic reference layer or in parallel or antiparallel configuration dependent on the imbalance in an AP-coupled reference layer system.

Another effect that is extremely important and requires attention is the so-called Neél- or Orange-peel-coupling which is caused by the roughness of the interfaces between the ferromagnetic layers and the non-magnetic spacer layer (Cu in GMR, tunnel barrier in TMR-stacks). [61] As is shown in fig. 6 the roughness at the surface of the ferromagnetic layers causes theformation of magnetic poles which induce a corresponding pole in the other ferromagnetic layer. This interaction is always ferromagnetic and

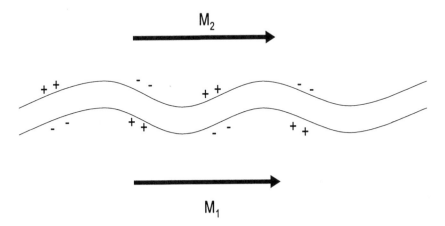

Figure 6: Schematic drawing showing the origin of the Neél coupling effect. A conformal roughness leads to a coupling between two ferromagnetic layers and will induce a preference to align the magnetization of the two layers to be parallel

hence results in a preferential parallel configuration. To reduce the offset caused by Neél-coupling one could introduce an appropriate imbalance in the AP-system. However, it much more preferable to optimize the growth conditions for the individual layers in the stack, particularly the ferromagnetic layers, in order to reduce their roughness.

For MRAM to become accepted as a viable memory technology, it is important that there is a clear technology path to future denser memories. Besides the issue of increased cell-to-cell interaction, there are a number of other important challenges that need to be addressed when scaling to higher densities. Perhaps the most important is the stability of the information stored in the cell against thermal fluctuations - $K_U V_{eff}$ (where K_U is the uniaxial magnetic anisotropy energy per unit volume and V_{eff} is the effective switching volume) may not be large in comparison to the thermal energy $k_B T$ with sufficiently high density and elements that are thin or narrow. The shape anisotropy also increases as the lateral dimensions decrease, for a given shape and film thickness, resulting in a higher switching field, higher currents and more power consumption. A decrease in the cell size will also lead to greater magnetostatic coupling between the free and pinned magnetic layers of the tunnel junction. There is thus a definite need to explore alternative approaches of writing to address the scaling-related challenges. We will look at one proposed novel approach in the following section.

4.6 SOME RECENT DEVELOPMENTS

In this section we cover some exciting new developments that have implications for significantly improving the read and write performance of MRAM devices necessary for scaling to higher densities. These advances have the potential of providing MRAM a strong competitive advantage over other technologies in the expanding market for non-volatile device applications. The first involves new materials development for MTJ stacks that promises to offer much higher MR and improved stack thermal stability. The other innovation addresses the much more serious issue of margin limitation caused by the dispersion in the switching thresholds and thermal relaxation suffered by the conventional MRAM cell design, as has been discussed in the previous section. A radically different switching concept - termed toggle-mode switching - has been proposed to address the scaling issues. The successful implementation of the toggle switching concept will require the use of a more complicated synthetic antiferromagnet (SAF) instead of a single ferromagnet for the free layer. Furthermore, the materials for this SAF will have to be carefully chosen to satisfy the requirement that the selected cell can be switched at the lowest possible field without disturbing the adjacent cells or be thermally activated.

4.6.1 MTJ Stacks with Significantly Higher Magnetoresistance

As mentioned in Section 4, recent theoretical calculations [45, 46] have predicted very high MR in epitaxial MTJs structures when the momentum of conduction electrons is conserved during the tunneling process, such as in (001) oriented Fe/MgO/Fe stacks. At the Fermi energy of Fe in the (100) direction, there is a state of Δ_1 symmetry for the majority spins, but not for the minority. Since states of Δ_1 symmetry decay much more slowly in the MgO barrier and also since, in an ordered system, the Δ_1 symmetry states cannot propagate in the minority channel, the tunneling conductance is dominated by the high conductance of the majority channel for parallel alignment of the electrodes. Because of this unique feature of the Δ_1 bands in Fe, an extremely high MR ratio is expected for coherent spin-polarized tunneling. Indeed, according to theoretical calculations, the MR is expected to increase with increasing MgO thickness, and reach a maximum value above 1000% for about 10 atomic planes of MgO.

Since the theoretical calculations by Butler *et al* and by Mathon and Umerski, fully epitaxial Fe(001)/MgO(001)/Fe(001) MTJ structures have been experimentally investigated by a number of groups. [62] While in principle the TMR can be extremely large in such structures, it is expected to be limited by any disorder that may be present at the barrier interfaces. With process optimization and better control of the interfaces, a MR ratio as high as 88% has recently been observed at room temperature (146 % at 20 K) in epitaxial stacks with MgO barrier grown by MBE as shown in fig. 7. [63]

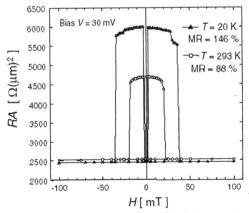

Figure 7: Variations in the *RA* versus magnetic field for an epitaxial Fe(001)/MgO (001) (2 nm)/Fe (001) junction measured at T = 293 and 20 K. [Ref. 63]

144

The bias dependence for these junctions is found to be considerably smaller than in tunnel junctions with AlO_x barrier and exhibits an asymmetric behavior. The bias voltage where the MR ratio reaches half the zero-bias value has been reported to be 1250 and 350 mV for the positive and negative bias directions, respectively. It is worth noting that the valuefor the positive bias voltage is significantly higher than that observed for AlO_x junctions.Parkin *et al* have shown that sputter-deposited polycrystalline MTJs grown on an amorphous underlayer, but with highly oriented (001) MgO tunnel barriers and CoFe electrodes, exhibit MR values of up to ~ 220% at room temperature and ~ 300% at low temperatures.[64] These textured MTJ stacks grown by standard sputtering appear more readily manufactureable, as compared to fully epitaxial structures grown by MBE, and also exhibit very good thermal stability at temperatures up to 400°C where MR even improves upon anneal (fig. 8).

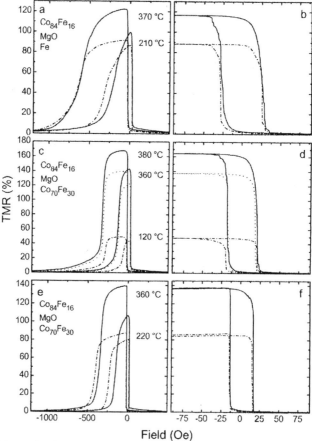

Figure 8: Major and minor loops for MgO-barriers with various electrode combinations. The measurements show very high MR for all cases which increases further upon annealing at high temperature. [Ref. 64]

An alternative approach that is also being actively pursued for enhancing the MR in MTJ stacks is to utilize non-traditional magnetic electrode materials that posses a much higher degree of spin polarization. In particular, half-metallic systems, which contain a gap in one spin band at the Fermi level and no gap in the other spin band, are expected to have a spin polarization value approaching 100%. Band structure calculations have shown that a number of magnetic oxide materials, such as the mixed-valence manganites ($La_{1-x}A_xMnO_3$, A=Ba, Sr, or Ca), magnetite (Fe_3O_4), and chromium dioxide (CrO_2), are half-metallic.[65] Additionally, a number half Heusler alloys (NiMnSb, PtMnSb) and full Heusler alloys (Co_2MnSi, Co_2MnGe, Co_2MnSn, Fe_2MnSi) are also predicted to be completely spin-polarized.[66] With the use of half-metals, the TMR should be effectively infinite in the absence of processes that mix the spin-channels. So far, the only experimental evidence for half-metallicity in tunnel junctions has come from results on manganatite-insulator-manganite junctions. These junctions have shown very large magnetotunneling effect at low temperatures, with MR values above 1000%.[67] The MR, however, decreases rapidly with temperature and at room temperature is only comparable with the standard AlO_x-based junctions and in the best case. Achievement of a huge room temperature MR in magnetic oxide heterostructures remains a big experimental challenge.

4.6.2 Toggle-Mode Operation for Achieving Improved Write Margin

In the past year, Motorola has disclosed a completely new mode of MRAM switching, termed the toggle mode configuration and named "Savtchenko switching" after its late inventor.[68] This new switching mode effectively eliminates the serious half-select and thermal activation problems that had been major impediments to MRAM scaling using the conventional switching concept. The dipole coupling between MTJ cells, which becomes increasingly important as they are packed closer together, is also effectively solved using this approach. This new switching mode relies on the use of a synthetic antiferromagnetic (SAF) as the free layer. The antiferromagnetic coupling energy, as well as the magnetostatic interaction between the layers in a patterned element, improves the thermal stability of the system against activated switching.

The SAF is formed from two ferromagnetic layers (e.g., NiFe or CoFe) with magnetization M_1 and M_2 separated by a non-magnetic coupling spacer such as Ru which induces AF coupling as shown in fig. 9. The properties of the SAF is qualitatively similar to that used in the pinned layer as discussed in section 3, although the requirements on the magnetic

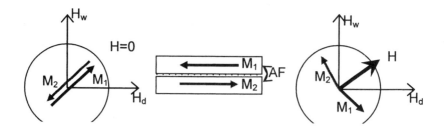

Figure 9: Schematic of the toggling operation of a synthetic antiferromagnet consisting of two magnetic layers with moment vectors M_1 and M_2 separated by a coupling layer. The magnetic moments of the two layers are equal for a perfectly balanced structure.

layers and the spacer for the toggle mode SAF will in general be very different. In the simpler case, the two layers are nominally identical in size, shape and magnetic properties. Each has a uniaxial anisotropy, with the same easy axis that is oriented 45° with the programming word and digit lines. Since an external field cannot switch this system to a known state (say the one with M_1 pointing to the left and M_2 to the right, opposite to the one in fig. 9) because the external field cannot affect M_1 and M_2 differently. However, an external field can toggle this element, leaving it in the opposite state to the original one. Thus the element must be read before being written. This is a disadvantage, but one that that is easily outweighed by its superior thermal stability.

As mentioned previously, in the preferred approach for the toggling operation the moments of the two ferromagnetic layers in the SAF are balanced. These coupled films act as the free layer switching in unison by continuous rotation driven by two sequential fields at 45° to the symmetry axes. The toggling is accomplished by applying the word and digit fields sequentially. The word field is applied first, and rotates M_1 and M_2 away from the easy axis direction as shown in fig. 10. In general, M_1 and M_2 bend toward the direction of the external field and the scheme is to rotate

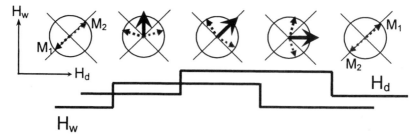

Figure 10: Schematic of the toggling operation of switching of the synthetic antiferromagnet (SAF). The word and digit line pulses are applied in sequence to rotate the SAF by 180° to the opposite resistance state.

the external field clockwise (and with it the magnetization) so that M_2 ends up pointing to the bottom-left and M_1 to the top-right, completing the toggle. This rotation is accomplished by turning on the digit field before the word field is turned off, so that the total field points to the right; then when the word field is turned off the field rotates again by 45° clockwise. Although this field rotation covers only 90°, not the 180° of a full toggle, it turns out to be sufficient to toggle the element provided the parameters are chosen properly.

Analytical modeling aided by numerical solutions of single domain switching elements have provided a good understanding of the toggle switching and the magnetic parameters that are need to be optimized for maximizing the operating field margins.[69, 70] The operation margin is principally determined by the difference between the saturation (H_{sat}) and spin-flop (H_{sf}) fields for the SAF. In the single domain limit, analytical expressions can be derived for H_{sat} and H_{sf} in terms of the intrinsic anisotropy of the layers, H_i, and their exchange coupling, J, and the magnetostatic coupling. It is desirable to make H_k as small as possible on the condition that the element exhibits sufficient robustness against thermal agitation. Similarly a very low value of J is preferred in order to reduce the switching field (H_w and H_d) needed for the toggle switching. The optimum exchange coupling needed for a SAF free layer will be determined by the size and shape of the cell, and can easily be about one to two orders of magnitude lower than the coupling for the standard SAF used for the pinned layer ($\sim 0.1 - 1$ erg/cm^2). With a reduction in cell size there will be increased magnetostatic interaction between the layers of the SAF, necessitating further reduction or even elimination of the exchange coupling.

4.7 SUMMARY

In this chapter we have attempted to provide a snapshot of the rapidly evolving field of materials research related to MRAM devices. This relatively new technology that involves the utilization of MTJs as the storage element is presently at the brink of becoming the primary contender for a universal memory device. There are a number of semiconductor companies that have already announced products or are working on product demonstrators. Still others have initiated internal development efforts in order to remain competitive in any future market. While there is a continuous flow of new ideas and concepts to improve the write and read performance, and also to enable scaling of the devices to smaller dimensions, the basic structure of the memory cell consisting of a reference layer, oxide tunnel barrier and a storage layer has remained more or less the same. The objective of MRAM materials development has now

148

expanded in several different directions. For MTJ-stacks using the established aluminum oxide barrier the focus now is on the ability to integrate this technology in existing semiconductor manufacturing process, with the goal of understanding and addressing any reliability issues. At the same time the scaling to smaller dimensions using advanced switching schemes, such as the toggle mode, is of importance to ensure the viability of MRAM for future generations. Similarly, new materials, such as those using MgO as a barrier layer, promise to provide higher MR equivalent to significantly larger read signals.

ACKNOWLEDGEMENTS

Some of the results reported in this article were obtained as part of a joint alliance between IBM and Infineon working on MRAM development. We express our gratitude to all the members of the development alliance for their support and help.

REFERENCES

[1] S. A. Wolf, D. D. Awschalom, R. A. Buhrman, J. M. Daughton, S. Von Molnár, M. L. Roukes, A. Y. Chtchelkanova and D. M. Treger, *Science* **294**, 1488 (2001); S. Parkin, X. Jiang, C. Kaiser, A. Panchula, K. Roche and M. Samant, *Proc. of IEEE* **91**, 661 (2003).

[2] *Semiconductor Spintronics and Quantum Computation*, edited by D. S. Awaschalom, D. Loss and N. Samarth (Springer, Berlin, 2003).

[3] M. N. Baibich, J. M. Broto, A. Fert, F. N. Vandau, F. Petroff, P. Eitenne, G. Creuzet, A. Friederich and J. Chazelas, *Phys. Rev. Lett.* **61**, 2472 (1988).

[4] For a review see J. S. Moodera, J. Nassar and G. Mathon, *Annu. Rev. Mater. Res.* **29**, 381 (1999).

[5] J. DeBrosse, C. Arndt, C. Barwin, A. Bette, D. Gogl, E. Gow, H. Hoenigschmid, S. Lammers, M. Lamorey, Y. Lu, T. Maffitt, K. Maloney, W. Obermeyer, A. Sturm, H. Viehmann, D. Willmott, M. Wood, W. J. Gallagher, G. Mueller and A. R. Sitaram, "A 16Mb MRAM Featuring Bootstrapped Write Drivers," *Symposium on VLSI Circuits*, p. 454 (2004).

[6] M. Durlap, P. J. Naji, A. Omair, M. DeHerrera, J. Calder, J. M. Slaughter, B. N. Engel, N. D. Rizzo, G. Grynkewich, B. Butcher, C. Tracy, K. Smith, K. W. Kyler, J. J. Ren, J. A. Molla, W. A. Feil, R. G. Williams and S. Tehrani, IEEE J. Solid-State Circuits 38, 769 (2003).

[7] S. Soeya, S. Tadokoro, T. Imagawa, M. Fuyama and S. Narishige, *J. Appl. Phys.* **74**, 6297 (1992).

[8] M. Lederman, *IEEE Trans. Magn.* **35**, 794 (1999).

[9] J. H. Lee, H. D. Jeong, C. S. Yoon, C. K. Kim, B. G. Park and T. D. Lee, *J. Appl. Phys.* **91**, 1431 (2002).

[10] V. Gehanno, P. P. Freitas, A. Veloso, J. Ferreira, B. Almeida, J. B. Sousa, A. Kling, J. C. Soares and M. F. da Silva, *IEEE Trans. Magn.* **35**, 4361 (1999).

[11] A. Wallash and Y. K. Kim, *IEEE Trans. Magn.* **34**, 1519 (1998).

[12] S. P. Bozeman, B. J. Daniels and D. J. Larson, *Mat. Res. Soc. Symp. Proc.* **615**, G5.6.1 (2000).

[13] K. Yagami, M. Tsunoda and M. Takahashi, *J. Appl. Phys.* **89**, 6609 (2001).

[14] For example: M. Mao, S. Funada, C.-Y. Hung, T. Schneider, M. Miller, H.-C. Tong, C. Qian and L. Miloslavsky, *IEEE Trans. Magn.* **35**, 3913 (1999); J. Fujikata, K. Hayashi and M. Nakada, *J. Appl. Phys.* **85**, 5021 (1999)

[15] V. S. Speriosu, B.A. Gurney, D. R. Wilhoit and L. B. Brown, *Proceedings of the INTERMAG '96.*

[16] H. A. M. van den Berg, W. Clemens, G. Gieres, G. Rupp, W. Schelter and M. Vieth, *IEEE Trans. Magn.* **32**, 4624 (1996).

[17] S. S. P. Parkin, R. Bhadra and K. P. Roche, *Phys. Rev. Lett.* **66**, 2152 (1991).

[18] H. A. M. van den Berg, J. Altmann, L. Baer, G. Gieres, R. Kinder, G. Rupp, M. Vieth and J. Wecker, *IEEE Trans. Magn.* **35**, 2892 (1999).

[19] H. A. M. van den Berg, G. Rupp, W. Clemens, G. Gieres, M. Vieth, J. Wecker and S. Zoll, *IEEE Trans. Magn.* **34**, 1336 (1998).

[20] R. R. Katti, *Proc. of IEEE* **91**, 687 (2003); M. Dax, *Semiconductor Int.* **20**, 84 (1997).

[21] M. Julliere, *Phys. Lett.* **54A**, 225 (1975).

[22] R. Jansen and J. S. Moodera, *Phys. Rev. B.* **61**, 9047 (2000) and *J. Appl. Phys.* **83**, 6682 (1998).

[23] J. S. Moodera, L. R. Kinder, T. M. Wong and R. Meservy, *Phys. Rev. Lett.* **74**, 5260 (1995).

[24] C. Fery, L. Hennet, O. Lenoble, M. Piecuch, E. Snoeck and J.-F. Bobo, *J. Phys.: Cond. Matter* **10**, 6629 (1998).

[25] E. Y. Chen, R. Whig, J. M. Slaughter, D. Cronk, G. Steiner and S. Tehrani, *J. Appl. Phys.* **87**, 6061 (2000).

[26] R. Bubber, M. Mao, T. Schneider, H. Hegde, K. Shin, S. Funada and S. Shi, *IEEE Trans. Magn.* **38**, 2724 (2002).

[27] Y. Ando, M. Hayashi, S. Iura, K.Yaoita, C. C. Yu, H. Kubota and T. Miyazaki, *J. Phys. D: Appl. Phys.* **35**, 2415 (2002).

[28] J. S. Bae, K. H. Shin and H. M. Lee, *J. Appl. Phys.* **91**, 7947 (2002).

[29] D. Song, J. Nowak and M. Covington, *J. Appl. Phys.* **87**, 5197 (2000).

[30] S. S. P. Parkin, K. P. Roche, M. G. Samant, P. M. Rice, R. B. Beyers, R. E. Scheuerlein, E. J. O'Sullivan, S. L. Brown, J. Bucchigano, D. W. Abraham, Y. Lu, M. Rooks, P. L. Trouilloud, R. A. Wanner and W. J. Gallagher, *J. Appl. Phys.* **85**, 5828 (1999).

[31] R. C. Sousa, J. J. Sun, V. Soares, P. P. Freitas, A. Kling, M. F. da Silva and J. C. Soares, *J. Appl. Phys.* **85**, 5258 (1999).

[32] X. Batlle, P. J. Cuadra, Z. Zhang, S. Cardoso and P. P. Freitas, *J. Magn. Magn. Mat.* **261**, L305 (2003).

[33] M. F. Gillies, W. Oepts, A. E. T. Kuiper, R. Coehoorn, Y. Tamminga, J. H. M. Snijders and W. M. A. Bik, *IEEE Trans. Magn.* **35**, 2991 (1999).

[34] S. Cardoso, V. Gehanno, R. Ferreira and P. P. Freitas, *IEEE Trans. Magn.* **35**, 2952 (1999).

[35] J. W. Freeland, D. J. Keavney, R. Winarski, P. Ryan, J. M. Slaughter and R. W. Dave, J. Janesky, *Phys. Rev. B* **67**, 134411 (2003).

[36] M. Tsunoda, K. Nishikawa, S. Ogata and M. Takahashi, *Appl. Phys. Lett.* **80**, 3135 (2002).

[37] M. M. Schwickert, J. R. Childress, R. E. Fontana, A. J. Kellock, P. M. Rice, M. K. Ho, T. J. Thompson and B. A. Gurney, *J. Appl. Phys.* **89**, 6871 (2001).

[38] P. Shang, A. K. Petford-Long, J.H. Nickel, M. Sharma and T. C. Anthony, *J. Appl. Phys.* **89**, 6874 (2001).

[39] B. G. Park and T. D. Lee, *Appl. Phys. Lett.* **81**, 2214 (2002).

[40] J. R. Childress, J.-S. Py, M. K. Ho, R. E. Fontana and B. A. Gurney, *J. Appl. Phys.* **93**, 6426 (2003).

[41] M. F. Gillies, A. E. T. Kuiper, J. B. A. van Zon and J. M. Sturm, *Appl. Phys. Lett.* **78**, 3496 (2001).

[42] H. Kyung, C.-S. Yoo, C. S. Yoon and C. K. Kim, *Mat. Chem. Phys.* **77**, 583 (2002).

[43] J. Wang, P. P. Freitas, E. Snoeck, P. Wei and J. C. Soares, *Appl. Phys. Lett.* **79**, 4387 (2001).

[44] M. Guth, V. Da Costa, G. Schmerber, A. Dinia and H. A. M. van den Berg, *J. Appl. Phys.* **89**, 6748 (2001).

[45] W. H. Butler, X.-G. Zhang, T. C. Schulthess and J. M. MacLaren, *Phys. Rev. B.* **63**, 054416 (2001).

[46] J. Mathon and A. Umerski, *Phys. Rev. B* **63**, R220403 (2001).

[47] J. M. De Teresa, A. Barthelemy, A. Fert, J. P. Contour, R. Lyonnet, F. Montaigne, P. Seneor and A. Vaures, *Phys. Rev. Lett.* **82**, 4288 (1999).

[48] A. Gupta, X. W. Li and G. Xiao, *Appl. Phys. Lett.* **78**, 1894 (2001).

[49] S. Parkin, X. Jiang, C. Kaiser, A. Panchula, K. Roche and M. Samant, *Proc. of IEEE* **91**, 661 (2003).

[50] For example, see D. Wang, J. Daughton, Z. Qian, M. Tondra and C. Nordman, *IEEE Trans Magn.* **39**, 2812 (2003).

[51] B.G. Park and T.D. Lee, *IEEE Trans. Magn.* **35**, 2919 (1999)

[52] H. Kano et al., *Digest of Intermag Conference*, BB-04 (2002).

[53] D. Wang, C. Nordman, J. Daughton, Z. Qian and J. Fink, *IEEE Trans. Mag.* **40**, 2269 (2004).

[54] K. Tsunekawa, Y. Nagamine, H. Maehara, D. D. Jjayaprawira, M. Nagai and N. Watanabe, *Abst. 9th Joint MMM-Intermag Conf.* 2004, BD-03.

[55] Y. Lu, X. W. Li, G. Xiao, R. A. Altman, W. J. Gallagher, A. Marley, K. Roche and S. S. P. Parkin, *J. Appl. Phys.* **83**, 6515 (1998).

[56] S. Zhang, P. M. Levy, A. C. Marley and S. S. P. Parkin, *Phys. Rev. Lett.* **79**, 3744 (1997).

[57] E. C. Stoner and E. P. Wohlfarth, *Phil. Trans. Roy. Soc.* **A240**, 599 (1948).

[58] For example, see C. K. Kim, J. Yi, J. N. Chapman, W. A. P. Nicholson, S. McVitie and C. D. W. Wilkinson, *J. Phys. D* **36**, 3099 (2003), and references therein.

[59] M. Motoyoshi, I. Yamamura, W. Ohtsuka, M. Shouji, H. Yamagishi, M. Nakamura, H. Yamada, K. Tai, T. Kikutani, T. Sagara, K. Moriyama, H. Mori, C. Fukumoto, M. Watanabe, H. Hachino, H. Kano, K. Bessho, H. Narisawa, M. Hosomi and N. Okazaki, *2004 Symposium on VLSI Technology*, Honolulu, Digest of Technical Papers, p. 22

[60] A. R. Sitaram et al, *2003 Symposium on VLSI Technology*, Kyoto, June 2003, Semiconductor Fabtech, 19th edition, p. 103.

[61] For example, see J. C. S. Kools, W. Kula, D. Mauri, and T. Lin, *J. Appl. Phys.* **85**, 4466 (1999); B. D. Schrag, A. Anguelouch, S. Ingvarrson, G. Xiao, Y. Lu, P. L. Trouilloud, A. Gupta, R. A. Wanner, W. J. Gallagher, P. M. Rice and S. S. P. Parkin, *Appl. Phys. Lett.* 77, 2373 (2000).

[62] M. Bowen, V. Cros, F. Petroff, A. Fert, C. M. Boubeta, J. L. Costa-Krämer, J. V. Anguita, A. Cebollada, F. Briones, J. M. de Teresa, L. Morellon, M. R. Ibarra, F. Güell, F. Piero and A. Cornet, *Appl. Phys. Lett.* **79**, 1655 (2001); J. Faure-Vincent, C. Tiusan, E. Jouguelet, F. Canet, M. Sajieddine, C. Bellouard, E. Popova, M. Hehn, F. Montaigne, A. Schuhl, *Appl. Phys. Lett.* **82**, 4507 (2003).

[63] S. Yuasa, A. Fukushima, T. Nagahama, K. Ando and Y. Suzuki, *Jpn. J. Appl. Phys.* **43**, L588 (2004).

[64] S. S. P. Parkin, C. Kaiser, A. Panchula, P. Rice, B. Hughes, M. Samant, and S. -H. Yang, Nature Materials (in press, 2004); S. S. P. Parkin, C. Kaiser, A. Panchula, P. Rice, B. Hughes, M. Samant and S. Yang, *49th MMM-Conference*, Jacksonville 2004, BD-09

65 A. Gupta and J. Z. Sun, *J. Magn. Magn. Mat.* **200**, 24 (1999); J. M. D. Coey and C. L. Chien, *MRS Bulletin* **28**, 720 (2003).

66 C. Palmstrøm, *MRS Bulletin* **28**, 725 (2003).

67 M. Bowen, M. Bibes, A. Barthelemy, J. P. Contour, A. Ananae, Y. Lemaitre and A. Fert,. *Appl. Phys. Lett.* **82**, 233 (2003).

68 B. N. Engel, J. Åkerman, B. Butcher, R. W. Dave, M. DeHerrera, M. Durlam, G. Grynkewich, J. Janesky, S. V. Pitambaram, N. D. Rizzo, J. M. Slaughter, K. Smith, J. J. Sun and S. Tehrani, *9th Joint MMM-Intermag Conf.* 2004, GE-05.

69 D. C. Worledge, *Appl. Phys. Lett.* **84**, 2847 (2004); D. C. Worledge, *Appl. Phys. Lett.* **84**, 4559 (2004).

70 H. Fujiwara, S.-Y. Wang and M. Sun, *J. Mag. Soc. Jpn.*, in print.

CHAPTER 5

MANGANITE, MAGNETITE, AND DOUBLE-PEROVSKITE THIN FILMS AND HETEROSTRUCTURES

S. B. Ogale, S. R. Shinde, and T. Venkatesan
Center for Superconductivity Research, Department of Physics, University of Maryland, College Park, MD 20740-4111, U.S.A.

R. Ramesh
Department of Materials Science and Engineering and Department of Physics, University of California, Berkeley CA 94720, U.S.A.

5.1 INTRODUCTION

The physical properties of oxide systems cover a broad range of interesting properties, which have attracted the attention of technologists for quite some time. However, oxides have not yet made the expected impact on the high technology sector, primarily due to a relatively poorer control on their growth and properties in thin film form. This situation appears to be changing quite rapidly with increased use of *in situ* growth characterization and control, and the growing recognition of the uniqueness and value of the novel electronic, magnetic, and optical responses of oxides in the context of modern micro and nanoelectronics. In this chapter we address the developments in this domain, focusing primarily on the exploration of thin films and device configurations involving ferromagnetic oxides with high spin polarization.

In a given ferromagnetic system, the number of carriers with spin up and spin down (w.r.t. the defining field direction) are unequal. The higher this differential the higher the spin polarization as per the definition: $P = (N_\uparrow - N_\downarrow)/(N_\uparrow + N_\downarrow)$, where, $N_\uparrow$ and $N_\downarrow$ are the numbers of spin up and spin down electrons. Materials with high spin polarization are considered key to the realization of a number of novel device concepts in the emerging field of spintronics [1-5]. This field envisages manipulation of the spin variable of electrons to control their transport in circuits and systems, in contrast to conventional electronics, which uses only the charge property. There are certain specific advantages of spintronics over electronics, some emanating from longer spin relaxation lifetimes and diffusion lengths as compared to charge momentum relaxation times and lengths, which can only be harnessed if the corresponding devices can be realized in practice. This realization not only requires materials with very high (preferably 100%) spin polarization, but also the actual realization of their high spin polarization in a real device configuration involving surfaces and interfaces with other materials. Such other materials may be insulators used as barrier layers in tunneling devices, metals used as contacts or property separating layers, or semiconductors used to configure the signal processing electronics. Highly spin-polarized materials are also of interest in magnetic sensing applications requiring high sensitivity. Another application area of interest is magneto-optics devices.

Amongst various ferromagnetic materials, which reportedly have or are predicted to have a high spin polarization, ferromagnetic oxides are of great interest because they can also support a host of interesting physical properties. If suitable oxides with very high spin polarization can be discovered/developed and their spin polarization is maintained in device architectures, a large number of possibilities could unfold for novel device concepts. One key advantage can be epitaxial integration of a magnetic oxide with semiconducting, metallic or superconducting oxides, which can be found within the same class of lattice/chemical systems. Moreover, successes have already been achieved in realizing epitaxial oxide layers on silicon[6,7] and GaAs[8,9], which can be used as templates for the integration of spintronic/oxide electronic devices with semiconductor electronics.

Over the years, a number of magnetic oxides have been predicted to have high spin-polarization (even 100%) based on the mechanism of ferromagnetism and transport therein. For example, the double exchange process in mixed-valent manganites suggests that these materials should have 100% spin polarization[10-13]. Similar predictions have also been made regarding magnetite (Fe_3O_4)[14-17] and CrO_2[18-24], with some evidence of experimental verifications. More recently, an exciting new development has occurred wherein high temperature ferromagnetism has been reported in some non-magnetic oxides dilutely doped with magnetic impurities. These so-called diluted magnetic semiconductor (DMS) oxides are currently the focus of intense scientific activity. Among the materials of interest are epitaxial films of Co and Fe doped TiO_2[25-28], SnO_2[29], LSTO[30], and Cu_2O[31], Mn doped ZnO[32-34], etc.

5.2 MATERIALS AND SCIENTIFIC ISSUES IN THE CONTEXT OF OXIDE ELECTRONICS DEVICES

As stated above, several families of oxide materials are believed to exhibit the property of complete or nearly complete spin polarization, including the rare-earth manganese perovskites such as $La_{1-x}Ca_xMnO_3$, 'double-perovskites' such as Sr_2FeMoO_6, magnetite (Fe_3O_4), CrO_2, and possibly the diluted magnetic semiconductors. This chapter reviews the current status of research and understanding regarding manganites, magnetite, and double-perovskite systems.

5.2.1 Mixed-Valent Manganites

In the case of manganites[39-43] two fundamental concepts seem to govern their behavior, namely tunability of kinetic energy and multiphase coexistence. The tunability of kinetic energy, which is basically the conduction electron bandwidth, emanates from the atomic physics and crystal chemistry of these oxides, and competes with other interactions in the system that act to localize electrons. Some of the d electrons (major contributors to the conduction band) in these oxides form an electrically inert 'core spin', while others can move from one transition metal site to another, subject to the constraint imposed by strong Hund's coupling, namely a d-level on a given site must have its spin parallel to the core spin on that site. Thus, spin correlations strongly influence the electron itinerancy from one site to another. This so-called 'double-exchange' implies the tunability of kinetic energy by altering the spin correlations by varying the magnetic field and/or temperature.

The mixed-valent manganites (and possibly double-perovskites too) exhibit the property of multiphase coexistence: the coexistence of two or more metallic, insulating, ferro-, ferri- and antiferromagnetic, charge- and orbitally-ordered phases, with typical domain sizes ranging from 10-100 nm[44-47]. This occurs because the state of the material can be easily perturbed and switched from kinetic-energy-dominated (i.e. metallic) to interaction-dominated (i.e. insulating, and charge/orbitally ordered) behavior by changing spin correlations. Indeed, the energy balance between different electronic/magnetic states of these materials is quite delicate. Multiphase coexistence is being increasingly looked upon as the primary cause of the high sensitivity of manganite properties to perturbations.

5.2.2 Magnetite

Magnetite or Fe_3O_4[17,36] is one of the most studied magnetic materials. It belongs to a family of inverse spinel ferrites. The interest in magnetite stems from its high $T_C \sim 850$ K, electrically conducting nature at room temperature (conductivity $\sim 250~\Omega^{-1}cm^{-1}$), and high spin polarization (half metallicity). At room temperature magnetite has f.c.c. lattice consisting of a close-packed face-centered cube of large O^{2-} anions. The smaller cations, Fe^{2+} and Fe^{3+}, are located in the interstitial sites of the anion lattice. There exist two kinds of cation sites in the crystal: (i) Tetrahedrally coordinated to oxygen (A-site), which is occupied only by Fe^{3+} ions; and (ii) Octahedrally coordinated to oxygen (B-site) and is occupied by equal numbers of Fe^{2+} and Fe^{3+} ions. At room temperature the electron hopping between the Fe^{2+} and the Fe^{3+} ions occupying the B site results in a high conductivity of manganite. Below ~ 120K, the conductivity of magnetite drastically reduces by over two orders of magnitude. This transition, known as the Verwey transition, is due to an ordering of the extra electron at the Fe^{2+} B site. There has been theoretical and experimental evidence that the magnetite is a half metallic ferromagnet[14-17,37,38]. As a consequence, the construction of efficient spintronic devices such as magnetic tunnel junctions utilizing this perfect spin-polarized material is conceivable.

5.2.3 Double-Perovskites

Double-perovskites[48-51] have $A_2BB'O_6$ type perovskite structure, where A is a divalent alkaline-earth ion and B and B' are the transition metal ions. In its ordered state, the transition-metal sites are occupied alternately by different B and B' cations. Intervening oxygen bridges every B and B' atom pair, thus forming alternating BO_6 and $B'O_6$ octahedra. Sr_2FeMoO_6 is one such example, in which Fe^{3+} (spin quantum number S = 5/2) and Mo^{5+} (S = 1/2) ions order alternately on the transition metal sites and the respective spins couple antiferromagnetically yielding a magnetic moment of $\sim 4~\mu_B$/f.u (formula unit). Sr_2FeMoO_6 has long been known as a conducting ferrimagnet with a fairly high magnetic transition temperature (T_C) of 410 - 450 K, i.e. much above the room temperature. These oxides are not as widely studied as manganites as potential device materials.

5.3 SURFACES/INTERFACES AND MATERIALS ENGINEERING

The surface properties of transition metal oxides are quite complex and difficult to control. The high level of difficulty in this respect has been the primary hurdle limiting the success in constructing magnetic tunnel junction (MTJ) or spin valve devices with desirable performance features as discussed later. The expected behavior is observed only at very low temperature. However, as the temperature is raised, the device response begins to degrade fairly rapidly even though the bulk magnetization is stable.

Another aspect which is to be considered, and which could be of critical significance in the fabrication of thin film based devices, is the strain in supported films, which has been addressed by us[52-54] and by others[55,56]. Interestingly, thin films (< 200 Å), grown on substrates with a 2% compressive strain, are insulating and apparently non-ferromagnetic, but may be driven metallic by application of a 5T magnetic field. Transmission electron microscopy (TEM)[57] and grazing-incidence x-ray diffraction (XRD) studies[58] have shown that substrate induced strain in manganite films is highly non-uniform and induces insulating behavior very similar to that observed in the 'multiphase' material (La,Pr,Ca)MnO$_3$. Chemical segregation at surfaces has also been established for some manganite materials[59] and should be of concern to the device development community. Since strain and surface/interface chemistry could influence the spin ordering and texture as well as the interface states, the success of oxide electronic devices is expected to depend largely on the ability to grow the spin polarized oxides with the desired high quality of surfaces and interfaces. Since many spintronics devices envisage multilayer configurations, their optimization also calls for an ability to probe buried interfaces.

In the following sections we discuss various attempts by different groups to grow and test spin valve, MTJ, and electric field effect devices bringing out the successes and failures *vis a vis* the desired goals. Thus far the device structures have mostly been attempted on manganites and hence those will be discussed in detail in this chapter, although summary of the work done on magnetite and double-perovskites will also be presented.

5.4 SPINTRONIC DEVICES: SPIN VALVES AND MAGNETIC TUNNEL JUNCTIONS

The discovery of a giant magnetoresistance (GMR) in metallic multilayers composed of ultra-thin magnetic (e.g. Fe) and nonmagnetic (e.g. Cu) metals sparked the field of spintronics[60,61]. It was observed that in these structures, the magnetic layers, which are separated by thin nonmagnetic layers, are coupled antiferromagnetically. In this antiferromagnetic arrangement of layers, the electrons traveling from one magnetic layer to the other experience a spin scattering from the second layer, as the magnetization direction of that layer is opposite. However, when a strong enough magnetic field is applied, the magnetization of all the magnetic layers becomes parallel. In this case, the electron can easily go from one layer to the other without experiencing spin scattering. As a result, the resistance of multilayer structure reduces in a magnetic field. The GMR effect received much importance because it is about 2 orders of magnitude larger than ordinary magnetoresistance and occurs at much smaller fields.

It was soon realized that read heads incorporating GMR materials would be able to sense much smaller magnetic fields, allowing the storage capacity of hard disks to increase substantially. GMR devices can also be made to act as switches by flipping the magnetization in one of the layers. This allows information to be stored as 0s and 1s as in a conventional transistor memory device. The advantage of such magnetic random access memory (MRAM) is that it is non-volatile. MRAM devices would be smaller, faster, cheaper, use less power and would be much more robust in extreme conditions such as high temperature or high-level of radiation. These potential applications led to the development of spin valve and MTJ devices.

A typical spin valve device, in principle, consists of four active layers (Fig. 1(a)). The first layer is an antiferromagnetic pinning layer (AF) whose function is to exchange bias the adjacent ferromagnetic layer and increase its coercivity. Such pinned ferromagnetic layer (F_1) requires larger field for switching its magnetization from one direction to other. F_1 is separated from the unpinned ferromagnetic layer (F_2) by a nonmagnetic metal. The pinning layer is not necessary if the top and bottom ferromagnetic layers have different coericivities. Thus application or removal of the magnetic field changes the magnetization direction of only the unpinned ferromagnetic layer. Such a device acts as a 'spin valve', letting through more electrons when the spin orientations in the two layers are the same and fewer when orientations are oppositely aligned. The electrical resistance of the device can therefore be changed dramatically. Such spin valves with metal electrodes have been reported to have GMR as high as ~ 20 % at room temperature.

Another version of magnetoresistive deivces, known as magnetic tunnel junction (MTJ, Fig. 1(b)), has shown greater promise for MRAM applications. The basic structure of MTJ devices is similar to the spin-valve device, except that in MTJ the ferromagnetic layers are separated by very thin insulating layer (I) rather than a metallic layer. The current through MTJ is the result of electron tunneling from one magnetic layer to the other. The tunneling probability depends on the relative orientation of magnetization in two ferromagnetic layers. The maximum tunneling magnetoresistance (TMR) one can obtain in MTJ[62] is:

$$TMR = \frac{2P_1P_2}{(1-2P_1P_2)}$$

where, P_1 and P_2 are spin-polarization of the ferromagnetic layers. Ideally, TMR could be infinite when P_1 and P_2 are 100 % (~ 20 – 35 %).

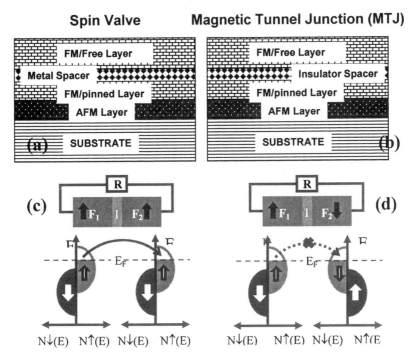

Figure 1: (a) Spin Valve, (b) MTJ heterostructures, (c and d) schematics showing operation of MTJ.

An oversimplified picture of MTJ operation is shown in Fig. 1 (c,d). Ferromagnetic materials have asymmetric spin bands, i.e. at Fermi level the number of electrons having spin up and spin down are not equal. For the case of 100 % spin polarized material, all the conduction electrons have the same spin, either up or down. Therefore, when magnetization of the two layers is aligned, the electrons tunneling from F_1 have available states in layer F_2, hence tunneling is possible (Fig.1c). On the other hand, when the magnetization of the two layers is anti-parallel, there are no states available for an electron tunneling to layer F_2. Therefore, tunneling is forbidden in this case (Fig.1d). The most successful MTJs are based on simple transition metal elements like Fe, Co, and their alloys. However, the spin polarization of these materials is much smaller than 100 %. Some magnetic oxides with 100% spin polarization thus offer more attractive alternatives in this project.

5.4.1 Manganites for Spintronics
5.4.1.1 Experimental Evidence of Half Metallicity of La$_{0.7}$Sr$_{0.3}$MnO$_3$

In 1998, by spin-resolved photoemission measurements, Park et al.[10] reported the half-metallic nature of La$_{0.7}$Sr$_{0.3}$MnO$_3$. For the majority spin, the photoemission spectrum clearly showed a metallic Fermi cut-off, whereas for the minority spin, it showed an insulating gap with disappearance of spectral weight at ~0.6 eV binding energy. The valence band spin-resolved photoemission spectra near E$_F$ measured at 40 K and 380 K are shown in Fig. 2. At 40 K (well below T$_C$), all the Mn 3d electron spins are aligned ferromagnetically and the spectra in Fig. 2a show considerable difference for the majority (↑) and minority (↓) spins. Moreover, the spectrum for the majority spin extends up to Fermi level and shows the metallic Fermi cut-off, whereas the one for the minority spin decreases rapidly and the spectral weight disappears before E$_F$, reflecting the insulating gap. The spin polarization was found to be ~100%, thereby revealing the half metallic nature of the compound. The difference spectrum shown in the bottom panel of Fig. 2a shows the metallic Fermi cut-off at E$_F$ and two peak features corresponding to the e$_g$ (1.2 eV) and t$_{2g}$ (2.2 eV) electron removal states.

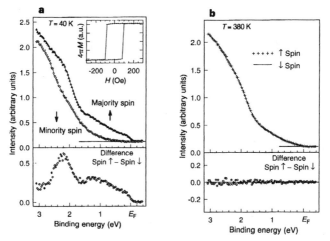

Figure 2: Valance band spin-resolved photoemission spectra for the LSMO film measured at (a) 40 K (below T$_C$) and 380 K (above T$_C$). [Ref. 10]

The observed spin anisotropy in the spin-resolved photoemission spectra at low temperature disappears above T$_C$, as expected. The spectra in Fig. 2b, which were measured at 380 K (above T$_C$), show no difference for the two different spins, and the spectral weight disappears at E$_F$ for both the spins, showing the pseudogap feature. These studies of La$_{0.7}$Sr$_{0.3}$MnO$_3$ show the half-metal feature well below T$_C$ and the occurrence of the metal to pseudogap state transition accompanied by loss of the ferromagnetic order on heating through T_C.

5.4.1.2 Brief Overview of Manganite Magnetic Tunnel Junctions

The first MTJ device using perovskite manganites was fabricated by Sun et al.[63]. They used $La_{0.7}Ca_{0.3}MnO_3$ (LCMO) or $La_{0.7}Sr_{0.3}MnO_3$ (LSMO) as ferromagnetic electrodes, and about 5 nm thick $SrTiO_3$ (STO) as a tunneling barrier. Using pulsed laser deposition technique, they grew LCMO/STO/LCMO and LSMO/STO/LSMO MTJs. Both of these devices showed magnetic field dependent resistance changes at temperatures below T_C. With application of a magnetic field of moderate strength (less than 200 Oe), which is of the order of coercivity of the material, they observed a resistance change of about factor of 2 at 4.2 K. Their results of LSMO/STO/LSMO MTJ are shown below in Fig. 3.

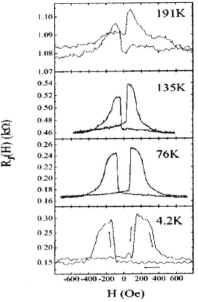

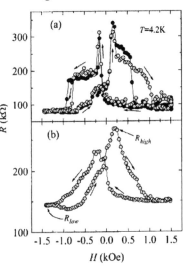

Figure 3: Magnetoresistance of LSMO/STO/LSMO junction [Reprinted with permission from A. Gupta, J.Z. Sun et al., *Appl. Phys. Lett.* **69**, 3266 (1996). Copyright by the American Institute of Physics. Ref. 63].

Figure 4: Magnetotransport data of LSMO/STO/LSMO MTJ. (a) Data from two measurements taken shortly after one another. (b) Average data of 25 measurements [Reprinted with permission from A. Gupta, J.Z. Sun et al., *Appl. Phys. Lett.* **70**, 1769 (1997). Copyright by the American Institute of Physics. Ref. 64].

The shape of the resistance versus applied field loop is similar to the metal–electrode based magnetic tunneling valves; however, the magnitude of resistance change in this case is much larger. The I-V characteristics of these junctions were found to be nonlinear, as expected for the case of tunneling conduction. Note here that the magnitude of the

162

magnetoresistance decreases with increase in temperature. These results demonstrated the successful operation of an all oxide MTJ.

Further improvement in oxide MTJs was made by growing these structures on LaAlO$_3$, which was found to be more stable than STO substrates[64]. Figure 4 shows the field dependent resistance data for such MTJ measured at 4.2 K. The low temperature transport properties of these devices were consistent with spin-dependent tunneling and a large magnetoresistive change of a factor of 5 was induced by an applied field only of the order of 200 Oe. Observation of such a large magnetoresistance indicates that the spin polarization of LSMO has to be ~81%. It was found that the disappearance of MR above 100 K was related to the defect states in the STO barrier, as was previously observed for α-Si barrier-based tunneling[65].

MTJs with substantially improved performance were obtained by Jo et al.[66]. They chose NdGaO$_3$ (NGO) as a substrate as well as a barrier layer separating LCMO layers because of excellent lattice matching between NGO and LCMO. These devices showed very large magnetoresistance (~ 86 %) at 77 K, with extremely sharp switching between the high and low resistance states. Figure 5 shows the magnetic field dependence of resistance at 77 K of a number of junctions from the same chip. Two striking features are: (i) the measured TMR is up to 86% (i.e., a factor of 7.3 resistance change); (ii) the switching between these states is extremely sharp with $R^{-1}(dR/dH) > 400\%/Oe$. Both these features were stable and reproducible for both magnetic history and thermal cycles.

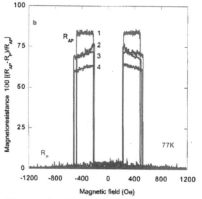

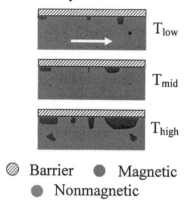

Figure 5: TMR data for LCMO/NGO /LCMO MTJ devices [Reprinted with permission from M. Blamire, M. Jo et al., *Phys. Rev. B* **61**, R14905 (2000). Copyright by the American Physical Society. Ref. 66].

Figure 6:. Schematics of the phase separation at one of the interfaces of manganites MTJ.

The large TMR at 77 K was seen to decrease to 40 % at 100 K and disappear above 150 K. They attributed this decrease to the presence of defective states in the NGO barrier. However, at lower temperatures they explain large MR by suggesting an active tunneling mechanism based on percolative phase separation. As shown schematically in Fig. 6, significant tunneling will occur only at low temperatures as the concentration of ferromagnetic regions increases with lowering the temperature. At higher temperatures, the system is largely phase separated[44-47] and hence has a smaller cross-section for tunneling, thereby causing a rapid fall in TMR with increasing temperature. The phase separation is known to be very sensitive to the strain[53]. The non-uniform distribution of strain leads to an enhanced phase separation in thin films. In this case, as the lattice parameters of NGO and LCMO are closely matched, the degree of strain at the interface is much smaller as compared to the other MTJs involving STO barrier (STO has larger lattice mismatch with manganites). Therefore, in this case one expects less phase separation and hence larger magnitude of TMR as indeed observed.

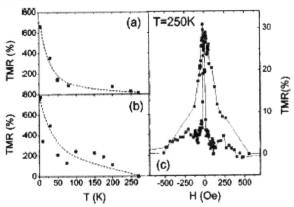

Figure 7: Temperature dependence of TMR observed by Bowen et al. for two devices is shown in (a) and (b). TMR measured at 250 K is shown in (c). [Reprinted with permission from A. Barthelemy, M. Bowen et al., *Appl. Phys. Lett.* **82**, 233 (2003). Copyright by the American Institute of Physics. Ref. 67]

Very recently, Bowen et al.[67] refuted this argument of strain being responsible for the lower TMR in manganite MTJs and rapid drop of TMR with increasing temperature by obtaining a large TMR in carefully fabricated LSMO/STO/LSMO MTJs. They argued that the key to achieving large TMR is careful photolithography pattering and ion milling. While ion milling the LSMO/STO/LSMO heterostructures, they monitored the etching process by secondary ion mass spectroscopy so as to stop the etching when entering the bottom LSMO layer. On these MTJs, they measured very large TMR with a resistance change of about a factor of 18 at 4.2 K. This TMR corresponds to a spin polarization of 95 % for LSMO. Remarkably, as shown in Fig. 7, the measurable TMR was present up to temperatures as high as 270 K in contrast to the previously reported data where the TMR disappeared at ~ 150 K.

164

5.4.1.3 Issues Involved in Manganite Magnetic Tunnel Junctions

The studies of MTJs involving manganites have revealed that while it is possible to obtain large TMR at very low temperatures, it reduces dramatically with increase in temperature. The temperature at which TMR disappears is much below the T_C of the used manganites. Hence, the reduction of TMR at high temperature cannot be attributed to the decrease in magnetization near T_C. In an interesting study Park et al.[68] measured and compared the temperature dependence of magnetization of the bulk, surface layer, and intermediate layer of $La_{0.7}Sr_{0.3}MnO_3$ epitaxial film. The magnetization of the film measured by SQUID magnetometry represents the bulk magnetization (M_B). The magnetization at the film surface boundary (M_{SB}) was obtained using spin-resolved photoemission spectroscopy (SPES), which has a ~5 Å probing depth. Magnetization at another length scale (M_{IM}) was determined from the Mn L-edge absorption magnetic circular dichroism (MCD) with ~50 Å of the probing depth.

Figure 8 shows the temperature dependence of magnetization at the above-mentioned length scales. The M_B showed the expected magnetization with temperature dependence very similar to that for a bulk. While the general nature of M_{IM} versus temperature curve is similar to that of M_B, for most of the temperature range M_{IM} is significantly smaller than M_B. Interestingly, the behavior of M_{SB} is totally different from that of M_B; it shows linear-like temperature dependence near T_C. Park et al.[68] concluded from these studies that the surface density of states at Fermi level deviates significantly from that of bulk, causing concomitant changes in the magnetization and the spin polarization at the surface. It should be noted here that while the temperature dependence of magnetization of LSMO film at three different length scales is vastly different, the T_C's at these scales appear to be the same.

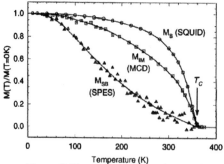

Figure 8: Temperature dependence of magnetization of LSMO thin film measured by SQUID magnetometry, magnetic circular dichroism, and spin-resolved photoemission spectroscopy. [Ref. 68]

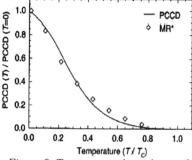

Figure 9: Temperature dependence of polarized charge carrier density and magnetoresistance observed in polycrystalline manganites. [Ref. 68]

In Fig. 9 is shown the polarized charge carrier density (PCCD), which is the difference in the charge carrier densities for up and down spins, at the surface boundary for LSMO as a function of normalized temperature. The PCCD drops down to zero before T_C and follows the temperature dependence similar to that observed for the magnetoresistance behavior for the polycrystalline manganites[69], which is known to be due to spin dependent tunneling (as in MTJ devices) between neighboring grains via insulating grain boundaries. These results throw some light on the issue of disappearance of TMR at higher temperatures.

It could be argued that the oxygen nonstoichiometry at the surface of the film may be responsible for the observed changes in M_{SB}. However, in manganites the T_C changes with oxygen content. Hence Park et al.[68] ruled out this possibility. In this context, Wad et al.[70] performed Monte Carlo computer simulation studies of LCMO. Their results show that an ideal termination of the manganite surface (without reconstruction and associated effects) cannot lead to a strong loss of surface magnetization, but a nonstoichiometry in the form of either oxygen vacancy gradient or a gradient in the La:Ca ratio does indeed lead to a temperature dependence of magnetization very similar to that observed by Park et al.[68]. Wad et al.[70] calculated the temperature dependence of magnetization for different cases: (i) the bulk magnetization without any oxygen vacancies (representing M_B) obtained with periodic boundary conditions along all axes; (ii) the magnetization of the top surface layer for the case of an ideally terminated surface without oxygen vacancies (M_{Ideal}), wherein the periodic boundary conditions are lifted for the z-axis; (iii) the top surface magnetization (M_S) for the case with oxygen vacancy profile (Bexp-l/3, l being the layer number from the surface, B=7%); and (iv) cumulative average magnetization of the top 5 Mn layers corresponding to a thickness of about 25 Å (representing M_{IM}). While calculating M_S, they introduced an oxygen vacancy concentration profile near the surface.

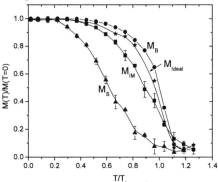

Figure 10: Monte Carlo simulation results of the temperature dependent magnetization of LCMO for different situations. [Ref. 70]

The calculated temperature dependence of magnetization for the above-mentioned cases is shown in Fig. 10. The magnetization, M_{Ideal}, of the top layer under an ideally terminated surface without oxygen vacancies is fairly close to the bulk magnetization, M_B. In the transition region, M_{Ideal} deviates slightly from M_B, which is because the top Mn atoms have one absent Mn-O-Mn bond above. This indicates that ideally, the magnetization of the surface of the manganites should be the same as that of the bulk. Interestingly, however, the results for the top surface (with oxygen vacancies) and top few layer magnetizations are remarkably similar in structure to those reported by Park et al.[68] for the LSMO case. With an increase in temperature, M_{IM} decreases faster than M_B; and M_S drops even faster than M_{IM}. While the calculated results are for LCMO, the phase diagram for LSMO near 0.7:0.3 La:Sr composition is principally similar to LCMO, although the temperature scales differ. Therefore the results for LCMO should also represent LSMO.

Practically, it is very likely to have either oxygen or La:A nonstoichiometry at the surface of manganites. Indeed, such nonstoichiometry at the surfaces and interfaces has been experimentally observed[71]. This could lead to a gradual viscous fingeringlike protrusion of demagnetization into the bulk, resulting in the typical behavior of the top surface magnetization observed in photoemission experiments. While viscous fingering type protrusions do leave fractional regions of device-worthy magnetic order up to the top surface, they also cause enhanced spin flip scattering in deeper layers offsetting that advantage. These studies strongly point out the need for a better control over fabrication process of oxide interfaces and surfaces for the realization of spintronics devices efficiently operating at room temperature.

Recently, Freeland et al.[72] observed that the bilayer manganite $La_{2-2x}Sr_{1+2x}Mn_2O_7$ exhibits a natural 1 nm thick insulating layer at the surface which does not have any long range ferromagnetic ordering. The next bilayer below this surface is ferromagnetic and possesses full spin polarization. They found that this abrupt localization of the surface effect is due to the quasi-two-dimensional nature of $La_{2-2x}Sr_{1+2x}Mn_2O_7$. They attributed the loss of ferromagnetism to the weakened double exchange in the surface bilayer that results in to an antiferromagnetic phase at the surface. Figure 11 shows the magnetization versus temperature data of the bulk (measured by neutron scattering), and the near-surface region that is dominated by the second bilayer (obtained by x-ray magnetic circular dichroism and x-ray resonant magnetic scattering). These data show identical temperature dependence of the near-surface magnetization that matches that of bulk except very close to T_C. These magnetization profiles match with that of the bulk except for a slight change near T_C. Further, through Au-tip point-contact scanning tunneling spectroscopy they concluded that the surface layer is indeed insulating. These findings of a naturally occurring surface insulator layer on top of fully spin polarized manganite promise the better fabrication of oxide MTJs and is expected to accelerate the developments in this field[72].

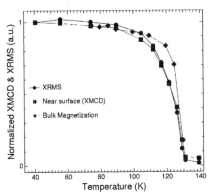

Figure 11: Temperature dependence of the bulk (diamonds, measured by neutron scattering), near-surface (squares, obtained by x-ray magnetic circular dichroism), and second bilayer (circles, measured by x-ray resonant magnetic scattering) of $La_{2-2x}Sr_{1+2x}Mn_2O_7$ single crystal [Reprinted with permission from J. Freeland et al., *Nature Materials* 4, 62 (2005). Copyright by the Nature Publishing Group. Ref. 72].

In another recent report, Yamada et al.[73] used magnetization-induced second-harmonic generation to probe the local magnetic properties at the interface between $La_{0.6}Sr_{0.4}MnO_3$ and $SrTiO_3(STO)$ or $LaAlO_3(LAO)$. By using PLD, they grew bilayer samples consisting of 120 nm thick $La_{0.6}Sr_{0.4}MnO_3$ with 5 unit cell thick STO or LAO on top. They found that antiferromagnetic spin canting occurs at the STO/ $La_{0.6}Sr_{0.4}MnO_3$ interface, while ferromagnetic spin arrangement is much less canted at the LAO/$La_{0.6}Sr_{0.4}MnO_3$ interface. As a consequence, a large magnetization-induced second harmonic generation (MSHG) is observed for the LAO/$La_{0.6}Sr_{0.4}MnO_3$ interface, while hardly any MSHG for the STO/$La_{0.6}Sr_{0.4}MnO_3$ interface. They further showed that by grading the doping profile on an atomic scale at the interface, robust ferromagnetism could be realized around room temperature. To enhance the interface magnetization without sacrificing the ferromagnetic and metallic characters of the film, they fabricated atomic level controlled compositionally graded $La_{0.6}Sr_{0.4}MnO_3$ interfaces with Sr doping level gradually changing from 0.4 to 0 toward the insulating layer. This was achieved by inserting a 2-unit cell thick layer of $LaMnO_3$ between STO and $La_{0.6}Sr_{0.4}MnO_3$, based on the anticipation that the originally antiferromagnetic $LaMnO_3$ layer can be hole-donated by $La_{0.6}Sr_{0.4}MnO_3$ and STO layers, thereby compensating for the interface effects. Such an atomically engineered interface not only showed a large MSHG signal but also when used in MTJ configuration

$(La_{0.6}Sr_{0.4}MnO_3/STO/LaMnO_3/La_{0.6}Sr_{0.4}MnO_3)$ showed an improved performance. These results improve our understanding about oxide MTJs and should lead to better performance.

5.4.1.4 Recent Breakthrough in Magnetic Tunnel Junction devices:

Recently, two papers were published in Nature Materials which represent dramatic improvements in tunneling magnetoresistance at room temperature for MTJs fabricated using crystalline MgO as a tunnel barrier[74,75]. In one study, Parkin et al.[74] fabricated MTJs with highly oriented (100) MgO tunnel barriers and CoFe electrodes. These MTJs exhibited magnetoresistance as high as ~220% at room temperature and ~300% at low temperatures. From superconducting tunneling spectroscopy data they found that the tunneling current has a very high spin polarization of ~85%. Further, these devices are highly stable up to a temperature of 400 °C.

In the same issue of Nature Materials, Yuasa et al.[75] reported similarly high values of magnetoresistance (180 % at room temperature) in single crystalline Fe/MgO/Fe MTJs. They found that the origin of such a high magnetoresistance is the symmetry of electron wave functions and accompanied coherent spin-polarized tunneling. In addition, they observed that the magnetoresistance oscillates as a function of tunnel barrier thickness, indicating that coherency of wave functions is conserved across the tunnel barrier.

The values of magnetoresistance reported in these papers are about three times higher than the previously observed magnetoresistance in any type of MTJ and clearly represents a breakthrough. This will accelerate the development of new families of spintronic devices, such as MRAM integrated circuits.

We would like to remark here that MgO has an excellent crystal lattice matching with magnetite (Fe_3O_4) and high quality epitaxial films have been grown on MgO substrates[76-78]. Therefore, one can expect high tunneling magnetoresistance in Fe_3O_4/MgO/ Fe_3O_4 MTJs.

5.4.2 Magnetite (Fe_3O_4) for Spintronics

Researchers have explored the possibility of using other materials having T_C much higher than room temperature, so that the magnetization and the spin polarization of the surface/interface would be substantial and stable at room temperature. One such magnetic oxide, for which 100 % spin polarization has been suggested, is Fe_3O_4[79]. The T_C for this compound is well above the room temperature (~850 K) and hence Fe_3O_4 is expected to have very high spin polarization at room temperature.

The first MTJ type heterostructure based on Fe_3O_4 was studied by Ghosh et al.[80] in the form of Fe_3O_4/STO/LSMO trilayer devices. The magnetotransport data, measured at different temperatures, is shown in Fig. 12. Significant MR was observed only below 200 K. The magnetoresistance was found to be composed of a sharp central positive MR feature at low field, and outer wings at higher fields reflecting a

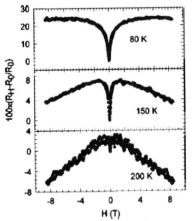

Fig. 12: Magnetotransport data of Fe_3O_4/STO/LSMO heterostructure measured at different temperatures. [Ref. 80]

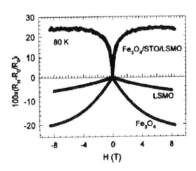

Fig. 13: Magnetoresistance of Fe_3O_4/STO/LSMO heterostructure measured at 80 K. For comparison similar data for LSMO and Fe_3O_4 films are also shown. [Ref. 80]

negative MR contribution to the total MR. The contribution of the positive MR increases with lowering the temperature. The MR data for the heterostructure measured at 80 K are plotted in Fig. 13 along with that for individual LSMO and Fe_3O_4 layers. Since the MR of LSMO and Fe_3O_4 is negative, the observed positive MR contribution has to be from the Fe_3O_4/STO/LSMO junction.

The MR at 80 K exhibits two distinct regions, one at lower field (H < 0.5 T) and the other at fields higher than 0.5 T. The magnetic hysteresis loop of the device measured at 80 K exhibited two values of coercivity: 0.04 and 0.1 T corresponding to LSMO and Fe_3O_4 layers, respectively. Thus the MR data do not reflect any specific correlation with these two coercivity values. This is in contrast to the clear correlation invariably observed in the case of the conventional spin valve/tunneling systems. Never the less, the observation of giant increase in junction resistance (positive MR) when the magnetization of the two electrodes is aligned parallel by applying magnetic field is very interesting.

In Fe_3O_4, the entire density of states near the Fermi energy is contributed by the minority spin band (carrier spin, S^C opposite to **M, H**), while in LSMO the primary contribution to the density of states near Fermi energy is from the majority spin band (S^C parallel to **M, H**). Therefore, the electrons contributing to conduction in Fe_3O_4 and LSMO have opposite spin orientations relative to the applied field, as shown in Fig. 14. Therefore, the resistance of the heterostructure for vertical transport is high at high field applied in the plane so that the magnetizations of Fe_3O_4 and LSMO are parallel. At low field, due to the canting of magnetization and

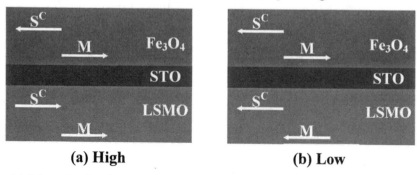

(a) High   (b) Low

Fig. 14: Schematics showing the magnetic states of the heterostructure for high and low R.

domain formation, the magnetization vectors in the two layers are no longer parallel and hence the resistance of the junction is lower.

In another study, Li et al.[81] fabricated $Fe_3O_4/MgO/Fe_3O_4$ MTJs on MgO substrates. In order to obtain different coercivities (Fig. 15) for the top and bottom Fe_3O_4 layers, the bottom layer was grown on $CoCr_2O_4$ buffer layer. At low temperature the hysteresis loop shows clear indication of two uncoupled magnetic layers. Figure 16 shows the field dependence of the tunneling resistance and the MR, $\Delta R/R_p$ (R_P being the peak resistance), at various temperatures. The switching fields for the increase and decrease in R at various temperatures correspond closely to the magnetic coercivities of the top and bottom electrodes (Fig. 15). At high fields, when

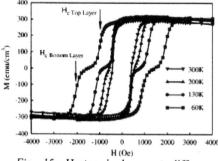

Fig. 15: Hysteresis loops at different temperatures for $Fe_3O_4/MgO/Fe_3O_4$ MTJ. [Reprinted with permission from A. Gupta, X. Li et al., *Appl. Phys. Lett.* 73, 3282 (1998). Copyright by the American Institute of Physics. Ref. 81]

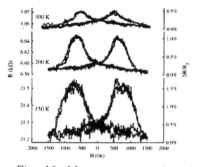

Fig. 16: Magnetotransport data for $Fe_3O_4/MgO/Fe_3O_4$ MTJ at different T. [Reprinted with permission from A. Gupta, X. Li et al., *Appl. Phys. Lett.* 73, 3282 (1998). Copyright by the American Institute of Physics. Ref. 81]

the magnetization of the two layers is parallel, R attains a low value. On the other hand, at fields between the coercivities of the two layers, R reaches the maximum value due to antiparallel orientation of the magnetization of the two layers.

While this study demonstrates the successful room temperature operation of $Fe_3O_4/MgO/Fe_3O_4$ MTJs, the magnitude of TMR is much smaller (~0.5% at 300 K and ~1.5% at 150 K) than expected. Several possibilities such as spin flip processes due to defective insulating barrier, formation of magnetic dead layer at the interface, and/or formation of antiferromagnetic oxide (e.g. $Fe_{1-x}O$) at the interface have been suggested.

A remarkably large TMR was observed by Seneor et al.[82] in very thin $Co/Al_2O_3/Fe_3O_4$ MTJs. They grew 15 nm thick Co electrode on glass followed by deposition of 1.5 nm thick Al film that was oxidized to form Al_2O_3. On top of this, Fe_3O_4 layer of thickness ~1.5 – 2 nm was grown and capped with 15 nm Al layer. The transmission electron microscopy analysis indicated that there might be a presence of small fraction of γ-Fe_2O_3 impurities in the Fe_3O_4 layer. At room temperature, these MTJs showed TMR of about 13 %, which increased to as high as 43 % at 4.2 K (Fig. 17).

The interesting observation of this study is that conductance of these MTJs depends on the bias voltage. This is shown in Fig. 18. The main feature is the abrupt drop of both the conductance and MR at very small bias voltage, between ± 10 mV. Such a drop in the conductance was not seen for the $Co/Al_2O_3/NiFe$ tunnel junction (bottom curve in Fig. 18). This indicates the existence of a gap of about 10 meV for the tunneling process in the $Co/Al_2O_3/Fe_3O_4$ tunnel junction. The similar behavior for positive and negative bias implies that these tunneling processes involve electronic levels on both sides of the Fermi level.

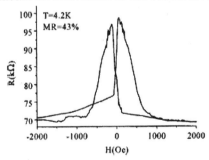

Fig. 17: TMR of $Co/Al_2O_3/Fe_3O_4$ MTJ at 4.2 K. [Reprinted with permission from P. Seneor et al., *Appl. Phys. Lett.* 74, 4017 (1999). Copyright by the American Institute of Physics. Ref. 82]

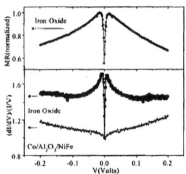

Fig. 18: Bias dependent MR and conductance of $Co/Al_2O_3/Fe_3O_4$ MTJ. For comparison, the conductance data for $Co/Al_2O_3/NiFe$ MTJ is also shown. [Reprinted with permission from P. Seneor et al., *Appl. Phys. Lett.* 74, 4017 (1999). Copyright by the American Institute of Physics. Ref. 82]

5.4.3 Double-Perovskites for Spintronics

Double-perovskites offer a broad parameter space for materials engineering; for example, Sr in Sr_2FeMoO_6 (SFMO) can be substituted by Ca, Ba, or their mixture and the combination of Fe–Mo in B site can be replaced by Fe–Re as well, thus providing tunability of the physical properties[83]. For example, among the series Ba_2MReO_6 (M = Mn, Fe, Co, Ni) the M=Fe compound is both metallic and ferrimagnetic, the M=Mn and Ni compounds are ferrimagnetic semiconductors, and M=Co compound is an antiferromagnetic semiconductor. Similarly, A_2CrMoO_6 (A=Ca, Sr), A_2CrReO_6 (A=Ca, Sr), and $ALaMnReO_6$ (A=Ca, Sr, Ba) are all ferrimagnetic but not metallic.

5.4.3.1 Half Metalicity of Double-Perovskites

The band structure calculations based on the generalized gradient approximation performed by Kobayashi et al.[84] brought out the half metallic nature of SFMO. Figure 19 shows the total density of states with majority `up' and minority `down' spins as well as the local density of states for individual elements. The down-spin band is mainly occupied by oxygen 2p states and around the Fermi level by both the Mo 4d t_{2g} and Fe 3d t_{2g} electrons, which are strongly hybridized with oxygen 2p states. Such a half-metallic nature results in 100% spin-polarized charge carriers in the ground state. In light of the fairly high T_C (410 - 450 K), an unusually high spin polarization is expected around room temperature, which makes this compound potentially useful for spintronics devices.

Owing to spin-polarized nature of the carriers, the polycrystalline samples of these compounds exhibit large low field magnetoresistance that is caused by spin-dependent tunneling[85]. A magnetoresistance in the range 10 – 40 % has been observed in different systems of double perovskite family. Such a large magnetoresistance was not found in single crystals[86], confirming the role of grain boundaries in magnetoresistance.

5.4.3.2 Anti-site Defects and Their Effects on the Physical Properties of Double-Perovskites

It has been observed that the electrical and magnetic properties of these compounds strongly depend on the degree of B-site cation ordering[87]. In order to examine the implications of octahedral cation (B-site) disorder, Ogale et al.[88] performed Monte Carlo simulation studies for the case of SFMO. They observed that the saturation magnetization and the T_C both decrease linearly with increase in the B-site disorder (Fig. 20). The

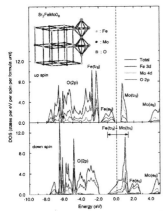

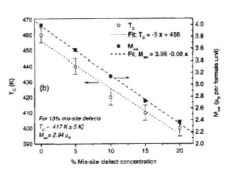

Figure 19: The density of states (DOS) of Sr_2FeMoO_6 [Reprinted with permission from Y. Tokura, Kobayashi et al., *Nature* 395, 677 (1998) Ref. 84].

Figure 20: Dependence of T_C and saturation magnetization on the mis-site defect concentration [Ref. 88].

presence of oxygen vacancies alone also reduces the saturation magnetization and the T_C. The simultaneous presence of oxygen vacancies and B-site cation disorder enhances the rate of decrease of the saturation magnetization.

Experimental studies[89-91] on thin films and bulk have verified that B-site disorder reduces the saturation magnetization and T_C. In an interesting study, Fontcuberta and coworkers studied the magnetotransport properties of SFMO samples with excess oxygen and found that excess oxygen mainly affects the chemical composition in the grain boundary region, As a result, the grain boundaries become insulating and gives rise to tunneling transport and increase in low field magnetoresistance. In another study, García-Hernández et al.[92] found that the presence of a moderate level of antisite disorder is responsible for the low field magnetoresistance. They found linear decrease in magnetoresistance with increase in antisite disorder. They explained this as follows: Antisite disorder implies the existence of Fe-O-Fe (antiferromagnetic) and Mo-O-Mo (paramagnetic) patches across a sample in between stoichiometric volumes, where the ordering of Fe and Mo is fulfilled, as illustrated in Fig. 21a.

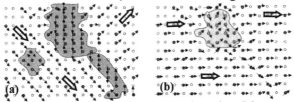

Figure 21: Schematics presenting antisite disordered patches (Mo atoms are open circles and Fe atoms are solid circles). (a) H= 0 and (b) H ≠ 0 [Reprinted with permission from M. García-Hernández et al., *Phys. Rev. Lett.* 86, 2443 (2001). Copyright by the American Physical Society. Ref. 92].

The antiferromagnetic Fe-O-Fe patches, and to a much lesser extent the paramagnetic Mo-O-Mo ones, act as barriers to the electronic transport between the ordered volumes. As the antisite disorder decreases, the average width of the antiferromagnetic islands decreases and the tunneling between stoichiometric patches becomes more probable, as soon as the orientation of the local magnetizations line up. As a side product of the alignment of the latter, provided the Fe-O-Fe patches are not very large, the antiferromagnetic interaction within the Fe-O-Fe patches has to compete with the FM ones imposed by the surrounding material. As a consequence, the antiferromagnetic order is partially broken, effectively thinning the barriers. Thus, the crossing of the spin-polarized electron is easier and the LFMR is enhanced (see Fig. 21b). For wide AF patches, at low fields, the AF interactions within the patch dominate within the core, and these paths remain closed to the electronic transport. Obviously, as antisite disorder increases and the saturation moment decreases, the magnetoresistance decreases.

5.4.3.3 Double-Perovskite Films Growth

Thin film growth of these materials has proved to be rather difficult as compared to that of other oxides, e.g. colossal magnetoresistive manganites. These difficulties are primarily due to fairly high growth temperature required for ordered phase formation, a very narrow window of growth parameters, and high probability of nucleation of impurity phases. Recently, Besse et al.[93] performed a comprehensive structural characterization of SFMO thin films. They evidenced the presence of Fe-rich parasitic phases nucleating during the early stage of the growth, which leads to a rough surface with L-shape valleys and outgrowths, thus demonstrating the need for a careful optimization of the growth parameters. However, despite these difficulties, high quality epitaxial films have been grown by researchers worldwide. In most of the cases, pulsed laser deposition technique has been used, the possible reason being its capability to transfer the exact stoichiometry from target to the film.

Manako et al.[94] fabricated thin films of SFMO on single crystalline $SrTiO_3$ substrates. Atomically smooth films with an atomic-scale step-and-terrace structure (Fig. 22a) have been realized on STO (100) substrates. On the other hand STO (111) substrates exhibited spiral structure of atomic-scale steps on individual grain corresponding to growth mediated by screw dislocation (Fig. 22b). Both these films showed B-site cation ordering and a Curie temperature above 400 K and also exhibited magnetoresistance of about 2 - 5 % even at room temperature.

Asano et al.[95] deposited epitaxial films of SFMO using a two-step growth process. First a very thin layer of 5 nm was deposited at 600 °C, and then a thicker layer was deposited at a higher temperature of 800 – 850

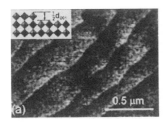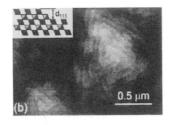

Figure 22: Atomic force microscope images epitaxial films grown on (a) SrTiO₃ (001) and (b) (111) substrates. Insets are schematic representations of step structures [Reprinted with permission from Y. Tokura, T. Manako et al., *Appl. Phys. Lett.* 74, 2215 (1999). Copyright by the American Institute of Physics. Ref. 94].

°C. Depending on the growth conditions, the films were either metallic with a residual resistivity of ~ 1μΩ-cm or semiconducting. The metallic films showed a large positive magnetoresistance (~35%) while the semiconducting ones showed small (~3%) and negative magnetoresistance. They observed formation of nano-sized clusters of Sr and Mo deficient phases in these films, which was claimed to be due to the volatility of the corresponding oxides. They also argued that presence of B-site disorder would lead to small clusters of $SrFeO_3$ and $SrMoO_3$ distributed within the film. These compounds have drastically different electrical and magnetic properties as compared to SFMO ($SrFeO_3$ has a resistivity of 1 μΩcm and is antiferromagnetic below 130 K while $SrMoO_3$ has a much lower resistivity of 10 μΩcm and exhibits Pauli paramagnetism). They concluded that the presence of such phases and corresponding interfaces could lead to the observed intriguing magnetoresistive properties.

Westerburg et al.[96] performed detailed transport and magnetotransport measurements of SFMO thin films grown by PLD. In semiconducting films where full magnetic order was not established, the temperature dependence of resistivity showed Kondo-like behavior below 30 K and variable range hopping like transport above 100 K. They observed electron-like ordinary Hall effect and a hole-like anomalous Hall contribution. Interestingly, these signs are reversed compared to the colossal magnetoresistive manganites. The value of the nominal charge carrier density at 300 K, obtained from Hall effect by a single band model, is about four electrons per formula unit. On the other hand, the metallic films[87] had a carrier concentration of 1.3 electrons per formula unit at 300 K in a single band model scenario. The increase of the nominal charge-carrier density in the semiconducting sample compared to the metallic one thus shows that a single band model is not appropriate for this compound and band-structure effects dominate the magnetotransport.

During PLD of SFMO films, Shinde et al.[97] observed that ordered and metallic films are realized in a very narrow window of substrate temperature. Under optimum growth conditions, the films grew epitaxially on STO substrates. These films exhibited XRD rocking curve width comparable to the substrate and also four peaks in XRD Φ-scan corresponding to the in-plane epitaxial relation with the substrate. The films grown at temperatures lower than the optimal temperatures were semiconducting. At higher (than optimal) growth temperature, the films were found to be highly conducting due to dissociation of Sr_2FeMoO_6 and formation of low resistivity $SrMoO_3$ phase. At optimal growth temperature, the films exhibited metallic conductivity with resistivity value close to that observed for single crystals. They observed that the films grown in vacuum were metastable and aged with time, while those grown in Ar atmosphere were stable and their temperature dependence of resistivity was similar to that of SFMO single crystal.

In a recent study, Venimadhav et al.[89] also observed a strong dependence of magnetic properties on the ambient used during PLD. Figure 23 shows the room temperature hysteresis loops obtained for the films grown in N_2+O_2, N_2+H_2, and Ar_2+O_2. The best films with a saturation magnetic moment of 3.8 μ_B were obtained in Ar_2+O_2 atmosphere. The films showed tunneling-type magnetoresistance, which they attributed to tunneling across the antiphase boundaries in films.

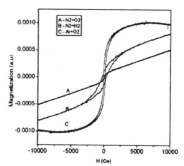

Figure 23: Magnetic hysteresis loops for the films deposited in different gases [Reprinted with permission from M. Blamire, A. Venimadhav et al., *J. Magn. Magn. Mat.* 269, 101 (2004). Ref. 89].

In order to gain some insight into the grain boundary magnetoresistance of SFMO, Shinde et al.[97] also studied polycrystalline films that were obtained by performing depositions on polycrystalline STO substrates. As expected, the polycrystalline films showed higher magnetoresistance as compared to the epitaxial films. However, no low-field behavior expected for the tunneling type grain boundary magnetoresistance was observed. They found that the polycrystalline films were metallic with a resistivity of the order of 1 mΩcm. Also, the current-voltage characteristics were found to be linear. Thus, in the as-grown films, the grain boundaries were conducting and no tunneling conduction existed. Shinde et al.[97] therefore performed a series of annealing experiments on these films and found interesting behavior. They annealed the films at different temperature in the range 400 – 500 °C in air. After annealing at temperature 475 °C or higher, the current-voltage characteristics were found to be non-linear and the resistivity showed semiconducting/insulating behavior. These effects were attributed to the changes occurred primarily at the grain boundaries. They observed that only the films that showed non-linear I-V behavior showed low-field magnetoresistance (Fig. 24). Thus, they concluded that in SFMO the low-field magnetoresistance corresponds to the magnetization dependent tunneling between two grains.

This observation of tunneling type magnetoresistance is consistent with the earlier experiments of Yin et al.[98]. They deposited high quality film of SFMO on STO bicrystal substrate by PLD and patterned it in the form of Wheatstone bridge as shown in Fig. 25. The magnetoresistance of the arm crossing the grain boundary showed tunneling-type low field magnetoresistance, whereas the arm that was not crossing the grain boundary did not show such behavior. Thus they concluded that the low-field MR is due to spin-dependent electron transfer across the grain boundary.

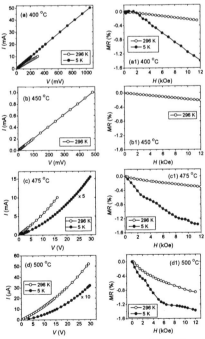

Figure 24: The I-V curves (a,b,c,d) and MR-H plots (a1,b1,c1,d1) for the polycrystalline films annealed at different temperatures [Ref. 97].

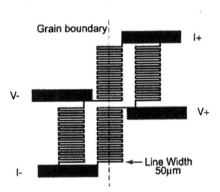

Figure 25: Diagram of the Wheatstone bridge configuration used for a Sr_2FeMoO_6 thin film grown on a $SrTiO_3$ bicrystal substrate with a 24° misalignment at the boundary. I1, I2, V1, and V2 are current and voltage terminals [Reprinted with permission from J. Goodenough, H. Yin et al., *Appl. Phys. Lett.* 75, 2812 (1999). Copyright by the American Institute of Physics. Ref. 98].

Among the other studies on double-perovskite films, Asano et al.[99] deposited SFMO films on STO substrates with and without $Ba_{0.4}Sr_{0.6}TiO_3$ buffer layer. The films with buffer layer showed higher magnetization that was close to the expected value. Philipp et al.[100] observed that PLD grown Sr_2CrWO_6 showed no B-site cation ordering presumably due to similar ionic sizes of Cr and W. Despite this, the films had a T_C well above 400 K, thereby questioning the previous theories about the requirement of sublattice ordering for high T_C.

Using ultrasonic spray pyrolysis deposition followed by high temperature postannealing in forming gas, Rager et al.[101] obtained highly oriented films of SFMO. These films showed metallic conductivity, small negative magnetoresistance that depended on the film growth temperature, and lower saturation magnetization ($\sim$ 2.5 μ_B/f.u.). The spin polarization for these films, as determined by fitting the degree of suppression of Andreev reflection at the sample surface was about 60 %. In the extension of this work, Branford et al.[102] doped films with 10 % La, Ca, or Ba. Compared to the undoped films, the doped films showed reduced saturation magnetization and magnetoresistance. However, the La-doped film showed a significant increase (40 K) in T_C. This is consistent with the reported increase in T_C by electron doping[103]. In their studies, Fontcuberta and co-workers[103] achieved this by replacing Sr^{2+} in Sr_2FeMoO_6 by La^{3+} and found increase in T_C of $\sim$70 K. These results indicate that the ferromagnetic coupling is mediated by itinerant carriers, and provide experimental support to the double-exchange picture.

5.4.3.4 Double-Perovskite Based Magnetic Tunnel Junction

In spite of a number of studies reported on thin films of double perovskite, not much work has been reported on devices based on these materials. This is partly due to the presence of inhomogeneities, surface roughness, and a very high sensitivity of these materials to different conditions (e.g. moisture) that makes handling and processing rather difficult. The first and probably the only MTJ device based on double perovskite was reported by Bibes and co-workers[104]. By using a nanolithography based on real time electrically controlled nanoindentation by a conducting tip atomic force microscopy they fabricated nano-sized SFMO/STO/Co MTJs (Fig. 26a). They observed a tunneling magnetoresistance of as high as 50 % at 4 K, as shown in Fig. 26b. This corresponds to a spin polarization of Sr_2FeMoO_6 of about 85 %, a value very close to the predicted value of 100 %.

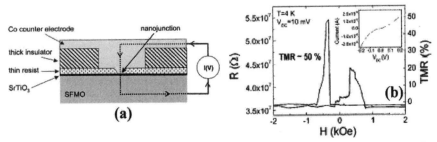

Figure 26: (a) Schematic of the final device. (b) Magnetoresistance curve of a $Sr_2FeMoO_6/SrTiO_3/Co$ tunnel junction defined by nanoindentation at 4 K and for a bias voltage of 10 mV; inset: current vs voltage curve obtained at 4 K. Positive bias corresponds to electron injection from Sr_2FeMoO_6 to Co [Reprinted with permission from M. Bibes et al., *Appl. Phys. Lett.* 83, 2629 (2003). Copyright by the American Institute of Physics. Ref. 104].

5.5 ELECTRIC FIELD EFFECT DEVICES

In most of the discussion above, the focus was on magnetic field related device actions and concepts. However, given the high sensitivity of the properties of the concerned materials to external perturbations, the effects of other perturbations could be fruitfully explored in the interest of oxide electronics. Amongst these, perhaps the most interesting possibility is to perturb the system with an electric field, much the same way as in the case of a semiconductor field effect transistor. In thin films, electric field of several MV/cm can be easily realized with several nanometer thick high quality dielectrics and a very practical applied voltage of a few volts. Electric field effects have been studied in high-T_C superconducting oxides, and most of these appear to be well explained by field-induced modulation of mobile carrier density. In the case of materials such as manganites, which display nanoscale electronic inhomogeneities, the strength of electric field effects as well as the corresponding physics could in fact be interestingly different.

5.5.1 Early Experiments

In the first experiment on this subject Ogale et al.[106] used mixed-valent manganite channels $Nd_{0.7}Sr_{0.3}MnO_3$ (NSMO) with STO dielectric. They observed several interesting effects: (i), Field-direction independent decrease of the resistance at temperatures above the resistivity peak temperature (T_P); (ii) Modulation proportional to E^2; (iii) A dramatic reduction in the magnitude and field-direction independent reversal of the sign of $\Delta R/R$ just below T_P; and (iv) A time response at high temperatures limited by RC time constant of a device but near T_P, the same being partly limited by intrinsic mechanism. They concluded that the observed changes result from the lattice distortions in the NSMO layer caused by electric field.

Following these initial studies[106], further experiments were performed on the manganite channel by Mathews et al.[107] using LCMO as a semiconductor and $PbZr_{0.2}Ti_{0.8}O_3$ as a ferroelectric gate, recognizing that the remnant polarization of ferroelectric thin films offers the possibility of nonvolatile memory elements. A ferroelectric field effect transistor (FET) permits nondestructive memory readout because the device can be interrogated by reading the conductance of the semiconductor channel. This all-perovskite ferroelectric field effect device made of a CMR channel showed at least 300% modulation with retention of at least on the order of several hours. Their capacitance versus voltage data were found to be consistent with the formation of accumulation and depletion regions in the oxide semiconductor.

5.5.2 Electric and Magnetic Field Effects: Electronic Phase Separation

Recently, Wu et al.[108] have investigated the electric field effects in mixed-valent manganites in many more details using a new, inverted device configuration (inset of Fig. 27a). This has lead to a new understanding of the phenomenon. The surface preparation procedure of Kawasaki et al.[109] was found to be crucial for obtaining high quality devices. The new geometry was a significant improvement over the previous designs and the strain in the channel layer was minimized.

Figure 27a shows the dependence of resistivity (ρ) of the $La_{0.7}Ca_{0.3}MnO_3$ (LCMO) channel with the PZT gate on temperature (T), for field biasing (V_{gate} from +6 to -6 V). The new features seen are as follows:

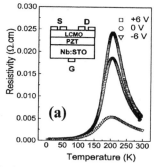 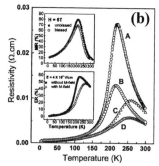

Figure 27: (a) Dependence of resistivity of the LCMO channel on temperature, for field biasing with a PZT gate over an applied gate voltage range from +6 to -6 V. Inset shows the device configuration. (b) LCMO channel resistivity on temperature for the unbiased (A) and electric-field-biased (B) channel, in the absence of magnetic field. A changes to C, and B changes to D under a magnetic field of 6 T. The insets show MR and ER data. [Ref. 108]

First, the effect is much stronger (electroresistance, $ER_{max} = [R(E) - R(0)]/R(0) = 76\%$ for $E = 4 \times 10^5$ V/cm near T_P) than reported previously[106]. Second, the ER changes sign with the field direction. Third, the magnitude of ER is not symmetric *vis a vis* the field direction, implying channel asymmetry for hole vs electron-type carrier modulation. Finally, the gate field leaves the temperature T_P, at which resistivity peaks, almost unchanged. In the PZT device the polarization dependence on field is nonlinear, but the device behavior was found to be similar to those based on dielectric material STO.

Results were also obtained for $Nd_{0.7}Ca_{0.3}MnO_3$ (NSMO) $La_{0.7}Ba_{0.3}MnO_3$ (LBMO) and $La_{0.5}Ca_{0.5}MnO_3$ channels by using the new device geometry. In NSMO, the sign of $-\Delta R/R$ was seen to be polarity independent and the magnitude of the effect was rather small, as seen previously[106]. For LBMO the behavior was found to be similar to that for LCMO, but the magnitude of the effect was much lower. The effect was also found to be negligibly small in the charge ordered insulator $La_{0.5}Ca_{0.5}MnO_3$. The fact that a large effect was seen only in the 0.7:0.3 LCMO samples, wherein metallic and insulating phase mixtures dominate in the transition region, brought out the significance of such phase separation in the context of the physics of electric field effect in manganites.

Figure 27b compares the E-and H-field effects on transport in LCMO, showing ρ vs T for the unbiased and field-biased (4×10^5 V/cm) channel, with and without 6T H-field. Wu et al.[108] observed that an E-field of 4×10^5 V/cm produces a change in ρ which is almost as large as that produced by an H-field of 6 T. Curves B and C in Fig. 27b show that, for same conductivity, the electroresistance [magnetoresistance (MR)] is lower (higher) for $T < T_P$ than that for $T > T_P$. Thus, the E-and H-fields have a complementary effect on the transport. Comparison of curves B and D, or of C and D, shows that the application of an H-field of 6 T to the E-field-biased channel, or an E-field to the channel subjected to a 6 T H-field, produces an additional large drop in ρ. The H-field shifts the temperature at which ρ and ER peak. The insets show MR vs T in 6 T, for the unbiased and E-field-biased CMR channel, and ER vs T, for the channel with and without an H-field of 6 T. The difference seen in the ER case arises from the peak shift characteristic of CMR.

The observations of Wu et al.[108] seem to be more naturally interpreted as a consequence of inhomogeneous charge ordering (CO) in the films. In an electronically inhomogeneous channel, partly metallic (M) and partly insulating (I), the E-field would change the relative volume fractions of the M and I regions by accumulation of charge between M and I phases, thereby causing the interface to move. The nature of accumulating charge, holes vs electrons, will depend on field polarity and thereby control the direction of interface movement. In this system, as T is lowered, "droplets" of ferromagnetic (FM) metal appear. Applying a negative gate voltage causes electrons to accumulate at I-M interfaces, which pushes the interfaces away from the electrodes, causing increased metallicity with a nonlinear V dependence. A positive voltage does the opposite, leading to a ρ which saturates at the intrinsic insulating value for the given T, which is not high. This picture also naturally explains the experimentally observed complementary character of the E-and H-field influence on the CMR channel. The H-field makes the intrinsic behavior of the insulating regions more metallic, while the E-field changes the connectivity of the metallic regions by modifying the volume fractions of the FM and CO components.

That the ER in LCMO is much larger than in NSMO, LBMO, or the 0.5-doped LCMO has a natural interpretation in the phase coexistence model. Essentially all of the 0.5-doped LCMO is in a strongly insulating state; essentially all of the NSMO is in a weakly insulating state; hence there is no ER. The lower resistivity and the higher T_C of LBMO show that it is mostly metallic. Only in the LCMO film are the two phases sufficiently closely balanced to give a truly large ER.

5.5.3 Manganite-Based Electric Field Effect Device on Silicon

Very recently, Zhao et al.[110] fabricated an all-perovskite ferroelectric field effect transistor (Fig. 28a) structure with a ferroelectric $Pb(Zr_{0.2}Ti_{0.8})O_3$ (PZT) gate and a CMR $La_{0.8}Ca_{0.2}MnO_3$ (LCMO) channel on Si and studied the effects of electric and magnetic fields on the LCMO electrical resistance. This study is important from the standpoint of oxide electronics and semiconductor microelectronics. The structure of the FeFET used by Zhao et al.[110] is shown in Fig. 28b,c. Following an

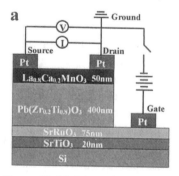
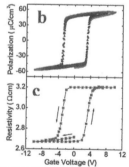

Figure 28: (a) Schematic structure of the Pt/LCMO/PZT/SRO/STO/Si FeFET. (b) The ferroelectric hysteresis loop of the PZT gate measured through the Gate and the Drain; (c) The resistivity of the LCMO channel as a function of gate voltage measured through the Source and the Drain at 200K without a magnetic field [Ref. 110].

approximately 20nm-thick STO template growth on Si by molecular beam epitaxy, a 75nm-thick SrRuO₃ (SRO) bottom electrode, a 400nm-thick PZT and a 50nm-thick LCMO layers were deposited by PLD. The final device structure was formed through standard photolithography, lift-off and chemical etching. A Pt layer was grown by PLD as the top electrode. The LCMO channel size was 20μm×200μm, while the Pt electrode size was 200μm×200μm.

Low frequency ferroelectric hysteresis measurements, Fig. 28, show a square hysteresis loop indicating the high quality of the PZT film. To measure the ferroelectric field effect on the LCMO channel resistivity, an electric voltage was applied to the PZT layer through the Gate and the Drain. The resistivity of the LCMO layer between Source and Drain was measured three seconds after the gate bias was turned off to allow any switching transients to die down. The LCMO resistivity at 200K as a function of applied gate voltage is plotted in Fig. 28c. A clear resistivity hysteresis loop was observed with sweeping of the gate voltage. The maximum channel resistivity modulation ($[(\rho_{max}-\rho_{min})/\rho_{min}]\times100\%$) seen was about 20%. The similarity in the shape of the hysteresis loop and the coercive voltage between the ferroelectric loop in Fig. 28b and the resistivity loop in Fig. 28c confirms that the resistivity change in the LCMO channel is attributed to the polarization change in the PZT gate resulting from the gate poling. The well-saturated resistivity hysteresis loop means that the channel resistance could be switched between two stable states with a difference of 20% by applying a +5 or –5 V pulse. In the initial tests, Zhao et al.[110] observed a change of less than 3% at each state for a retention time of 20 hours.

5.5.4 Other Interesting Current and Electric Field Induced Effects

Tokura and coworkers[111] have shown that the switching of resistive states in manganites can be achieved not only by a magnetic field, but also by injection of an electric current. For manganites of the form Pr₁₋ₓCaₓMnO₃, they reported that an electric current triggers the collapse of the low-temperature, electrically insulating charge-ordered state to a metallic ferromagnetic state, thereby causing drastic (more than ten orders of magnitude) decrease in resistivity. This current switching phenomenon is expected to open up new channels for fabrication of electromagnets on a submicrometer scale, as the switching between the low- and high-resistivity states in manganites is expected to be accompanied by a metamagnetic transition.

Liu et al.[112] discovered a large electric-pulse-induced reversible resistance change (active at room temperature and under zero magnetic field) in colossal magnetoresistive (CMR) $Pr_{0.7}Ca_{0.3}MnO_3$ thin films. They observed electric field-direction-dependent resistance changes of more than 1700% under applied pulses of ~100 ns duration and as low as ~5 V magnitude. The resistance changes were found to be reversible and nonvolatile. They grew $Pr_{0.7}Ca_{0.3}MnO_3$ (PCMO) films, partially on a conducting layer (bottom electrode) of either $YBa_2Cu_3O_{7-x}$ (YBCO) or Pt. Silver contact pads were sputtered on top of the PCMO and on the YBCO layers in PCMO/YBCO bilayer samples. Electrical pulses of 100 ns duration were applied across the PCMO film through these contacts and the resistance between the two electrodes was measured after each pulse. At room temperature, and under zero applied magnetic field, the measured resistance of the PCMO film was observed to change after the peak voltage of the applied pulse was increased above a threshold value (~ 5 V). It was found that the resistance of the PCMO film could either be decreased or increased according to the pulse polarity. The sample resistance was seen to decrease precipitously when positive pulse went from PCMO layer to YBCO layer, reaching a saturation low value under accumulating pulses of the same polarity and peak voltage. Upon reversing the direction of the pulsed field, the resistance was seen to increase to a high saturation value with increasing number of pulses. Subsequent changes of the pulsed field direction lead to reversal of the resistance change of the PCMO film resulting in an electrically reversible variable resistance for PCMO. At 300K a given resistance state was found to be stable with time, exhibiting a resistance variation of less than 1.5 % over 2×10^5 s.

5.6 INVESTIGATIONS OF BURIED INTERFACES

In the variety of oxide electronics devices discussed above, the device performance is defined and controlled by the quality of interfaces involved. The electrical and magnetic quality of an interface depends on the structural and chemical compatibility of materials as well as the growth parameters used in developing the corresponding layered film configuration. A search for the right kind of interface for a particular application should thus involve their post-formation characterization. Idzerda and coworkers[71,113] have shown that x-ray absorption spectroscopy (XAS) and x-ray magnetic circular dichroism (XMCD) techniques can be used very effectively to investigate the electronic structure and magnetic properties of the interfaces.

5.6.1 La$_{0.7}$Sr$_{0.3}$MnO$_3$/YBa$_2$Cu$_3$O$_7$ Interface

In their first study on the LSMO/YBCO interface using Mn L$_{2,3}$ XAS and XMCD Stadler et al.[71] brought out that the magnetic properties and electronic structure of the LSMO are adversely affected by the YBCO overlayer caused by cation displacement/exchange that effectively reduces the concentration of the La atoms in the LSMO near the YBCO/LSMO interface[71,113]. Two total electron yield spectra were recorded for each case, one with the incident light helicity (circular polarization) oriented parallel to the remnant magnetization of the sample, and one antiparallel. The XAS is the average of these two spectra, while the XMCD is the difference between the two. Here, the helicity of the light was kept constant while the remnant magnetization of the sample was reversed by reversing a magnetic field, which was applied in the plane of the sample. All XMCD spectra were corrected for incomplete photon polarization and the 45° sample orientation with respect to the incoming beam. With this experimental configuration, the measurements were sensitive to the top 50–100 Å of the LSMO film. The presence of the additional overlayer only marginally modifies the Mn L-edge spectra by both increasing the sensitivity to the YBCO/LSMO interfacial region and decreasing the signal intensity. Nonetheless, the results reported are not due to the variation of the probed interfacial volume as confirmed with spectra collected at a very grazing angle of incidence.

The Mn L$_{2,3}$ XAS and XMCD spectra for YBCO(t)/LSMO(1000 Å)/STO with thickness, t=0, 20, 50, and 80 Å are shown in Fig. 29.

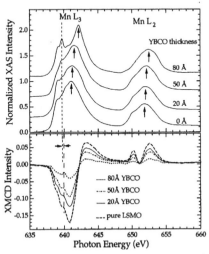

Figure 29. Mn L$_{2,3}$ XAS and XMCD spectra for LSMO capped with various thicknesses of YBCO. [Refs. 71]

The XAS spectral changes represent interface while the XMCD monitors the magnetic behavior of the LSMO at that interface. It is clear that the deposition of the YBCO strongly affects the electronic and magnetic properties of the LSMO in the interfacial region, identifying the source of the poor spin-polarized transport properties observed for these structures. The spectra exhibit four distinct changes as the YBCO overlayer thickness is increased; the change in the L_3 peak shape with the development of a second peak at fixed lower binding energy, the continuous energy shift of the maximum of the L_3 peak to higher energy, the narrowing of the L_3 distribution, and the decrease in the magnitude of the XMCD spectrum.

These XAS spectral variations are very similar to the results obtained for the Mn $L_{2,3}$ XAS of $La_{1-x}Sr_xMnO_3$ as a function of changing Sr concentration. As the composition of the LSMO is varied from $La_{0.7}Sr_{0.3}MnO_3$ to $La_{0.1}Sr_{0.9}MnO_3$, Abbate et al.[114] have reported that the Mn $L_{2,3}$ XAS spectra shape exhibits precisely the same spectral changes (narrowing, doubling, and continuous energy shift) as that of Mn XAS spectra for increasing YBCO coverage. This strongly suggests that the YBCO overlayer facilitates a cation displacement or interchange whose net effect is the out-diffusion of La (interchange of La with Ba) or by exchange of Sr from deeper in the bulk with La at the interface (a Mn–Cu interchange, although unlikely, also cannot be ruled out). It is known that, as the concentration changes in favor of Sr, the unit cell volume of LSMO decreases and becomes a better match for the overlying YBCO lattice. Additional evidence for this conclusion is the observed diminishing of the dichroism signal as a function of increasing YBCO thickness. The $La_{0.7}Sr_{0.3}MnO_3$ is ferromagnetic and as the relative Sr concentration is increased, it becomes antiferromagnetic. As more interfacial Mn is transformed into the antiferromagnetic state, the XMCD will diminish, as is indeed observed. It is known that the existence of oxygen vacancies also influences the transport properties of LSMO by reducing the hole concentration. However, Idzerda and co-workers[71,113] eliminated this possibility for their interface region by further controlled experiments.

5.6.2 Fe_3O_4/TiN and Fe_3O_4/$SrTiO_3$ Interfaces

In another study involving magnetite (Fe_3O_4), a magnetic oxide with high (possibly 100%) spin polarization of potential interest to spintronics, Idzerda and coworkers examined the interface formation between different thicknesses of strontium titanate ($SrTiO_3$) or titanium nitride (TiN) with a 2000 Å Fe_3O_4 film by using XAS and XMCD[115]. These interface systems are of interest for magnetic tunnel junction and spin valve type applications, respectively. Their results showed that deposition of 10–50 Å of TiN results in an immediate and substantial removal of oxygen from the near interface region; resulting in the formation of spin-randomizing FeO interlayers. Indeed, Fig. 30a shows the spectral weight of the L_3 edge

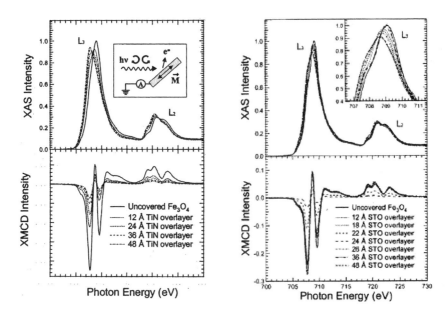

Figure 30: Peak area normalized Fe $L_{2,3}$ XAS (top) and XMCD (bottom) spectra as a function of (a) TiN and (b) SrTiO$_3$ overlayer thickness. The inset in (a) is the experimental geometry. The inset in (b) is an enlargement of the L_3 peak. [Ref. 115]

shifting from the Fe$_3$O$_4$ position to the FeO position as the thickness of the TiN overlayer is increased from 0 to 48 Å, suggesting that there is a continuous decrease of magnetic Fe$_3$O$_4$ in the interfacial region via the formation of antiferromagnetic FeO. For the case of deposition of SrTiO$_3$ (thin layers) on Fe$_3$O$_4$, these measurements showed only a small deviation from the Fe$_3$O$_4$ characteristic XAS signature, suggesting the limited formation of perhaps only one monolayer of another Fe oxide at the interface. The persistent XMCD signal, however, confirmed the preservation of Fe$_3$O$_4$ in its ferromagnetic state.

Figure 30b shows the peak area normalized Fe $L_{2,3}$ XAS and XMCD spectra as a function of STO coverage. Unlike the behavior from TiN overlayers, thin STO overlayers generated only minor changes in the XAS spectra, and no change in the XMCD spectra. However, for thicker overlayers (22 Å and 24 Å thicknesses), a substantially reduced XMCD intensity was seen. The corresponding XAS spectra also exhibited a peculiar trend. Thus thin STO layers can serve as good barrier layers for tunnel structures involving magnetite.

5.7 CONCLUSIONS

Oxide electronics is a new frontier, which holds considerable promise for the future. Some basic device concepts (magnetic tunnel junctions, spin valve, electric field effect transistor etc.) have been demonstrated, in principle. However, practical device development with the desirable level of device performance is hampered by our limited ability to control and manipulate the surfaces of magnetic oxides, and their interfaces with other materials including oxides, metals and semiconductors. High quality *in situ* characterization of film growth and use of novel techniques to probe the electronic and magnetic properties of buried interfaces hold the key to the success of the device fabrication process.

ACKNOWLEDGEMENT

The authors gratefully acknowledge support under NSF-MRSEC (DMR-96-32521 and DMR-00-80008) and DARPA Spin-S (N000140210962, Dr. S. Wolf) programs. We also acknowledge fruitful discussions and collaborations with R. L. Greene, H. D. Drew, S. Das Sarma, T. Zhao, Y. Zhao, T. Wu, E. Li, A. Orozco, M. Robson, R. Godfrey, M. Johnson, S. Lofland, S. M. Bhagat, H. D. Drew (University of Maryland) James Buban, Nigel Browning (University of Illinois at Chicago), A. J. Millis (Columbia University), S. W. Cheong (Rutgers University), R. Droopad, K. Eisenbeiser (Motorola Labs), Yves Idzerda, Dario Arena, Joe Drovak, Alexander Lussier (Montana State University). The authors would also like to thank researchers and research journals who gave us permission to use specific figures from published papers.

REFERENCES

1. S. D. Sarma, *Am. Sci* .**89**, 516 (2001).
2. S. D. Sarma, *Nature* **2**, 292 (2003).
3. M. Johnson, *Nature* **416**, 809 (2002).
4. S. A. Wolf, D. D. Awschalom, R. A. Buhrman, J. M. Daughton, S. Von Molnar, M. Roukes, A. Y. Chtchelkanova and D. M. Treger, *Science* **294**, 1488 (2001).
5. H. Ohno, *Science* **281**, 951 (1998).
6. S. J. Liu, J. Y. Juang, K. H. Wu, T. M. Uen, Y. S. Gou and J.-Y. Lin, *Appl. Phys. Lett.* **80**, 4202 (2002).
7. D. Kumar, S. Chattopadhyay, W. M. Gilmore, C. B. Lee, J. Sankar, A. Kvit, A. K. Sharma, J. Narayan, S. V. Pietambaram and R. K. Singh, *Appl. Phys. Lett.* **78**, 1098 (2001).
8. D. K. Fork and G. B. Anderson, *Appl. Phys. Lett.* **63**, 1029 (1993).
9. L. D. Chang, M. Z. Tseng, E. L. Hu and D. K. Fork. *Appl. Phys. Lett.* **60**, 1753 (1992).
10. J. H. Park, E. Vescovo, H. J. Kim, C. Kwon, R. Ramesh and T. Venkatesan, *Nature* **392**, 794 (1998).

11. Y. Okimoto, T. Katsufuji, T. Ishikawa, A. Urushibara, T. Arima and Y. Tokura, *Phys. Rev. Lett.* **75**, 109 (1995).

12. H.Y. Hwang,, S.-W. Cheong, N.P. Ong and B. Batlogg, *Phys. Rev. Lett.* **77**, 2041 (1996).

13. P.K. de Boer, H. van Leuken, R.A. de Groot, T. Rojo and G.E. Barberis, *Solid State Commun.* **102**, 621 (1997).

14. K. Ghosh, S. B. Ogale, S. P. Pai, M. Robson, Eric Li, I. Jin, Zi-wen Dong, R. L. Greene, R. Ramesh, T. Venkatesan and M. Johnson, *Appl. Phys. Lett.* **73**, 689 (1998).

15. P. Seneor, A. Fert, J. L. Maurice, F. Montaigne, F. Petroff and A. Vaure`s, *Appl. Phys. Lett.* **74**, 4017 (1999).

16. A. Yanase and K. Siratori, *J. Phys. Soc. Jpn.* **53**, 312 (1984).

17. Z. Zhang and S. Satpathy, *Phys. Rev. B* **44**, 13319 (1991).

18. D. J. Huang, L. H. Tjeng, J. Chen, C. F. Chang, W. P. Wu, S. C. Chung, A. Tanaka, G. Y. Guo, H.-J. Lin, S. G. Shyu, C. C. Wu and C. T. Chen, *Phys. Rev. B* **67**, 214419 (2003).

19. E. Z. Kurmaev, A. Moewes, S. M. Butorin, M. I. Katsnelson, L. D. Finkelstein, J. Nordgren and P. M. Tedrow, *Phys Rev B* **67**, 155105 (2003).

20. Y.S. Dedkov, M. Fonine, C. Konig, U. Rudiger, G. Guntherodt, S. Senz and D. Hesse, *Appl. Phys. Lett.* **80**, 4182 (2002).

21. S.P. Lewis, P.B. Allen and T. Sasaki, *Phys. Rev. B* **55**, 10253 (1997).

22. A. Anguelouch, A. Gupta, G. Xiao, D.W. Abraham, Y. Ji, S. Ingvarsson and C.L. Chien, *Phys. Rev. B* **64**, 180408 (2001).

23. Y. Ji, G.J. Strijkers, F.Y. Yang, C.L. Chien, J.M. Byers, A. Anguelouch, G. Xiao and A. Gupta, *Phys. Rev. Lett.* **86**, 5585 (2001).

24. A. Gupta, X.W. Li, and G. Xiao, *Appl. Phys. Lett.* **78**, 1894 (2001).

25. Y. Matsumoto, M. Murakami, T. Shono, T. Hasegawa, T. Fukumura, M. Kawasaki, P. Ahmet, T. Chikyow, S.-Y. Koshihara and H. Koinuma, *Science* **291,** 854 (2001).

26. S. A. Chambers, S. Thevuthasan, R. F. C. Farrow, R. F. Marks, J. U. Thiele, L. Folks, M. G. Samant, A. J. Kellock, N. Ruzycki, D. L. Ederer and U. Diebold, *Appl. Phys. Lett.* **79**, 3467 (2001).

27. S. R. Shinde, S. B. Ogale, S. Das Sarma, J. R. Simpson, H. D. Drew, S. E. Lofland, C. Lanci, J. P. Buban, N. D. Browning, V. N. Kulkarni, J. Higgins, R. P. Sharma, R. L. Greene and T. Venkatesan, *Phys. Rev. B* **67**, 115211 (2003).

28. S. Guha, K. Ghosh, J.G. Keeth, S. B. Ogale, S.R. Shinde, J. R. Simpson, H. D. Drew and T. Venkatesan, *Appl. Phys. Lett.* Accepted.

29. S. B. Ogale, R. J. Choudhary, J. P. Buban, S. E. Lofland, S. R. Shinde, S. N. Kale, V. N. Kulkarni, J. Higgins, C. Lanci, J. R. Simpson, N. D. Browning, S. Das Sarma, H. D. Drew, R. L. Greene and T. Venkatesan, *Phys. Rev. Lett.* **91**, 077205 (2003).

30. Y. G. Zhao, S. R. Shinde, S. B. Ogale, J. Higgins, R. J. Choudhary, V.N. Kulkarni, R. L. Greene, T. Venkatesan, S. E. Lofland, C. Lanci, J. P. Buban, N. D. Browning, S. Das Sarma and A. J. Millis, *Appl. Phys. Lett.* in press.

31. S. N. Kale, S. B. Ogale, S. R. Shinde, M. Sahasrabuddhe, V. N. Kulkarni, R. L. Greene and T. Venkatesan, *Appl. Phys. Lett.* **82**, 2100 (2003).

32. D. P. Norton, S. J. Pearton, A. F. Hebard, N. Theodoropoulou, L. A. Boatner and R. G. Wilson, *Appl. Phys. Lett.* **82**, 239 (2003).

33. S-J. Han, J. W. Song, C.-H. Yang, S. H. Park, J.-H. Park, Y. H. Jeong and K. W. Rhie, *Appl. Phys. Lett.* **81**, 4212 (2002).

34. W. Prellier, A. Fouchet, B. Mercey, C. Simon and B. Raveau, *Appl. Phys. Lett.* **82**, 3490 (2003).

35. J.M.D. Coey, A.E. Berkowitz, L. Balcells, F.F. Putris and F.T. Parker, *Appl. Phys. Lett.* **72**, 734 (1998).

36. G.Q. Gong, A. Gupta, G. Xiao, W. Qian and V.P. Dravid, *Phys. Rev. B* **56**, 5096 (1997).

37. A. Yanase and K. Siratori, *J. Phys. Soc. Jpn.* **53**, 312 (1984).
38. Y.S. Dedkov, U. Rüdiger, and G. Güntherodt, *Phys. Rev B* **65**, 064417 (2002).
39. A.P. Ramirez, *J. Phys.: Cond. Mat.* **9**, 8171 (1997).
40. T. Hotta, A.L. Malvezzi, and E. Dagotto, *Phys. Rev. B* **62**, 9432 (2000).
41. Y. Tokura and Y. Tomioka, *J. Mag. Magn. Mater.* 200, 1 (1999).
42. M.B. Salamon and M. Jaime, *Rev. Mod Phys.* **73**, 583 (2001).
43. T. Venkatesan, M. Rajeswari, Z.W. Dong, S.B. Ogale and R. Ramesh, *Phil. Trans. Royal Soc. A* **356**, 1661 (1998).
44. L. Zheng, X. Xu, Li Pi and Y. Zhang, *Phys. Rev. B* **62**, 1193 (2000).
45. V. Podzorov, M. E. Gershenson, M. Uehara and S-W. Cheong, *Phys. Rev. B* **64**, 115113 (2001).
46. T. Wu, S. B. Ogale, J. E. Garrison, B. Nagaraj, A. Biswas, Z. Chen, R. L. Greene, R. Ramesh, T. Venkatesan and A. J. Millis, *Phys. Rev. Lett.* **86**, 5998 (2001).
47. M. Uehara, S. Mori, C. H. Chen and S.-W. Cheong, *Nature* **399**, 560 (1999).
48. J. Navarro, J. Fontcuberta , M. Izquierdo, J. Avila and M. C. Asensio, *Phys. Rev. B* **70**, 054423 (2004).
49. K.-I. Kobayashi, T. Kimura, H. Sawada, K. Terakura and Y. Tokura, *Nature* **395**, 677 (1998).
50. Abhijit S. Ogale, S. B. Ogale, R. Ramesh and T. Venkatesan, *Appl. Phys. Lett.* **75**, 537 (1999).
51. D. Niebieskikwiat, A. Caneiro, R. D. Sa´nchez, and J. Fontcuberta, *Phys. Rev. B* **64**, 180406 (2001).
52. S. Kale, S. M. Bhagat, S. E. Lofland, T. Scabarozi, S. B. Ogale, A. Orozco, S. R. Shinde, T. Venkatesan, B. Hannoyer, B. Mercey and W. Prellier, *Phys. Rev. B* **64**, 205413 (2001).
53. T. Wu, S. B. Ogale, S. R. Shinde, A. Biswas, T. Polletto, R. L. Greene, T. Venkatesan and A. J. Millis, *J. Appl. Phys.* **93**, 5507 (2003).
54. S.R. Shinde, R. Ramesh, S.E. Lofland, S.M. Bhagat, S.B. Ogale and T. Venkatesan, *Appl. Phys. Lett.* **72**, 3443 (1998).
55. S. Valencia, LI Balcells, J. Fontcuberta and B. Martınez, *Appl. Phys. Lett.* **82**, 4531 (2003).
56. Mandar Paranjape, A. K. Raychaudhuri, N. D. Mathur and M. G. Blamire, *Phys. Rev. B* **67**, 214415 (2003).
57. O.I. Lebedev, G. Van Tendeloo, S. Amelinckx, B. Leibold and H. –U. Habermeier, *Phys. Rev. B* **58**, 8065 (1998).
58. B. Vengalis, A. Maneikis, F. Anisimovas, R. Butkute, L. Dapkus and A. Kindurys, *J. Magn. Mag. Mater.* **211**, 35 (2000).
59. S. Stadler, Y. U. Idzerda, Z. Chen, S. B. Ogale and T. Venkatesan, *J. Appl. Phys.* **87**, 6767 (2000).
60. M. N. Baibich, J. M. Broto, A. Fert, F. N. Van Dau, F. Petroff, P. Eitenne, G. Creuzet, A. Friederich and J. Chazelas, *Phys. Rev. Lett.* **61**, 2472 (1988).
61. B. Dieny, V. S. Speriosu, S. S. P. Parkin, B. A. Gurney, D. R. Wilhoit, and D. Mauri, *Phys. Rev. B* **43**, 1297 (1991).
62. M. Julliere, *Phys. Lett. A* **54**, 225 (1975).
63. J. Z. Sun, W. J. Gallagher, P. R. Duncombe, L. Krusin-Elbaum, R. A. Altman, A. Gupta, Yu Lu, G. Q. Gong and G. Xiao, *Appl. Phys. Lett.* **69**, 3266 (1996).
64. J. Z. Sun, L. Krusin-Elbaum, P. R. Duncombe, A. Gupta and R. B. Laibowitz, *Appl. Phys. Lett.* **70**, 1769 (1997).
65. Y. Xu, D. Ephron and M. R. Beasley, *Phys. Rev. B* **52**, 2843 (1995).
66. Moon-Ho Jo, N. D. Mathur, N. K. Todd and M. G. Blamire, *Phys. Rev. B.* **61**, R14905 (2000).
67. M. Bowen, M. Bibes, A. Barthe´ le´my, J. P. Contour, A. Anane, Y. Lemaı̂tre and A. Fert, *Appl. Phys. Lett.* **82**, 233 (2003).

68. J. H. Park, E. Vescovo, H. J. Kim, C. Kwon, R. Ramesh and T. Venkatesan, *Phys. Rev. Lett.* **81**, 1953 (1998).

69. H. Y. Hwang, S-W. Cheong, N. P. Ong and B. Batlogg, *Phys. Rev. Lett.* **77**, 2041 (1996).

70. U. P. Wad, Abhijit S. Ogale, S. B. Ogale and T. Venkatesan, *Appl. Phys. Lett.* **81**, 3422 (2002).

71. S. Stadler, Y. U. Idzerda, Z. Chen, S. B. Ogale and T. Venkatesan, *Appl. Phys. Lett.* **75**, 3384 (1999).

72. J. W. Freeland, K. E. Gray, L. Ozyuzer, P. Berghuis, Elvira Badica, J. Kavich, H. Zheng and J. F. Mitchell, *Nature Materials* **4**, 62 (2005), See also, M. Coey, *Nature Materials* **4**, 62 (2005)

73. H. Yamada, Y. Ogawa, Y. Ishii, H. Sato, M. Kawasaki, H. Akoh and Y. Tokura, *Science* **305**, 646 (2004).

74. S. S. P. Parkin, C. Kaiser1, A. Panchula1, P.M. Rice, B. Hughes, M. Samant and S.-H. Yang, *Nature Materials* **3**, 862 (2004).

75. S. Yuasa, T. Nagahama, A. Fukushima, Y. Suzuki and K. Ando, *Nature Materials* **3**, 868 (2004).

76. Y.X. Chen, C. Chen, W.L. Zhou, Z.J. Wang, J. Tang, D.X. Wang and J.M. Daughton, *J. Appl. Phys.* **95**, 7282 (2004).

77. R.J. Kennedy and P.A. Stampe, *J. Magn. Magn, Mater.* **195**, 284 (1999).

78. B. Aktas, *Thin Solid Films* **307**, 250 (1997).

79. P. A. Cox, Transition Metal Oxides (Oxford University Press, Oxford, 1995).

80. K. Ghosh, S. B. Ogale, S. P. Pai, M. Robson, Eric Li, I. Jin, Zi-wen Dong, R. L. Greene, R. Ramesh, T. Venkatesan and M. Johnson, *Appl. Phys. Lett.* **73**, 689 (1998).

81. X. W. Li, A. Gupta, Gang Xiao, W. Qian and V. P. Dravid, *Appl. Phys. Lett.* **73**, 3282 (1998).

82. P. Seneor, A. Fert, J. L. Maurice, F. Montaigne, F. Petroff and A. Vaure`s, *Appl. Phys. Lett.* **74**, 4017 (1999).

83. J. Gopalakrishnan, A. Chattopadhyay, S.B. Ogale, T. Venkatesan, R. L. Greene, A. J. Millis, K. Ramesha, B. Hannoyer and G. Marest, *Phys. Rev. B* **62**, 9538 (2000).

84. K.-I. Kobayashi, T. Kimura, H. Sawada, K. Terakura, and Y. Tokura, *Nature* **395**, 677 (1998).

85. K.-I. Kobayashi, T. Kimura, Y. Tomioka, H. Sawada, K. Terakura and Y. Tokura, *Phys. Rev. B* **59**, 11159 (1999).

86. H. Yanagihara, M. B. Salamon, Y. Lyanda-Geller, Sh. Xu and Y. Moritorno, *Phys. Rev. B* **64**, 214407 (2001).

87. W. Westerburg, D. Reisinger, and G. Jakob, *Phys. Rev. B* **62**, R767 (2000).

88. A. S. Ogale, S.B. Ogale, R. Ramesh and T. Venkatesan, *Appl. Phys. Lett.* **75**, 537 (1999).

89. A. Venimadhav, F. Sher, J. P. Attfield and M. G. Blamire, *J. Magn. Magn. Mater.* **269**, 101 (2004).

90. Ll. Balcells, J. Navarro, M. Bibes, A. Roig, B. Martı´nez and J. Fontcuberta, *Appl. Phys. Lett.* **78**, 781 (2001).

91. J Navarro, Ll Balcells, F Sandiumenge, M Bibes, A Roig, B Mart´ınez and J Fontcuberta, *J. Phys.: Condens. Matter* **13**, 8481 (2001).

92. M. García-Hernández, J. L. Martínez, M. J. Martínez-Lope, M. T. Casais and J. A. Alonso, *Phys. Rev. Lett.* **86**, 2443 (2001).

93. M. Besse, F. Pailloux, A. Barthélémy, K. Bouzhouane, A. Fert, J. Olivier, O. Durand, F. Wyczisk, R. Bisaro and J.-P. Contour, *J. Crystal Growth* **241**, 448 (2002).

94. T. Manako, M. Izumi, Y. Konishi, K.-I. Kobayashi, M. Kawasaki and Y. Tokura, *Appl. Phys. Lett.* **74**, 2215 (1999).

95. H. Asano, S.B. Ogale, J. Garrison, A. Orozco, Y. H. Li, E. Li, V. Smolyaninova, C. Galley, M. Downes, M. Rajeswari, R. Ramesh and T. Venkatesan, *Appl. Phys. Lett.* **74**, 3696 (1999).

96. W. Westerburg, F. Martin and G. Jakob, *J. Appl. Phys.* **87**, 5040 (2000).

97. S. R. Shinde, S. B. Ogale, R. L. Greene, T. Venkatesan, K. Tsoi, S.-W. Cheong and A. J. Millis, *J. Appl. Phys.* **93**, 1605 (2003).

98. H. Q. Yin, J.-S. Zhou, J.-P. Zhou, R. Dass, J. T. McDevitt and J. B. Goodenough, *Appl. Phys. Lett.* **75**, 2812 (1999).

99. H. Asano, Y. Kohara and M. Matsui, *Japanese J. Appl. Phys. Part II: Lett.* **41**, L1081 (2002).

100. J. B. Philipp, D. Reisinger, M. Schonecke, M. Opel, A. Marx, A. Erb, L. Alff and R. Gross, *J. Appl. Phys.* **93**, 6853 (2003).

101. J. Rager, A. V. Berenov, L. F. Cohen, W. R. Branford, Y. V. Bugoslavsky, Y. Miyoshi, M. Ardakani, and J. MacManus-Driscoll, *Appl. Phys. Lett.* **81**, 5003 (2001).

102. W. R. Branford, S. K. Clowes, Y. V. Bugoslavsky, Y. Miyoshi, L. F. Cohen, A. V. Berenov, J. L. MacManus-Driscoll, J. Rager and S. B. Roy, *J. Appl. Phys.* **94**, 4714 (2003).

103. J. Navarro, C. Frontera, Ll. Balcells, B. Martínez and J. Fontcuberta, *Phys. Rev. B* **64**, 092411 (2001).

104. M. Bibes, K. Bouzehouane, A. Barthélémy, M. Besse, S. Fusil, M. Bowen, P. Seneor, J. Carrey, V. Cros, A. Vaurès, J.-P. Contour and A. Fert, *Appl. Phys. Lett.* **83**, 2629 (2003).

105. P. Schiffer, A. P. Ramirez, W. Bao and S.-W. Cheong, *Phys. Rev. Lett.* **75**, 3336 (1995).

106. S. B. Ogale, V. Talyansky, C. H. Chen, R. Ramesh, R. L. Greene and T. Venkatesan, *Phys. Rev. Lett.* **77**, 1159 (1996).

107. S. Mathews, R. Ramesh, T. Venkatesan and J. Benedetto, *Science* **276**, 238 (1997).

108. T. Wu, S. B. Ogale, J. E. Garrison, B. Nagaraj, Amlan Biswas, Z. Chen, R. L. Greene, R. Ramesh, T. Venkatesan and A. J. Millis, *Phys. Rev. Lett.* **86**, 5998 (2001).

109. M. Kawasaki et al., *Appl. Surf. Sci.* **107**, 102 (1996).

110. T. Zhao, S. B. Ogale, S. R. Shinde, R. Ramesh, R. Droopad, J. Yu, K. Eisenbeiser and J. Misewich, *Appl. Phys. Lett.* **84**, 750 (2004)

111. A. Asamitsu, Y. Tomioka, H. Kuwahara and Y. Tokura, *Nature* **388**, 50 (1997).

112. S. Q. Liu, N. J. Wu and A. Ignatiev, *Appl. Phys. Lett.* **76**, 2749 (2000).

113. S. Stadler, Y. U. Idzerda, Z. Chen, S. B. Ogale and T. Venkatesan, *J. Appl. Phys.* **87**, 6767 (2000).

114. M. Abbate, F. M. F. de Groot, J. C. Fuggle, A. Fujimori, O. Strebel, F. Lopez, M. Domke, G. Kaindl, G. A. Sawatzky, M. Takano, Y. Takeda, H. Eisaki and S. Uchida, *Phys. Rev. B* **46**, 4511 (1992).

115. A. Lussier, Y. U. Idzerda, S. Stadler, S. B. Ogale, S. R. Shinde and T. Venkatesan, *J. Vac. Sci. Technol. B* 20, 1609 (2002).

Diluted Magnetic Semiconductors and Insulators

CHAPTER 6

DILUTED MAGNETIC OXIDE SYSTEMS

S.B. Ogale, S.R. Shinde, Darshan C. Kundaliya and T. Venkatesan

Center for Superconductivity Research, Department of Physics, University of Maryland, College Park, MD 20742

6.1 INTRODUCTION

In recent years there is considerable emphasis in materials research on tailoring the properties of different materials to the required specifications, the latter being dictated by the ever growing demands of modern technology for new materials. Such "designer" approach to materials synthesis also holds the key to novel device architectures, which may not necessarily suffer from the limitations imposed by the property-set of the standard hosts. Magnetic materials are well known vehicles for a number of interesting applications in both low and high technology sectors. In the past, such materials were considered primarily for storage and sensor applications, but in the emerging high-tech scene they are needed to perform various other interesting tasks as well. Indeed such newly identified tasks generally focus on the property of spin polarization and its manipulation across interfaces in heterostructures and other modulated architectures by electric and magnetic fields as well as electromagnetic radiation. The efforts along these lines have come to be recognized and qualified as a new discipline of advanced magneto-opto-electronics.

A scientifically challenging and technologically rewarding approach in this field is to explore the possibilities of magnetizing functional non-magnetic materials by introducing a dilute concentration of magnetic impurities therein, thereby harnessing the benefits of magnetic response without completely affecting the application bearing attractive properties of the non-magnetic host. Clearly, the non-magnetic host in this context could be any functional materials such as Si, Ge, III-V or II-VI compound semiconductors, oxides or nitrides and as expected, work is being actively pursued on all these materials. The materials class of interest to this article is oxides, which offers the broadest structure-property space for design and manipulation. In the following we discuss the background, experimental knowledge, and theoretical issues related to some interesting oxide based diluted magnetic systems.

6.2 DILUTED MAGNETIC OXIDES (DMO) SEMICONDUCTORS, DMO METALS, AND DMO INSULATORS:

A comprehensive magneto-opto-electronic device effort embodies development of magnetic materials with diverse properties such as semiconductors (field effect devices, spin injection devices, optical elements), metals (contacts, magnetotransport channels), and insulators (magneto-optic devices, tunnel materials for spin filter). Fortunately oxides provide the broadest range of such properties of interest. In this section we discuss some of the DMOs reported in the recent literature.

6.2.1 Mn, Co doped ZnO

Early work on the development of DMO focused on epitaxial films of Mn doped ZnO.[1,2] Carrier densities in excess of 10^{19} /cm^3 were achieved. However, only spin glass behavior was observed, but with high Curie Weiss temperature corresponding to a strong antiferromagnetic coupling. In a subsequent study on Co doped ZnO a huge optical magnetic circular dichroism (O-MCD) signal, as shown in Fig. 1, was observed.[3] In yet another study on the same system, ferromagnetism above room temperature was also seen but with less than 10% reproducibility.[4] O-MCD spectra of ZnO doped with various transition metals have also been examined and compared. No discernable magneto-optical effect was observed for Sc, Ti, V and Cr doped samples, but Mn, Co, Ni and Cu doped samples did show pronounced negative MCD peaks near 3.4 eV, which corresponds to the band gap of the host ZnO semiconductor (Fig. 2).[5] These results suggested that ZnO alloyed with Mn, Co, Ni, and Cu is a diluted magnetic semiconductor with strong exchange interaction between sp-band carriers and localized d electrons. Other reports on Mn doped ZnO either showed paramagnetic nature[6,7] or suggested ferromagnetism at very low temperature.[8] In an interesting study on Mn implantation in ZnO codoped with Sn, ferromagnetism with Curie temperature of ~250 K was suggested.[9]

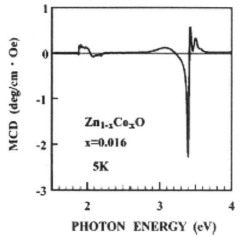

Figure 1: Magnetic circular dichroism spectrum of $Zn_{1-x}Co_xO$ (x=0.016) at 5 K [Reprinted with permission from K. Ando et al., Appl. Phys. Lett. **78**, 2700 (2001). Copyright by the American Institute of Physics. Ref. 3].

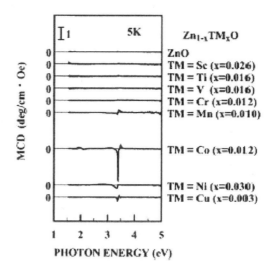

Figure 2: MCD spectra of $Zn_{1-x}TM_xO$ (TM: Sc, Ti, V, Cr, Mn, Co, Ni, or Cu) films at 5 K [Reprinted with permission from K. Ando et al., *J. Appl. Phys.* **89**, 7284 (2001). Copyright by the American Institute of Physics. Ref. 5].

In a recent[10] report on low temperature (~500 °C) processed mixtures of MnO_2 and ZnO powders ferromagnetism above room temperature was demonstrated and the authors of this work suggested that Mn incorporates uniformly in ZnO under their processing conditions and therefore this is a diluted magnetic semiconductor. Other works including some of our own recent studies have contested this claim.[11] Our study in fact suggests that

198

ferromagnetism in this system originates in a vacancy stabilized metastable phase ($Mn_{2-x}Zn_xO_{3-\delta}$) rather than via carrier induced interaction between separated Mn atoms in ZnO. We also have shown that ferromagnetism in low temperature processed $Zn_{1-x}Mn_xO$ persists up to ~ 900 K and further heating transforms this metastable phase and kills ferromagnetism.

Thermo-Gravimetric Analysis (TGA) was used to bring out the role of thermal transformations of Mn-oxide in the process. X-ray diffraction (XRD) data suggest that subsequent to the low temperature processing which leads to ferromagnetism, MnO_2 grains remain mostly unaffected, implying limited interdiffusion and ferromagnetism only in the surface shell. As shown in Fig. 3, in the presence of ZnO the oxygen loss begins at a temperature as low as 450–500 K, which could be due to Zn incorporation into Mn^{4+} sites in MnO_2 causing oxygen release and on further heating weight loss stabilizes at Mn_3O_4 level at around ~780K rather than at Mn_2O_3 level. This tends to suggest that in the presence of Zn, manganese oxide is skipping a phase transformation from $MnO_2 \Rightarrow Mn_2O_3$. But the qualitative graphical similarity between both the curves lends itself to the following possible interpretation, especially as the temperature of transformation to the plateau phase (780 K) is much lower than the temperature for the transformation to the Mn_3O_4 phase (over 1,000 K). In Mn_2O_3, the valence of Mn is 3+, and with incorporation of Zn^{2+} in the matrix, the system will lose half oxygen atom per Zn atom. Also, if one

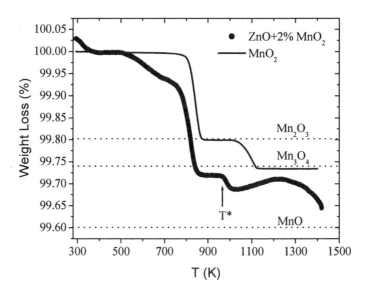

Figure 3: TGA data for MnO_2 (line) and ZnO + 2% MnO_2 powder mixture (circles) shown as percentage weight loss. The data for the mixture are slightly up-shifted as explained in the text. The weight levels for total conversion of MnO_2 into other stoichiometric forms are indicated by horizontal dashed lines. [Ref. 11]

removes one oxygen per three molecules of Mn_2O_3, the corresponding stoichiometry looks like Mn_3O_4. Thus, one may actually have an oxygen-vacancy-stabilized Zn-incorporated phase of the form $Mn_{2-x}Zn_xO_{3-\delta}$. At T* this metastable phase could then transform into Mn_3O_4 with oxygen uptake for valence control, albeit at somewhat lower temperature. If the ferromagnetism is identified with the phase $Mn_{2-x}Zn_xO_{3-\delta}$, it is understandable that the moment vanishes at ~900 K.

The XRD analysis of these samples (Fig 4) reveals that the contributions corresponding to the minority (2% MnO_2) phase are present even after the low-temperature thermal treatment of the nonmagnetic unsintered mixture leading to the thermally processed ferromagnetic phase. This implies that the ferromagnetic phase resides only in the thin region near the interfaces between the powder particles, and refutes the suggestion that Mn gets uniformly incorporated in the ZnO matrix. We also studied interface diffusion and reaction between ZnO and manganese oxide thin film bilayers and found that a uniform solution of Mn in ZnO does not form under *low temperature processing*; instead a metastable ferromagnetic phase develops with Zn diffusing into Mn oxide. Direct low-temperature film growth of Zn-incorporated Mn oxide by pulsed laser deposition also shows ferromagnetism at low Zn concentration for an optimum oxygen growth pressure.

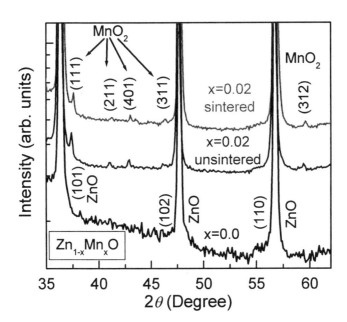

Figure 4: XRD pattern for ZnO (bottom), sintered 2 at.% MnO_2–ZnO (top) mixture, and unsintered 2 at.% MnO_2–ZnO (middle) mixture. The intensity is plotted on a log scale. [Ref. 11]

Recently, Swartz and Gamelin[12] have discovered reversible 300 K ferromagnetic ordering in the diluted magnetic semiconductor Co:ZnO. They have demonstrated switching of room-temperature ferromagnetism between "on" and "off" states in Co^{2+}: ZnO by lattice incorporation and removal of the native n-type defect, interstitial Zn. Based on spectroscopic and magnetic data they argue in favor of a double-exchange mechanism for ferromagnetism. Although the story of ZnO based DMS systems continues to unfold awaiting finality regarding materials issues, these results present new opportunities for integrating magnetism and conductivity in semiconductor sensor or spin-based electronics devices.

6.2.2 Co doped TiO_2

With the report of room temperature ferromagnetism in Co doped anatase TiO_2 thin films by Matsumoto et al.[13] brisk activity ensued in many laboratories around the world to confirm this significant claim and to establish the microscopic nature of the sample. While some groups initially argued that this is indeed a uniform DMS system (suggesting also that oxygen plasma assisted molecular beam epitaxy OPA-MBE gives better results than laser deposition)[14], many reports followed[15,16,18] wherein cobalt clustering, either in the form of cobalt metal clusters or clusters of Co-Ti-O complex, was suggested to occur and possibly be the cause of the observed ferromagnetism. It was argued that since the processing conditions cause significant changes in the properties of this system, disorder or defects states might have an important role in this context. Under a specific processing condition of high temperature annealing discussed in our own work[16] the clusters were seen to dissolve with a concomitant relaxation of lattice parameter (suggesting matrix incorporation of cobalt and RBS ion channeling for cobalt).[16] Recently we have used high temperature growth instead of high temperature annealing and have obtained samples with no apparent clusters.[19] Interestingly such high temperature grown or annealed samples are fairly insulating thereby ruling out the possibility of RKKY type ferromagnetism initially suggested for the rather conducting or semiconducting anatase TiO_2 films grown at relatively lower temperatures. Other mechanisms such as the polaronic mechanism[20, 21] (discussed below) appear more likely in such a case.

Most studies on Co doped TiO_2 have focused on thin films grown by different techniques such as laser Molecular Beam Epitaxy (LMBE)[14,22], Pulsed Laser Depostion (PLD)[15-19], sputtering [23], spray pyrolysis[24], etc. Some studies have also been performed on bulk polycrystals and nanoparticles.[25] In addition to the basic magnetic and structural characterization techniques such as SQUID and vibrating sample magnetometry, x-ray diffraction, transmission electron microscopy, Electron energy-loss spectroscopy (EELS), etc. other methods such as optical and x-ray magnetic circular dichroism (O-MCD, X-MCD)[26,27], x-ray absorption (XAS)[14,28-30], photoluminescence, Raman spectroscopy, optical spectroscopy[31], Hall effect [18,32,33] etc. have also been employed. The XAS data have shown that in controlled samples the valence of Co is 2+.[28,29,30] The bulk-sensitive Co L-edge XAS spectrum for a $Ti_{0.93}Co_{0.07}O_2$ film, along with analogous spectra for relevant reference compounds is described in ref. 14. Comparison of these spectra reveals that Co is in the +2 formal oxidation.

It has been observed that the cobalt distribution in TiO_2 films strongly depends on the growth conditions, especially on the growth temperature and the oxygen pressure during growth. At a growth temperature of 700 °C, which has been widely used in most of the reported studies, we observe a limited solubility (up to ~ 2%) of cobalt in anatase TiO_2. This limit was established through our structural examination by XRD and TEM, photoluminescence studies, and optical band-edge shift analysis. [16,31,33] Figure 5 shows that a film with x=0.01 has a T_C ~ 700 °C and no indication of clusters is found in TEM analysis (Fig. 5b)

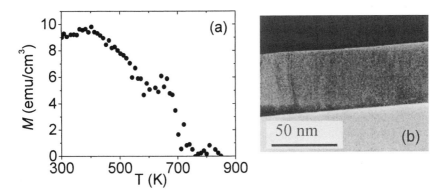

Figure 5: (a) M-T data above room temperature by VSM and, (b) TEM image for the $Ti_{0.99}Co_{0.01}O_{2-\delta}$ film. [Ref. 16]

Above x ~ 2 %, cobalt was seen to segregate and form cobalt metal clusters. For example, for a concentration of 7 %, cobalt clusters of size 20 – 50 nm were found. It is important that one should also look on a coarser scale rather than just the high-resolution scale while performing the TEM analysis since the clusters are separated from each other and the TiO_2 matrix between the clustered regions is of very high crystalline quality. The growth conditions chosen for these studies were optimal for TiO_2 growth as is evident from Fig. 6(a), where the ion channeling minimum yield is of ~5%, an indication of very good crystalline quality. Yet, the cobalt doped

TiO$_2$ films grown under identical conditions exhibited poor channeling for Ti and no channeling for cobalt due to cluster formation (Fig. 6(b)).

The measured Curie temperature, T_C, for these films was found to be ~1200 K, which in fact corresponds to cobalt clusters rather than Co:TiO$_2$. However, after a high temperature treatment, the very same film showed a T_C ~ 650 K (Fig. 7(a)), close to that of a film with lower cobalt concentration, where no clusters are formed. No clusters were observed for such a high temperature annealed film (Fig.7(b)) and the cobalt sub-lattice also showed ion channeling (Inset of Fig. 7(a)). Thus, our studies showed that under the employed growth conditions (substrate temperature =700 °C and oxygen pressure ~10^{-5} Torr) cobalt has a limited solubility in TiO$_2$ at low temperature, and a high temperature treatment is needed for incorporation of higher cobalt concentrations.

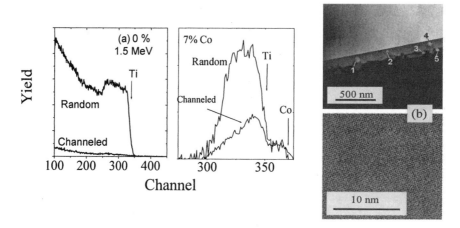

Figure 6: (a)1.5-MeV He$^+$ RBS random and channeled spectra for TiO$_2$ and Ti$_{0.93}$Co$_{0.07}$O$_2$ films. (b) TEM images of a Ti$_{0.93}$Co$_{0.07}$O$_2$ film. [Ref. 16]

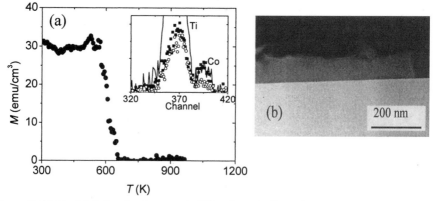

Figure 7: (a) The M-T data for the annealed Ti$_{0.93}$Co$_{0.07}$O$_2$ film. The inset shows channeling of cobalt signal. (b) TEM image of the same films. [Ref. 16]

Optical spectroscopy in these samples has shown a blue shift of the band edge due to Co doping but no mid-gap states as suggested by theoretical calculations. [34] With increasing Co concentration x, the band edge shifts to higher frequencies, showing a maximum blue shift of 100 meV for x=0.02 (Fig. 8).

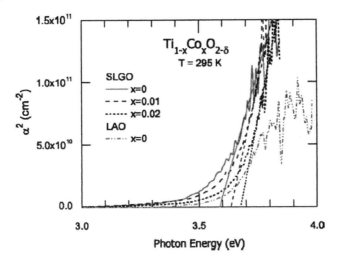

Figure 8: Room temperature frequency dependence of the square of the absorption coefficient of $Ti_{1-x}Co_xO_{2-\delta}$ films grown on SrLaGaO3 substrates for x=0, 0.01, and 0.02. Straight lines represent linear fits. An undoped TiO_2 sample grown on a $LaAlO_3$ substrate is shown for comparison. [Ref. 34]

Anomalous Hall effect (AHE) has been suggested as one of the signature tests to confirm the intrinsic nature of DMS in these new materials.[32] Unfortunately, for quite sometime no *unambiguous* demonstration of the AHE could be made either for the anatase phase of Co:TiO₂ or, for any of the DMOs. This brought into question the intrinsic nature of ferromagnetism in these systems. Given other markers of substitututional dopant incorporation, however, was suggested that non-observation of AHE in this case may presumably be due to the contribution being masked by the strong contribution of the normal Hall signal for the given resistivity range. This led three groups to explore growth under even more reducing conditions than used before so as to achieve higher carrier concentration and reduction in the normal Hall contribution.[18,32,35] Interestingly, all the three groups (one working with Fe instead of Co, and two others including ours using Co) have seen AHE in their highly reduced samples. Wang et al. [35] observed unexpected p-type conduction; while in our case the expected n-type behavior was observed in reduced TiO₂ (shallow donor states due to O vacancies) concurrently with the AHE signal (Fig. 9).

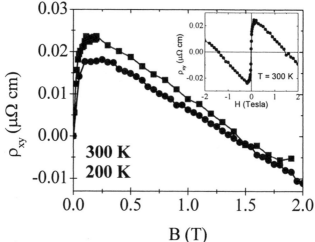

Figure 9: Hall resistivity data for highly reduced Co:TiO$_2$ sample at 300 K (square symbols) and (200 K) circle symbols. Inset shows the Hall resistivity as a function of field. [Ref. 18]

Unfortunately however, these samples grown under highly reduced conditions are rutile phase (in fact with Magnéli phase characteristics) and are not anatase. It is observed that the AHE signal increases with increase in carrier concentration, a scenario consistence with carrier-induced ferromagnetism. However, in our detailed magnetic characterization of the films grown under nominally similar conditions of Toyosaki et al.[32] and Wang et al., [35] we observed co-existence of superparamagnetism and AHE.[18] Subsequent analysis of these films by TEM further showed that tiny cobalt metal clusters organize at the film-substrate interface (Fig. 10). This further implies that AHE can't really be regarded as a strict test of intrinsic DMS.

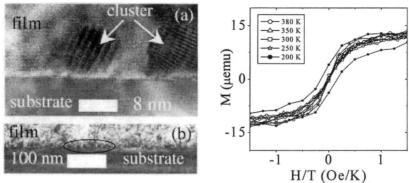

Figure 10: Two TEM images of Ti0.98Co0.02O2-δ film at different magnifications.Some of the clusters are marked in image (b) by a black loop. (c) Magnetization vs H/T loops at for different temperatures, indicating the onset of superparamagnetism above ~250K. [Ref. 18]

More recently, in order to establish the possible role of carriers in the ferromagnetism in this anatase Co:TiO$_2$ system, we have examined the possibility of tuning the magnetism therein by external electric field applied through a gate in field effect transistor (FET) geometry. In order to obtain nominally Co cluster free channel in the *in situ* growth of the heterostructure, however, we grew the channel at high temperature (875 C), which showed no apparent clustering by TEM. As stated earlier this channel is insulating in nature. We have seen a fairly strong modulation of saturation magnetization in this case, suggesting that this may indeed be an intrinsic DMO material. These results may also help resolve the questions about the precise origin of ferromagnetism in this *insulating* anatase Co:TiO$_2$.

Taking clue from the earlier observation[16] that the high temperature annealing leads to dispersion of Co in TiO$_2$ matrix, recently Shinde et al.[19] employed a new growth procedure wherein the films were grown at 875 °C in oxygen environment. The TEM and magnetic characterization studies revealed that Co did not segregate in these films. Interestingly, these films exhibit a strong Co concentration dependence of the magnetic properties. The saturation magnetization was found to decrease systematically as the Co concentration was reduced. Also the hysteresis loop shape was found to change with Co concentration. These data agree very well with the recently proposed models of F-center based exchange interaction[21] (Fig. 11) and percolation of associated polarons.[20,36] A computer simulation of polaron

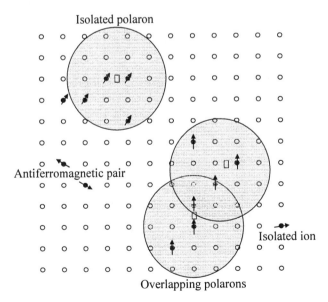

Figure 11: F-center Polaron Percolation Model by J. M. D. Coey. [Reprinted with permission from Coey et al., *Nature Materials* **4**, 173 (2005). Copyright by Nature Publishing Group. Ref 21]

percolation was also performed and it was shown that the results agree with the experimental data. Shinde et al.[19] have also noted a remarkable Co concentration dependent metal-insulator transition. In such Co:TiO$_2$ films, an electric field induced modulation of ferromagnetism (magnetization and coercivity) was also been observed.[37]

6.2.3 Co, Fe, Mn doped SnO$_2$

SnO$_2$ is a rather unique material with peculiar set of physical properties, most notably the concurrent high electrical conductivity and high optical transparency. It configures in rutile structure and has n-type conduction. The origin of simultaneous conductivity and transparency has not yet been fully understood, although recent theoretical studies point to the combined role of tin interstitial and oxygen vacancy defects and related lattice relaxations. SnO$_2$ has widespread applicability in fields such as gas sensing (even for flammable and toxic gases),[38] transparent conducting electrodes in flat-panel displays and solar cells, IR detectors, optoelectronic devices etc. It also has superior chemical stability in comparison with other wide gap semiconductors. Magnetizing this oxide by dilute doping is thus an attractive research goal.

In our work on Co doped SnO$_2$, occurrence of room temperature ferromagnetism was demonstrated in pulsed laser deposited thin films of Sn$_{1-x}$Co$_x$O$_{2-\delta}$ (x<0.3).[39] Interestingly, films of Sn$_{0.95}$Co$_{0.05}$O$_{2-\delta}$ grown on R-plane sapphire not only exhibit ferromagnetism with a Curie temperature close to 650 K, but also a giant magnetic moment of 7 μ_B/Co (Fig. 12).

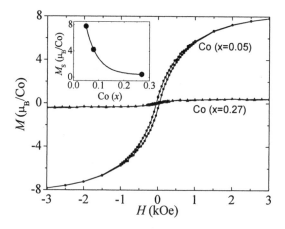

Figure 12: 300 K hysteresis loops for Sn$_{0.95}$Co$_{0.05}$O$_2$ and Sn$_{0.73}$Co$_{0.27}$O$_2$ films. The inset shows the dependence of saturation moment M$_S$ on Co concentration x. [Ref. 39]

The films were characterized by different techniques to verify the crystalline quality, microstructure and cobalt uniformity. The results are summarized in Fig. 13. Figure 13(a) shows the x-ray diffraction (XRD) pattern for a $Sn_{0.95}Co_{0.05}O_{2-d}$ thin film grown on R-plane sapphire. Only the (101) family of rutile phase film peaks is seen. The rocking curve (inset) full width at half maximum of 0.23 signifies excellent film orientation. Similar good XRD signatures were obtained even for films with Co concentrations as high as 27 at.%. At even higher Co content, a decrease in the XRD intensity was encountered [Fig. 13(b)]. The inset of Fig. 13(b) shows a scanning transmission electron microscopy (STEM) image for a $Sn_{0.73}Co_{0.27}O_{2-d}$ film, indicating that even for such high Co content the film microstructure is uniform. The electron energy loss spectroscopy (EELS) data recorded at various points spread over the TEM image [Fig. 13(c)] clearly show that cobalt is distributed uniformly in the film.

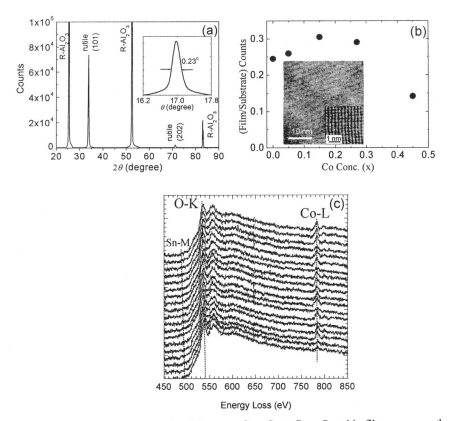

Figure 13: (a) X-ray diffraction (XRD) pattern for a $Sn_{0.95}Co_{0.05}O_{2-d}$ thin film grown on the R-plane sapphire substrate. The inset shows the rocking curve (b) normalized (101) XRD peak intensity as a function of the Co content. The inset shows high resolution STEM images for a $Sn_{0.73}Co_{0.27}O_{2-d}$ film. (c) EELS data recorded at various points spread over the TEM image domain of the image shown in (b). The spectra show the Sn-M, O-K, and Co-L edges. They are shifted on the y scale for clarity. [Ref. 39]

The films are also semiconducting and optically highly transparent. Similar giant moments and film properties have also been recently confirmed by Coey et al.[21] The origin of the observed giant moment is still unclear. But it is useful to make a few remarks on such giant magnetic. This moment is much larger than the value of ~1.67 μ_B/Co for cobalt metal, or that for small Co clusters ~2.1 μ_B/Co, or that of any of the standard Co oxides wherein the orbital moment is quenched. One possibility is that the atoms surrounding the cobalt atoms have acquired a moment through electronic effects, or the orbital moment of cobalt remains unquenched. Magnetic moments significantly larger than the spin-only moments (~6–16 μ_B/Co) have indeed been reported in transition metal atoms doped in or spread on the surfaces of alkali metal solids such as Cs,[40-46] and these have been attributed to the unquenched orbital contributions. In such cases, an increase in the dopant concentration has been found to cause a rapid decrease in the moment (as observed in Co doped SnO_2 sample) due to enhanced dopant-dopant associations leading to progressive orbital moment quenching. The decreased moment observed in higher Co-doped SnO_2 samples, and the drop in the moment after a high temperature treatment in low Co-doped samples, possibly caused by enhanced associations, suggest that the scenario may be similar in Co doped SnO_2 case. Interestingly, these films are also seen to exhibit aging effects suggesting either the metastability of defects in SnO_2 or contribution of spin glass state. These matters need further explorations.

Recently, Coey et al.[47] have shown that thin films grown by pulsed-laser deposition from targets of $Sn_{0.95}Fe_{0.05}O_2$ are transparent ferromagnets with Curie temperature of 610 K and spontaneous magnetization of 2.2 Am^2kg^{-1}, respectively. Mössbauer spectra show that iron is all high-spin Fe^{3+} but the films are magnetically inhomogeneous on an atomic scale, with only 23 % of the iron exhibiting magnetic ordering. The net ferromagnetic moment per ordered Fe ion is found to be ~1.8 μ_B, which is greater than that for any other iron oxide. These authors have proposed a mechanism of ferromagnetic coupling of ferric ions via an electron trapped in a bridging oxygen vacancy (F center) to explain the occurrence of ferromagnetism and the high Curie temperature.

Coey and coworkers[48] have also found room temperature ferromagnetism in $(Sn_{1-x}M_x)O_2$ (M = Mn, Fe, x = 0.05) ceramics (bulk) having a rutile-structure phase. Room temperature saturation magnetic moments of 0.11 or 0.95 μ_B per Mn or Fe have been realized, and the Curie temperatures are found to be 340 K and 360 K, respectively. This magnetization could not be attributed to any identified impurity phase. Mössbauer spectra of the Fe-doped SnO_2 samples showed that about 85 % of the iron is in a magnetically ordered high spin Fe^{3+} state, the remaining being paramagnetic.

The results discussed above collectively suggest that SnO_2 is an interesting functional oxide matrix for exploration of novel magnetic effects by dilute magnetic doping.

6.2.4 Co doped (La,Sr)TiO₃

$La_{1-x}Sr_xTiO_{3-\delta}$ is the mixed state of the end compounds $LaTiO_3$ and $SrTiO_3$. In cobalt doped end compounds no ferromagnetism was observed.[49] It is suggested that the mixed-valent strongly correlated character of the compound should provide favorable ground for ferromagnetic coupling between the magnetic ions. Therefore, the magnetic properties of this compound doped with different transition metal impurities in dilute concentration were explored. Indeed, ferromagnetism was observed at and above room temperature by Zhao et al.[49] in pulsed laser deposited epitaxial films of Co-doped Ti-based oxide perovskite ($La_{0.5}Sr_{0.5}TiO_{3-\delta}$) (Fig. 14). The system has the characteristics of an intrinsic diluted magnetic semiconductor (metal) with a Curie temperature of ~ 450 K at low Co concentrations (< ~ 2 %), but develops inhomogeneity at higher cobalt concentrations. The inhomogeneity is in the form of tiny clusters of LSCO and Co as reflected by two additional magnetic transitions near 210 K (LSCO) and above 1000K (Co metal). The films range from being opaque metallic to transparent semiconducting depending on the oxygen pressure during growth and are yet ferromagnetic. This remarkable tunability is interesting from applications standpoint because it enables fabrication of active layers as well as layers for contacts in an epitaxial configuration. The dependence of the magnetic

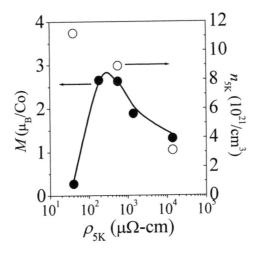

Figure 14: The dependence of magnetization and carrier concentration on resistivity for $La_{0.5}Sr_{0.5}Ti_{0.985}Co_{0.015}O_{3-\delta}$ films. [Ref. 49]

and transport properties of epitaxial $La_{0.5}Sr_{0.5}Ti_{1-y}Co_yO_{3-\delta}$ perovskite films on y and δ suggests that the ferromagnetism at low (<2%) cobalt concentration is carrier-induced.

6.2.5 Other systems

Ferromagnetism has also been reported in Co doped Cu_2O films. In this work, thin films of 5% Co doped Cu_2O were grown on single-crystal (001) MgO substrates by PLD, without and with 0.5% co-doping with Al, V, or Zn.[50] The films showed phase-pure character under the optimized growth conditions. In 'Co doped' films without any codopant, spin-glass-like behavior was observed. Ferromagnetism at room temperature was found only in the case of Co:Cu_2O films co-doped with Al.

In an interesting study performed on bulk Mn doped CuO samples (3.5–15 at. % of Mn) prepared by chemical co-precipitation method, Yang et al.[51] observed ferromagnetism with the transition (T_C) at 80 K. The transition is also associated with a metal–insulator transition. A weakly negative magnetoresistance is noted in the vicinity of the transition, but the same becomes positive in a wide range of temperatures below T_C. These experimental results suggest a possibility of interpretation in terms of the Zener double-exchange mechanism and strong electron–phonon interactions. Very recently, Wei et al. have also realized ferromagnetism in Mn doped Cu_2O with a moment of ~ 0.4 μ_B/Mn[52].

Apart from doping transition metals in binary oxide systems, researchers from University of Florida have also tried implantation of Mn, Co etc. in bulk single crystals of ternary oxide systems e.g. $BaTiO_3$, $SrTiO_3$ and $KTaO_3$[53]. Ferromagnetism with $T_C \geq 250K$ or in some cases above room temperature, with a coercive field of the order of few hundred Oe is observed in all these systems by implanting 3% or 5% of Co or Mn.

There are a few recent reports of the observation of room temperature ferromagnetism by doping transition metals (V, Cr, Ni, Fe) in TiO_2 apart from Co.[54] As discussed in the Co:TiO_2 section, clustering of Co in the host matrix restricted its use as a true DMO semiconductor unless prepared under very specific set of parameters.[19,] Recent presentations in various meetings by Chambers and coworkers suggests that Cr is suitable for doping in TiO_2 to get ferromagnetism and there are no Cr clusters or Cr-Ti-O complexes seen in this case.

Most recently, as shown in Figure 15, Coey and coworkers have reported anisotropic ferromagnetism in Zinc oxide substituted by Sc, Ti, V, Fe, Co and Ni.[55] A large moment of 2.6 μ_B/Co was observed for the case of $Zn_{0.95}Co_{0.05}O$. Ferromagnetism in Sc doped ZnO (0.3 μ_B/Sc) is a surprise

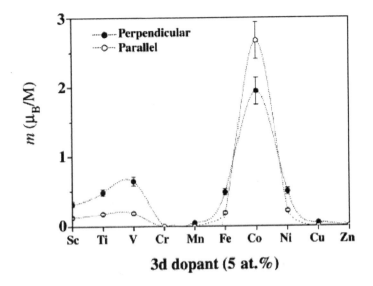

Figure 15: Magnetic moment of $Zn_{1-x}M_xO$ films, M =Sc; Ti; . .Cu; Zn, measured at room temperature. Solid circles are for the field applied perpendicular to the film plane and open circles are for the field applied in the plane of the film. [Reprinted with permission from J.M.D. Coey, *Phys. Rev. Lett.* **93**, 177206 (2004). Copyright by the American Physical Society Ref. 55]

since none of these elements is magnetic. A new source of magnetism is suggested for such case, namely, triplet states associated with two-electron defects. The same group has also observed an unexpected magnetism in a high-k dielectric and wide band gap Hafnium oxide (HfO_2),[56] which attracted a tremendous attention because of its outstanding potentials for replacing silicon dioxide as a gate dielectric in semiconductor devices and the new spin functionality due to the observed ferromagnetism. Unfortunately, there are no reports yet about the reproducibility of this significant result by other groups. In a very systematic approach, we examined the possibility of occurrence of ferromagnetism in undoped and dilutely Co and Fe doped HfO_2 films grown by pulsed laser deposition (PLD) using a variety of characterization techniques. Contrary to the recent report, we did not observe ferromagnetism in undoped HfO_2 films grown under a wide variety of conditions. However we did observe ferromagnetism in lightly Co and Fe doped HfO_2 films grown under a specific temperature and oxygen partial pressure (O_2pp) parameter space.[57]

In a recent interesting and detailed article, Coey et al.[21] have discussed the observations of ferromagnetism in materials that are high-*k* dielectrics supporting degenerate or n-type thermally activated transport. They argue that conventional super-exchange or double-exchange interactions could not lead to long-range magnetic order at a few percent doping with magnetic cations. They have proposed a model for ferromagnetic exchange mediated by shallow donor electrons that form bound magnetic polarons (BMP), which overlap to create a spin-split impurity band. They have also shown that within this model the Curie temperature in the mean-field approximation varies as $(x\delta)1/2$ where x and δ are the concentrations of magnetic cations and donors, respectively. When empty minority-spin or majority-spin d states lie at the Fermi level in the impurity band high Curie temperatures are suggested to be possible. They have provided a magnetic phase diagram that includes regions of semiconducting and metallic ferromagnetism, cluster paramagnetism, spin glass and canted antiferromagnetism. In addition to the role of the cations with d-electron character in ferromagnetism, these authors have also made a case for the key role of the defects introduced at film/substrate interface, the possible candidates being the two-electron defects that have a triplet ground state or low-lying triplet excited state are the V_0 centre (a cation vacancy) and the F_t centre (adjacent F+ centres). They argue that in an ultra-thin defective interface layer the triplet molecular orbitals may themselves form an impurity band, taking the role of the magnetic polarons in inducing ferromagnetism. Such a model is suggested particularly for cases wherein ferromagnetism is observed at low dopant concentrations, $x \leq 1\%$.

6.3 THEORETICAL APPROACHES

T. Dietl and coworkers[58] have provided a Zener model description of ferromagnetism in Zinc-Blend type magnetic semiconductors. This theory considers ferromagnetic correlation mediated by holes originating from shallow acceptors in the ensemble of the localized spins in doped magnetic semiconductors. It explicitly takes into account the complex valence-band structure of such ferromagnetic semiconductors. The corresponding results bring out the important effects of the spin-orbit coupling in the valence band in determining the magnitude of the T_C as well as the direction of the easy axis in p-type ferromagnetic semiconductors. Considerably high T_C values are predicted for materials containing greater concentrations of holes and magnetic ions or consisting of lighter elements.

Chattopadhyay et al.[59] have formulated a theory of doped magnetic semiconductors in the dynamical mean field approximation and have found the T_C as a function of magnetic coupling strength J, carrier density n, and Mn density x . A subtle interplay between carrier density and magnetic coupling is shown to determine the value of T_C. The theory is applicable to both the weak and intermediate coupling limits, and allows calculation of the resistivity and optical conductivity.

Kaminski and Das Sarma[20] have examined the development of spontaneous magnetization in diluted magnetic semiconductors as arising from a percolation of bound magnetic polarons. Their analytical theory takes into account both disorder and strong magnetic interaction starting with localized carriers and employs the physically appealing magnetic polaron percolation picture. Analytic expressions are derived for the Curie temperature and the magnetization in the limit of low carrier density. Excellent quantitative agreement is obtained with Monte Carlo simulation results. These percolation-theory-based calculations agree well with the existing data in strongly insulating materials, such as $Ge_{1-x}Mn_x$. The issues of non-mean-field like magnetization curves and the observed incomplete saturation magnetization values in diluted magnetic semiconductors have also been addressed. Carrier density is found to be the crucial parameter determining the magnetization behavior. Their calculated dependence of magnetization on external magnetic field is also in excellent agreement with the existing experimental data.

Das Sarma and coworkers[60] have also developed a disordered lattice mean field theory for DMS ferromagnetism. It incorporates spatial fluctuations associated with random lattice locations of the impurity moments, the finite carrier mean free path, and the Mn-Mn nearest-neighbor antiferromagnetic coupling. The RKKY form is assumed for the carrier-mediated indirect exchange interaction between the impurity local moments, based on the support from first principles band structure calculations and numerical calculations. All the magnetic properties, including T_C, are seen to depart significantly from the predictions of the continuum Weiss mean-field theory. A strong theoretical correlation is noted between T_C and the metallicity of the system (i.e. r_0), consistent with experimental observations. The theory allows estimation of magnetic properties as a function of five independent physical parameters (J_0, J_{AF}, n_c, n_i, r_0).

Dagotto and coworkers[61] have numerically studied a lattice spin-fermion model for diluted magnetic semiconductors (DMS), improving on previously used mean-field approximations. Curie temperatures are obtained varying the Mn spin and hole densities, and the impurity-hole exchange J in the units of the hopping t. Broad guidelines to improve T_C are discussed. It is suggested that the optimum must be intermediate between the itinerant and localized limits. The number of antisite defects must be controlled. The simplest procedure to increase T_C is suggested to rely on increasing the scale t. This work also suggests formal analogies between DMS and manganite models, with similar T_C s.

Bruno and coworkers[62] have also examined carrier-induced ferromagnetism in diluted magnetic semiconductors. They have shown that the effect of a proper treatment of disorder, which includes all single-site, multiple scattering seems to play a crucial role. A standard RKKY calculation that neglects disorder is shown to strongly underestimate the Curie temperature and is suggested to be inappropriate to describe magnetism in DMS. It is further shown that an antiferromagnetic exchange favors higher Curie temperature.

Sullivan and Erwin[63] have performed first-principles microscopic calculations of the formation energy, electrical activity, and magnetic moment of Co dopants and a variety of native defects in TiO_2 anatase. Equilibrium thermodynamics is used to predict the resulting carrier concentration, the average magnetic moment per Co, and the dominant oxidation state of Co. The predicted values are found to be in good agreement with experiments on samples grown under O-poor growth conditions. In this regime, the calculations suggest that a substantial fraction of Co dopants occupy interstitial sites as donors, and incomplete compensation of these donors by substitutional Co acceptors leads to the experimentally observed n-type behavior.

6.4 SUMMARY AND CONCLUSIONS

During the past few years, the field of diluted magnetic oxides has witnessed many interesting studies, results and excitements. At the same time the field is also clouded by concerns regarding materials issues, leading to suspicions about extrinsic Vs intrinsic nature of ferromagnetism. Efforts are being made to eliminate all the potential sources of extrinsic ferromagnetism (including specific sample handling tools) which could cause experimental artifacts and variable/irreproducible results. New discoveries are still being made and questioned. Researchers appear to be converging on a few specific cases as intrinsic ferromagnets. Devices based on such materials are likely to be fabricated and tested in the near future. On the whole, this field promises an exciting and mature phase of research in the next few years.

ACKNOWLEDGEMENTS

The authors gratefully acknowledge support under NSF-MRSEC (DMR-96-32521 and DMR-00-80008) and DARPA Spin-S (N000140210962, Dr. S. Wolf) programs. The authors also acknowledge fruitful discussions and collaborations with R. L. Greene, H. D. Drew, S. Das Sarma, T. Zhao, Y. Zhao, S. E. Lofland, R. Choudhary, J. Higgins, J. Simpson (University of Maryland) James Buban, Nigel Browning (University of Illinois at Chicago), R. Ramesh (University of California, Berkeley) A. J. Millis (Columbia University), S. W. Cheong (Rutgers University), Yves Idzerda, Alexander Lussier (Montana State University). The authors would also like to thank researchers and research journals who gave us permission to use specific figures from published papers.

REFERENCES

1. T. Fukumura, Z. Jin, A. Ohtomo, H. Koinuma and M. Kawasaki, *Appl. Phys. Lett.* **75**, 3366 (1999).

2. T. Fukumura T. Fukumura, Zhengwu Jin, M. Kawasaki ,T. Shono, T. Hasegawa, S. Koshihara and H. Koinuma. *Appl. Phys. Lett.* **78**, 958 (2001).

3. K. Ando, H. Saito, Zhengwu Jin, T. Fukumura, M. Kawasaki,Y. Matsumoto and H. Koinuma, *Applied Phys. Lett.* **78**, 2700 (2001).

4. K. Ueda, H. Tabata and T. Kawai, *Appl. Phys. Lett.* **79**, 988 (2001).

5. K. Ando, H. Saito Zhengwu Jin, T. Fukumura, M. Kawasaki Y. Matsumoto and H. Koinuma, *J. Appl. Phys.* **89**, 7284 (2001).

6. A. Tiwari et al., *Solid State Commn.* **121**, 371 (2002).

7. X.M. Cheng & C.L. Chien, *J. Appl. Phys.* **93**, 7876 (2003).

8. S.W. Jung et al., *Appl. Phys. Lett.* **80**, 4561 (2002).

9. D. P. Norton, S. J. Pearton, A. F. Hebard, N. Theodoropoulou, L. A. Boatner, R. G. Wilson, *Appl. Phys. Lett.* **82**, 239 (2003).

10. P. Sharma, A. Gupta, K.V. Rao, Frank J. Owens, R. Sharma, R. Ahuja, J. M. Osorio Guillen, B. Johansson and G. A. Gehring, *Nature Materials* **2**, 673 (2003).

11. D. C. Kundaliya, S. Ogale, S. Lofland, S. Dhar, C. J. Metting, S. R. Shinde, Z. Ma, B. Varughese, K. V. Ramanujachary, L. Salamanca-Riba and T. Venketesan, "On the origin of high temperature ferromagnetism in low temperature processed Mn-Zn-O system," *Nature Materials* **3**, 709 (2004).

12. D.A. Schwartz and D.R. Gamelin, *Advanced Materials* **16**, 2115 (2004).

13. Y. Matsumoto, M. Murakami, T. Shono, T. Hasegawa, T. Fukumura, M. Kawasaki, P. Ahmet, T. Chikyow, S.-Y. Koshihara and H. Koinuma, *Science* **291**, 854 (2001).

14. S. A. Chambers, S. Thevuthasan, R. F. C. Farrow, R. F. Marks, J. U. Thiele, L. Folks, M. G. Samant, A. J. Kellock, N. Ruzycki, D. L. Ederer and U. Diebold, *Appl. Phys. Lett.* **79**, 3467 (2001).

15. J.-Y. Kim, J. H. Park, B.-G. Park, H.-J. Noh, S.-J. Oh, J. S. Yang, D.-H. Kim, S. D. Bu, T. W. Noh, H.-J. Lin, H. H. Hsieh and C. T. Chen, *Phys. Rev. Lett.* **90**, 017401 (2003).

16. S. R. Shinde, S. Ogale, S. Das Sarma, J. R. Simpson, H. D. Drew, S. Lofland, C. Lanci, J. P. Buban, N. D. Browning, V. N. Kulkarni, J. Higgins, R. P. Sharma, R. L. Greene and T. Venketesan, *Phys. Rev. B* **67**, 115211 (2003).

17. S. Guha, K. Ghosh, J. G. Keeth, S. Ogale, S. R. Shinde, J. R. Simpson, H. D. Drew and T. Venketesan, *Appl. Phys. Lett.* **83**, 3296 (2003).

18. S. R. Shinde, S. Ogale, J. Higgins, H. Zheng, A. J. Millis, R. Ramesh, R. L. Greene and T. Venketesan, *Phys. Rev. Lett.* **92**, 166601 (2004).

19. S.R. Shinde, S. B. Ogale, A.S. Ogale, H. Zheng, S. Dhar, D. C. Kundaliya, M.S.R. Rao, T. Venkatesan, *Unpublished* (2005).

20. A. Kaminski and S. Das Sarma, *Phys. Rev. Lett.* **88**, 247202 (2002).

21. J. M. D. Coey, M. Venkatesan and C. B. Fitzgerald, *Nature Materials* **4**, 173 (2005).

22. S. A. Chambers, T. Droubay, C. M. Wang, A. S. Lea, R. F. C. Farrow, L. Folks, V. Deline and S. Anders, Appl. Phys. Lett. 82, 1257 (2003).

23. A. Punnoose, M.S. Seehra, W.K. Park and J.S. Moodera, *J. Appl. Phys.* **93**, 7867 (2003).

24. A. Manivannan, M.S. Seehra, S.B. Majumder and R.S. Katiyar, *Appl. Phys. Lett.* **83**, 111 (2003).

216

25. Y. L. Soo, G. Kioseoglou, S. Kim, Y. H. Kao, P. Sujatha Devi, John Parise, R. J. Gambino and P. I. Gouma, *Appl. Phys. Lett.* **81**, 655 (2002).

26. T. Fukumura, Y. Yamada, K. Tamura, K. Nakajima, T. Aoyama, A. Tsukazaki, M. Sumiya, S. Fuke, Y. Segawa, T. Chikyow, T. Hasegawa, H. Koinuma and M. Kawasaki, *Jpn. J. Appl. Phys. Part 2-Lett.* **42**, L105 (2003).

27. J. R. Simpson, H. D. Drew, S.R. Shinde, S. B. Ogale and T. Venkatesan, *Bull. Am. Phys. Soc.* (2005).

28. A. Lussier, Y. Idzerda, S.B. Ogale, S. R. Shinde, R.J. Choudhary and T. Venkatesan, *J. Appl. Phys.* **95**, 7190 (2003).

29. A. Lussier, J. Dvorak, Y. Idzerda, S. R. Shinde, S.B. Ogale and T. Venkatesan, *Physica Scripta* (In press).

30. S. A. Chambers, S. M. Heald and T. Droubay, *Phy. Rev. B.* **67**, R100401 (2003).

31. S. Guha, K. Ghosh, J.G. Keeth, S. B. Ogale, S.R. Shinde, J. R. Simpson, H. D. Drew and T. Venkatesan, *Appl. Phys. Lett.* **83**, 3296 (2003).

32. H. Toyosaki, T. Fukumura, Y. Yamada, K. Nakajima, T. Chikyow, T. Hasegawa, H. Koinuma and M. Kawasaki, *Nature Materials* **3**, 221 (2004).

33. J. Higgins, S. R. Shinde, S. Ogale, T. Venketesan and R. L. Greene, *Phys. Rev. B* **69**, 073201 (2004).

34. J. R. Simpson, H. D. Drew, S. R. Shinde, R. J. Choudhary, Y. G. Zhao, S. Ogale and T. Venketesan, *Phys. Rev. B* **69**, 193205 (2004)

35. Zhenjun Wang, Wendong Wang, Jinke Tang, Le Duc Tung, Leonard Spinu and Weilie Zhou, *Appl. Phys. Lett.* **83**,518 (2003).

36. M. Mayr, G. Alvarez and E. Dagotto, *Phys. Rev. B* **65**, 241202 (2002).

37. T. Zhao, S. R. Shinde, S. Ogale, H. Zheng, T. Venketesan, R. Ramesh and S. Das Sarma, *Phys. Rev. Lett.* **94**, 126601 (2005)

38. Cetin Kilic and Alex Zunger, *Phys. Rev. Lett.* **88**, 95501 (2002).

39. S. B. Ogale, R. J. Choudhary, J. P. Buban, S. Lofland, S. R. Shinde, S. N. Kale, V. N. Kulkarni, J. Higgins, C. Lanci, J. R. Simpson, N. D. Browning, S. Das Sarma, H. D. Drew, R. L. Greene and T. Venketesan, *Phys. Rev. Lett.* **91**, 077205 (2003).

40. P. Gambardella *et al.*, *Phys. Rev. Lett.* **88**, 047202 (2002).

41. G. Bergmann and M. Hossain, *Phys. Rev. Lett.* **86**, 2138 (2001).

42. S. K. Kwon and B. I. Min, *Phys. Rev. Lett.* **84**, 3970 (2000).

43. H. Beckmann and G. Bergmann, *Phys. Rev. Lett.* **83**, 2417 (1999).

44. Gerd Begmann, *Phys. Rev. B* **23**, 3805 (1981).

45. J. Flouquet, O. Taurian, J. Sanchez, M. Chapellier and J. L. Tholence, *Phys. Rev. Lett.* **38**, 81 (1977).

46. R.M. Bozorth, P. A. Wolff, D. D. Davis, V. B. Compton and J. H. Wernick, *Phys. Rev.* **122**, 1157 (1961).

47. J.M.D. Coey, A.P. Douvalis, C.B. Fitzgerald, M. Venkatesan, *Appl. Phys. Lett.* **84**, 1332 (2004).

48. C.B. Fitzgerald, M. Venkatesan, A.P. Douvalis, S. Huber, J.M.D. Coey and T. Bakas, *J. Appl. Phys.* **95**, 7390 (2004).

49. Y. G. Zhao, S. R. Shinde, S. Ogale, J. Higgins, R. J. Choudhary, V. N. Kulkarni, R. L. Greene, T. Venketesan, S. Lofland, C. Lanci, J. P. Buban, N. D. Browning, S. Das Sarma and A. J. Millis, *Appl. Phys. Lett.* **83**, 2199 (2003).

50. S. N. Kale, S. Ogale, S. R. Shinde, M. Sahasrabuddhe, V. N. Kulkarni, R. L. Greene and T. Venketesan, *Appl. Phys. Lett.* **82**, 2100 (2003).

51. S.G. Yang, T. Li, B.X. Gu, Y.W. Du, H.Y. Sung, S.T. Hung, C.Y. Wong, A.B. Pakhomov, *Appl. Phys. Lett.* **83**, 3746 (2003).

52. M. Wei, N. Braddon, D. Zhi, P. A. Midgley, S. K. Chen, M. G. Blamire, and J. L. MacManus-Driscoll, *Appl. Phys. Lett.* 86, 072514 (2005).

53. S.J. Pearton, W.H. Heo, M. Ivill, D.P. Norton and T. Steiner, *Semi. Sci. Tech.* **19**, R59 (2004).

54. N.H. Hong, J. Sakai and W. Prellier, *Phys. Rev. B* **70**, 195204 (2004).
55. M. Venkatesan, C.B. Fitzgerald, J.G. Lunney and J.M.D. Coey, *Phys. Rev. Lett.* **93**, 177206 (2004).
56. M. Venkatesan, C.B. Fitzgerald and J.M.D. Coey, *Nature* **430**, 630 (2004).
57. M.S.R. Rao et al., (Unpublished).
58. T. Dietl, H. Ohno, F. Matuskura, J. Cibert and D. Ferrand, *Science* **287**, 1019 (2000).
59. A. Chattopadhyay, S. Das Sarma and A. J. Millis, *Phys. Rev. Lett.* **87**, 227202 (2001).
60. S. Das Sarma, H. Y. Hwang and A. Kaminski, *Phys. Rev. B* **67**, 155201 (2003).
61. G. Alvarez, M. Mayr and E. Dagotto, *Phys. Rev. Lett.* **89**, 277202 (2002).
62. G. Bouzerar, J. Kudrnovsky and P. Bruno, *Phys. Rev. B* **68**, 205311 (2003).
63. J.M. Sullivan and S.C. Erwin, *Phys. Rev. B* **67**, 144415 (2003).

CHAPTER 7

EPITAXIAL GROWTH AND PROPERTIES OF MAGNETICALLY DOPED TiO$_2$

Scott A. Chambers, Timothy C. Droubay and Tiffany C. Kaspar

Fundamental Science Directorate
Pacific Northwest National Laboratory
Richland WA

7.1 INTRODUCTION

Spin electronics and photonics represent exciting frontiers in which new paradigms are envisioned for signal processing and computing.[1-4] Using quantum mechanical spin states associated with charged particles to propagate information offers potential advantages over the use of charge alone, as in present-day technology. These advantages include the ability to combine logic and memory on the same chip, lower power consumption, decrease switching times in device architectures similar to those used today, and incorporate entirely new functionalities.

Some years ago, Datta and Das[5] proposed an electronic analog to an electro-optic modulator which has since been dubbed "spin-FET". Although the ability of this device to function as a useful transistor has been seriously questioned,[6] the spin-FET is a useful heuristic. In a spin-FET, spin polarized carriers are injected from a ferromagnetic (FM) source into a high mobility channel. The spins are made to precess by an electric field applied to the associated gate via the Rashba effect.[7] A 180° spin precession as the carrier traverses the channel will result in a high resistance state at the interface to a FM drain, for which the magnetization is oriented the same as that of the source. The channel thus goes from "on" to "off" by rotating the spin of the carrier. In order for a spin-FET to work as envisioned, several physical processes must occur. First, a spin polarized current must be injected from the FM source into the channel. Second, the spin polarization must remain as high as possible in the channel in order to maximize the signal, which is given by $\Delta R/R_p = (R_{ap} - $

$R_p)/R_p$, where R_p and R_{ap} are the channel resistances for parallel and antiparallel magnetic alignment of carriers with the drain magnetization, respectively. Third, the interaction with the electric field must be sufficiently strong that a $180°$ degree rotation is attained over the channel length, which must be kept short (i.e. on the nanometer scale) for fast switching times. Fourth, generation and propagation of the spin polarized current must occur at or above room temperature in order to avoid the need for cryogenic cooling.

These stringent device requirements place significant demands on the associated materials. The obvious first choice for the magnetic source and drain materials is a FM metal. However, the conductivity mismatches between metals and appropriately doped semiconductors are sufficiently large that the spin injection efficiency is typically very low, $<\sim1\%$, due to conductivity mismatch,[8-12] unless a cleverly designed Schottky barrier is used, in which case efficiencies as high as $\sim30\%$ have been achieved.[13] However, the best choice for spin injection into a nonmagnetic semiconductor is a diluted magnetic semiconductor (DMS) that is FM, and for which the conductivity can be matched to that of the nonmagnetic semiconductor through selective doping. DMSs are semiconductors doped with a few to several atomic percent of some magnetic dopant. The magnetic dopant substitutes for one of the host ions and is randomly distributed throughout the lattice. The prototypical DMS, Mn-doped GaAs, has been investigated for many years.[14, 15] The FM interaction between spins associated with Mn ions, mediated by free carriers (holes), is rather weak in this material, resulting in a Curie point well below room temperature.

In 2000, Dietl et al.[16] performed mean field calculations employing the Zener model of ferromagnetism to predict which semiconductors might exhibit Curie points at or above room temperature when doped p-type with 5 at.% Mn. These calculations predict that GaN and ZnO are candidates for room temperature ferromagnetism. Since then, these materials have been actively investigated and interesting results have been achieved.[17] At nearly the same time as the Dietl calculation, Matsumoto et al.[18, 19] performed combinatorial growths using pulsed laser deposition (PLD) in which epitaxial TiO_2 films of both rutile and anatase (not included in Dietl's calculations) were doped with first-row transition metals ranging from Sc to Zn to see if any exhibited room temperature ferromagnetism. Of these many combinations, only Co-doped anatase was observed to be FM at and above room temperature. This result was unprecedented in the world of magnetically doped III-V semiconductors because of the exceedingly high Curie point. Shortly thereafter, Chambers et al.[20] confirmed this result and showed that using oxygen plasma assisted molecular beam epitaxy (OPAMBE) as the growth method results in considerably higher saturation magnetization and remanence than does PLD.

Since its initial discovery in 2001, Co-doped TiO_2 anatase has been the center of controversy. The significantly higher Curie point relative to more traditional DMS materials has prompted skepticism. The skepticism intensified when claims of intrinsic ferromagnetism in poorly characterized material were made. It eventually became clear that incomplete oxidation of Co (and the resultant formation of embedded Co metal clusters) results if the oxygen partial pressure is too low during growth.[21] This finding continues to fuel speculation in some circles that any and all FM behavior in Co-doped TiO_2 anatase is due to "hidden" Co metal nanoparticles, although such particles are superparamagnetic at room temperature if the particle size is less than ~10 nm.[22] Since 2001, other magnetic dopants and different semiconducting oxide host lattices have been investigated, both experimentally and theoretically. Although many papers make claims that are not defensible, enough good research has been done to establish that room temperature ferromagnetism in magnetically doped TiO_2 *can* occur under certain conditions. However, the mechanism of magnetism is not yet established. Nevertheless, several useful clues have recently been uncovered that may allow the mechanistic details to be fully elucidated in the near future. Then and only then will it be possible to say with any level of confidence how this new generation of high-T_c ferromagnetic oxide semiconductors will be useful for spintronics technology.

Although it is not yet clear what form spin electronics will take, a "tool box" of materials will be required to bring device concepts to reality. Two of the necessary classes of materials are thermally robust FM semiconductors and insulators. These are needed for spin injection and spin filtering at and above room temperature. The purpose of this chapter is to examine the detailed growth and properties of one of the more promising class of materials -- magnetically doped TiO_2.

7.2 FUNDAMENTAL CONSIDERATIONS

The ideal FM semiconductor or insulator would be based upon a host matrix that can be easily doped, exhibits high electron mobility (for semiconductor applications), and is well lattice matched to and "interface compatible" with some high-mobility nonmagnetic semiconductor. Spin-based devices that use electrons as the majority carrier are most attractive because spin coherence times are typically much longer for electrons than for holes in all common nonmagnetic semiconductors.[23] Moreover, minimizing the hole population in an n-type spin device is essential because electron-hole exchange interaction can greatly accelerate the rate of electron spin decoherence.

The magnetic dopant in an ideal material should have a high solid solubility in the host matrix. The highest solid solubility typically occurs when the formal oxidation state and ionic radius of the magnetic dopant closely match those of one of the ions in the host. Additionally, the magnetic dopant should have a high magnetic moment, which typically results from being in a high-spin state associated with unpaired d electrons, and high remanence. Finally, the material system is simplified if the magnetic dopant can also act as an *electronic* dopant, providing free carriers to ferromagnetically couple the dopant spins. The dopant must of course have a different formal charge than the host cation in order to be an electronic dopant, and this property may limit its solid solubility. It is typically found for carrier-mediated FM exchange interaction that the Curie point goes approximately linearly with the mole fraction of the magnetic dopant, at least up to the point at which the average distance between magnetic dopants becomes sufficiently small that antiferromagnetic interaction, mediated by superexchange, becomes energetically more favorable than FM coupling. [14, 24-26]

Interface compatibility means simply that the DMS lattice nucleates on the nonmagnetic semiconductor surface without disruption of the latter, resulting in an interface of high crystallographic perfection and very low defect density. Interfacial defects, particularly edge dislocations, have been shown to be potent spin-flip scattering centers in the case of the MnZnSe/GaAs interface.[27]

7.3 Co:TiO$_2$

7.3.1 Anatase

TiO$_2$ occurs naturally as three polymorphs – rutile, anatase, and brookite. All are stable at room temperature. The relative stabilities of the three are rutile > anatase > brookite. Anatase undergoes an irreversible phase transition to rutile above 575°C.[28] The electron mobility is greater in *n*-type anatase than in *n*-type rutile.[29, 30] The crystal structures of anatase and rutile are shown in Fig. 1. The space groups for anatase and rutile are $I4_1/amd$ and $P4_2/mnm$, respectively. The local structural environment for Ti in the two lattices is somewhat different; both exhibit octahedral coordination, but the octahedral cage is heavily distorted in anatase, while only slightly distored in rutile. The Ti-O bond lengths are 1.89Å (4x) and 1.96Å (2x) in anatase, and 1.95Å (4x) and 1.98Å (2x) in rutile.[31] Anatase (001) is reasonably well lattice matched to Si(001) ($\Delta a/a_{Si}$ = +1.6%, after performing a 45° rotation about [001] such that $[100]_{ana} \parallel [110]_{Si}$), whereas rutile is not lattice matched to Si in any orientation. From the points of view of electron mobility and epitaxy, anatase is preferable to rutile for spin injection applications with Si.

As mentioned above, Matsumoto *et al.*[18, 19] used combinatorial PLD to survey the magnetic properties of epitaxial TiO_2 anatase and rutile films doped with variable quantities of transition metals ranging from Sc to Zn. Of these, only Co-doped anatase with concentrations of 8 at. % Co or less was found to be FM. Interestingly, the saturation moment in PLD-grown films, while not particularly high, was found to persist up to 400K. The moment was found to be ~0.28 μ_B per Co at a field of 4000 G in $Co_xTi_{1-x}O_2$ anatase films for which x = 0.07. Subsequent work in our lab has confirmed these results, in addition to showing that the saturation moment in OPAMBE grown material is considerably higher, ~1.25 μ_B per Co.[20, 32] We have also demonstrated that Co substitutes for Ti in the lattice, that the formal oxidation state of Co is +2, and that no detectable Co metal is present in OPAMBE films grown on oxide substrates such as $LaAlO_3(001)$ and $SrTiO_3(001)$.[20, 33] A consequence of Co(II) substituting for Ti(IV) is that one O atom vacancy must be generated for each substitutional Co in order to maintain charge neutrality, making the correct empirical formula $Co_xTi_{1-x}O_{2-x}$.[34] Detailed analysis of Co K-shell extended x-ray absorption fine structure (EXAFS) data reveals that these O vacancies are adjacent to substitutional Co(II) ions.[33]

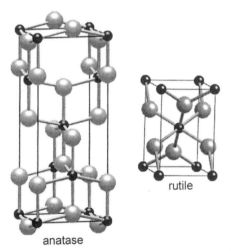

rutile

anatase

Figure 1: Unit cells of anatase and rutile. The small black circles are Ti, and the large gray circles are O.

The standard enthalpies of formation for CoO and TiO_2 are -238 and -944 kJ/mol, respectively.[35] Therefore, Ti is thermodynamically favored for oxidation to the +4 state relative to Co(II) formation, and Co oxidation is not expected to occur unless there is ample oxygen present at the growth front. This conclusion has been reached in PLD studies which have shown that below a certain threshold O_2 background pressure in the chamber during growth, $Co_xTi_{1-x}O_{2-x}$ anatase films invariably contain small Co metal precipitates which often segregate at the interface.[21,36] The undetected presence of such particles can mire the magnetic properties investigation because Co metal particles with diameters in excess of ~10 nm are FM at room temperature.[22] In the limit of complete phase separation, the saturation moment per Co is typically close to ~1.7 μ_B per Co, the value for bulk Co metal.

It has also been reported that epitaxial $Co_xTi_{1-x}O_{2-x}$ anatase films with no detectable Co(0) can be grown on $LaAlO_3(001)$ by reactive RF sputtering when H_2O is used as the reactive gas.[37] These films exhibit saturation moments of ~0.6 μ_B per Co. This result is encouraging for eventual technological deployment because sputtering is much more amenable to a manufacturing environment than OPAMBE or PLD.

Although OPAMBE[38, 39] and reactive sputtering[37] do not result in $Co:TiO_2$ with Co metal inclusions when grown on oxide substrates, the Co(II) is typically *not* uniformly distributed. On very rare occasions we grow OPAMBE films in which the Co is uniformly distributed, but such growths are not reproducible. Rather, Co tends to segregate to a Co-enriched $Co_xTi_{1-x}O_{2-x}$ anatase phase (x as large as ~0.3) that nucleates as islands on the surface of a relatively Co-poor, but continuous epitaxial $Co_xTi_{1-x}O_{2-x}$ film (x $\leq$ ~0.01). The same may also be true for PLD-grown films under certain conditions, although Co metal particle formation tends to prevail.[21, 40, 41] The areal density of these particles is at most ~5% of the total film surface area in OPAMBE-grown material, but higher in sputtered films. Magnetic force microscopy (MFM) images of these surfaces reveal the presence of perpendicular magnetic dipoles on the surface particles, but not on the film proper, indicating that the magnetization detectable by MFM is concentrated in the Co-enriched anatase particle phase.[38] The continuous Co-poor film may also be FM, but the magnetization measured by MFM is dominated by that of the Co-enriched anatase particles. Matsumoto *et al.*[18] have seen magnetic domains throughout PLD-grown $Co_xTi_{1-x}O_2$ films by scanning SQUID magnetometry. The tendency toward Co-enriched $Co_xTi_{1-x}O_{2-x}$ anatase cluster formation is consistent with the observation that Co is extremely mobile in anatase. We have found that the post-growth annealing of $Co_xTi_{1-x}O_{2-x}$ anatase films in ultrahigh vacuum at temperatures up to 550°C results in the coalescence of these rather small, dispersed Co-enriched anatase islands into larger islands of the same phase. There is no measurable reduction of Co(II) to Co(0). There is also no measurable change in the macroscopic magnetic properties as measured with vibrating sample magnetometry (VSM).[42] In addition, exposing $Co_xTi_{1-x}O_{2-x}$ anatase films to activated oxygen from an electron cyclotron resonance plasma source at room temperature results in the formation of secondary phase CoO and the loss of ferromagnetism.

There is an interesting relationship between conductivity and magnetism in $Co_xTi_{1-x}O_{2-x}$ anatase films grown by OPAMBE. The conductivity can be controlled via the activated oxygen flux.[34] When grown under oxygen-rich conditions, the films are insulating. If we reduce the O flux during growth, we can controllably introduce *additional* O vacancies, which act as shallow donors in anatase.[43] We show in Fig. 2 the dependence of coercive field and magnetic remanence on resistivity for $Co_xTi_{1-x}O_{2-x}$ epitaxial films grown on $LaAlO_3(001)$. The coercive field falls off rapidly with increasing resistivity. The remanence also falls off with resistivity, but at a lower rate, and with more scatter in the data. Moreover, the coercivity and remanence do *not* go to zero at high resistivity. Several films with resistivities of a few thousand Ω-cm exhibit coercivities of ~100 Oe and remanences of ~5%. These results suggest that free carriers (electrons) effectively mediate FM interaction between isolated particles of Co-enriched anatase, but that even in the absence of carriers, some residual FM coupling remains. This finding is relevant to understanding the anomalous Hall effect in this material, as explained below.

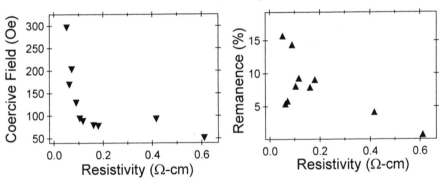

Figure 2: In-plane coercive field (left) and magnetic remanence (right) vs. resistivity for $Co_xTi_{1-x}O_2$ anatase films on $LaAlO_3(001)$.

Interestingly, there is no difference between in-plane and out-of-plane magnetization for $Co_xTi_{1-x}O_{2-x}$ thin films grown on $LaAlO_3(001)$ when the Co segregates to the Co-enriched anatase surface particle phase. No magnetic anisotropy is expected if the Co-enriched anatase particles carry the bulk of the magnetism, as we have found by MFM. In contrast, there is a strong in-plane anisotropy when the Co is uniformly distributed throughout the film, as we have observed in those rare instances when uniform Co distribution is achieved.

The driving force behind the Co-enriched anatase surface particle phase may be related to a lowering of the total lattice energy when the distance between substitutional Co(II) ions is small. Yang *et al.*[44] have performed density functional theory (DFT) calculations in which they varied the Co-Co distance and found a lattice energy stabilization when this distance was reduced to values expected for a more concentrated phase. These theoretical results have been corroborated by DFT calculations in which the oxygen vacancies described above have been included.{Jaffe, 2005 #136} No such vacancies were included in the calculations by Yang et al.

The charge state and local structural environment of Co in Co:TiO$_2$ anatase, and the relationship of these properties to growth method and specimen preparation, is currently the topic of some debate. As mentioned above, OPAMBE and reactive sputtering result in Co being in the +2 charge state for both growth methods. Moreover, Co(II) substitutes for Ti(IV) at cation sites, at least for OPAMBE grown material. Such films are FM as grown, even when insulating, exhibiting a saturation moment of 1.1-1.2 μ_B/Co and a Curie point of ~700K. In contrast, PLD grown material typically contains Co metal nanoparticles at the interface unless a high O$_2$ ambient pressure is used, in which case Co is in the +2 charge state. When Co metal precipitates are present, the films are also FM, but exhibit saturation moments and Curie points close to those of Co metal (1.5-1.7 μ_B/Co and ~1200K, respectively). Against this backdrop, Shinde et al.[36] have performed experiments in which PLD grown films containing Co nanoparticles were annealed in pure Ar at 1 atm. This procedure resulted in a drop in Curie point to ~650K, and expanded d_{004} spacings [$d_{004} = (\frac{1}{4})c$ for anatase], which is consistent with, but not exclusively the result of, Co(II) substituting for Ti(IV).[46] These results were interpreted as resulting from Co(0) being driven from Co metal nanoparticles into a Co$_x$Ti$_{1-x}$O$_{2-x}$ solid solution. However, just the opposite result was obtained by Kim et al.[41] These investigators subjected a film grown by PLD in which Co was in the +2 charge state to an anneal at 673K and 10^{-6} torr O$_2$, and found that Co(II) was driven out of solid solution and into a Co metal nanoparticle phase. We have obtained similar results to those of Kim et al. for OPAMBE grown material. However, a higher temperature (~1200K) in UHV was required to reduce the substitutional Co(II) to Co metal.[42] Given the difficulty with which Co is oxidized to the +2 state, as revealed by the relatively high free energy of formation of CoO, it is hard to visualize how Co could be oxidized to Co(II) in an environment that is oxygen poor (i.e. TiO$_{2-\delta}$, where δ is relatively large being annealed in Ar), as Shinde et al. posit.

An interesting question that arises in the case of Co$_x$Ti$_{1-x}$O$_{2-x}$ anatase films is that in light of the unusual film morphology, to what extent are free carriers spin polarized throughout the material? This question can be at least partially addressed by looking for an anomalous Hall effect (AHE) in the material. The AHE occurs when free carriers interact with magnetic moments in the host material via the spin-orbit interaction and, as a result, exhibit a Hall resistivity in excess of that associated with the normal Lorentz force interaction. The total Hall resistivity in a ferromagnetic material is given by

$$\rho_H = R_0 B + R_e \mu_0 M \qquad (1)$$

Here, R_0 is the ordinary Hall coefficient, B is the transverse magnetic field, R_e is the anomalous Hall coefficient, and M is the specimen magnetization. For homogeneous magnetic materials such as DMSs, R_e is proportional to ρ^n, where ρ is the longitudinal resistivity in zero field, and n depends on the details of the interaction of free carriers with the magnetic moments. Asymmetric or skew scattering, in which carriers of opposite spin polarization scatter asymmetrically from moments via the spin-orbit interaction[47] results in n = ~1, and dominates in low-resistivity, dilute systems, such as DMSs. The "side jump" phenomenon occurs when carriers undergo lateral displacements as a result of spin-orbit interaction with local moments, and n is typically ~2 for this interaction.[48] Side jump is dominant in more concentrated and conductive homogeneous specimens such as pure ferromagnetic metals.

Of particular relevance to the case of $Co_xTi_{1-x}O_2$ anatase is the fact that the AHE has been observed in a number of granular magnetic materials, such as Ni and Co in SiO_2 and Co in Ag. These systems consist of magnetic particles embedded in a nonmagnetic host, not unlike the Co-enriched anatase or Co metal particle phases in a more-or-less pure anatase film phase. Although all three systems exhibit an AHE, a simple ρ^n scaling law for R_e is *not* followed for Ni/SiO_2 and Co/SiO_2.[49] (This scaling relationship has not yet been investigated in Co-doped anatase.) The scaling parameter n was found to be as high as 3.7 for Co/Ag.[50]

We show in Fig. 3 ρ_H vs. magnetic field at 380K for ~150Å of OPAMBE-grown $Co_{\sim0.05}Ti_{\sim0.95}O_{\sim1.95}$ anatase on $LaAlO_3(001)$. The sample exhibited the characteristic low density of Co-enriched TiO_2 anatase particles atop a continuous epitaxial TiO_2 anatase film, as seen in the AFM image in the upper left corner of the figure. All detectable Co is in the +2 oxidation state. The resistivity of the film was 0.19 Ω-cm at 300K. The Hall curve is clearly nonlinear, indicating an AHE well above room temperature. In contrast, Higgins *et al.*[51] report that Co:TiO$_2$ anatase grown by PLD on $LaAlO_3(001)$ does *not* exhibit an AHE, again illustrating the difference between Co:TiO$_2$ produced by PLD and OPAMBE. However, the mere appearance of an AHE is not sufficient to conclude that this material is a true DMS, as we know from studies of Co and Ni in SiO_2 and Co in Ag discussed above. It is essential to systematically vary the dopant and carrier concentrations to see how the Hall resistivity depends on these parameters, and to determine the scaling parameter n. The observation of an AHE in semiconducting $Co_xTi_{1-x}O_{2-x}$ grown by OPAMBE is consistent with the increased coercive field and remanence at low resistivity (Fig. 2). Together, these results suggest that conduction band electrons are spin polarized throughout the film, and that free electrons mediate the ferromagnetic coupling within and between Co-enriched $Co_xTi_{1-x}O_{2-x}$ particles. The residual coercivity and remanence at resistivities higher than those that permit AHE measurement suggest FM coupling in semi-insulating material by some other mechanism. Possibilities include dipolar interaction of closely spaced $Co_xTi_{1-x}O_{2-x}$ particles, a magnetocrystalline effect made possible by the epitaxial orientation of the $Co_xTi_{1-x}O_{2-x}$ particles on the Co-poor continuous epitaxial film, or F-center mediated exchange, as discussed in more detail in Section 4.2. However, the tendency of Co to phase segregate to either Co metal nanoparticles in the case of PLD, or Co-enriched $Co_xTi_{1-x}O_{2-x}$ anatase particles in the cases of OPAMBE and reactive sputtering, makes $Co:TiO_2$ anatase a less than desirable candidate for a high-temperature DMS.

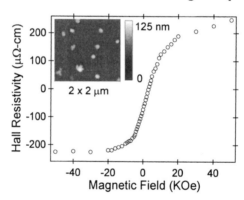

Figure 3: Hall resistivity vs magnetic field at 380K for ~150Å of OPAMBE grown $Co_{-0.05}Ti_{-0.95}O_{-1.95}$ anatase on $LaAlO_3(001)$. AFM of the film surface shows the presence of Co-enriched anatase particles.

7.3.2 Rutile

Shortly after the discovery of room-temperature feromagnetism in Co-doped TiO_2 anatase,[18] it was shown that heteroepitaxial $Co_xTi_{1-x}O_2$ rutile grown on r-cut α-Al_2O_3 sapphire by PLD exhibited ferromagnetism at room temperature for x up to ~0.05.[52] The saturation magnetization was reported to be ~1 μ_B per Co. However, in a subsequent publication by the same group, it was stated that of the many candidate dopants used in both rutile and anatase, *only* Co-doped anatase exhibited room temperature ferromagnetism.[19] This ambiguity nonwithstanding, other groups have investigated this phase with interesting results. Polycrystalline, phase-pure FM $Co_xTi_{1-x}O_2$ rutile was grown by reactive co-sputtering on quartz and unetched Si(001) substrates for x as high as ~0.12.[53] The saturation moment was reported to be ~0.9 μ_B per Co in this material. AFM images revealed large grain growth across the film surface. No attempt was made to determine the Co oxidation state. Polycrystalline Co-doped TiO_2 rutile was also grown on unetched Si(001) by PLD from a single ceramic target.[54] The resulting films were weakly ferromagnetic at room temperature, exhibiting a maximum saturation moment of ~0.3 μ_B per Co. AFM images revealed the presence of particles on the film surface, and Rutherford backscattering (RBS) showed strong segregation of Co in the top few hundred Å of the film. Although no experiments were done to determine the Co oxidation state in this work either, the authors argue that the relatively low moment per Co suggests that no Co(0) is present.

Higgins et al.[51] found that PLD growth of Co:TiO_2 on $LaAlO_3$(001) at oxygen pressures less than 10^{-6} torr resulted in rutile formation, whereas anatase preferentially nucleated above 10^{-6} torr. The rutile films were heavily reduced, as revealed by low (mΩ range) room-temperature resistivities, and were weakly FM, with saturation magnetizations of ~0.3 μ_B/Co at room temperature. No spectroscopic measurements to determine the charge state of Co throughout the film were reported, although TEM results were mentioned which reveal the presence of ~10 nm Co metal clusters at the interface. Similar PLD growth conditions on r-cut α-Al_2O_3 sapphire substrates also produced heavily reduced Co:TiO_2 rutile.[55]

Ion implanted $Co_{\sim 0.02}Ti_{\sim 0.98}O_2$(110) rutile that was free of any measurable Co(0) was shown to be FM at room temperature.[56] Here, the saturation magnetization at room temperature was shown to be ~0.4 μ_B per Co, and all Co was found by Co K-edge x-ray absorption near edge spectroscopy (XANES) to be in the +2 oxidation state, although some CoO was detected. The rutile crystals exhibited a blue tint characteristic of partial reduction and n-type semiconducting behavior as a result of implanting at high temperature (800°C). Depth profiles revealed that the Co was uniformly distributed throughout the depth of the implantation region (~3000Å). The reduced moment per Co is in all likelihood associated with the formation of a secondary antiferromagnetic CoO phase.

Toyosaki et al.[30] have used PLD to grow heteroepitaxial $Co_xTi_{1-x}O_2$ rutile on r-cut α-Al_2O_3 sapphire. The absence of surface particles was taken as evidence for uniform distribution of the Co. If the Co is truly distributed in a uniform fashion, and substitutes for Ti in the lattice, $Co_xTi_{1-x}O_2$ rutile is amenable to a detailed study of the magnetic properties as a function of Co doping level and carrier concentration to check for DMS candidacy. To this end, Toyosaki et al.[30] varied the free carrier concentration by changing the oxygen pressure from 10^{-8} torr to 10^{-4} torr during growth. The resulting room temperature conductivities ranged from ~0.005 to ~20 Ω-cm. This film set exhibited weak magnetoresistance and an AHE. However, more significantly, a scaling law intermediate between skew scattering (n=1) and side jump (n=2) was found to consistently hold over six orders of magnitude in conductivity. We show in Fig. 4 the anomalous Hall conductivity vs. longitudinal conductivity for several values of x and a range of carrier concentrations. The near constant scaling parameter (n=1.5-1.7) derived from these plots support the conclusion that spin-polarized conduction band electrons in the semiconductor are responsible for the ferromagnetism in this material. This result is potentially significant because it is suggestive of carrier mediated exchange interaction.

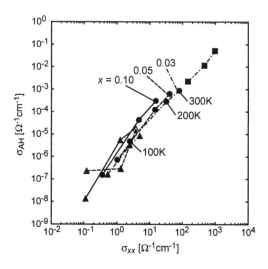

Figure 4: The anomalous Hall effect for $Co_xTi_{1-x}O_2$ rutile films grown by PLD on r-cut α-Al_2O_3, measured as a function of x, carrier concentration, and temperature. The rectangles, circles, and triangles represent data taken from films for which the O_2 partial pressure during growth was 10^{-8}, 10^{-7}, and 10^{-6} Torr, respectively. Lines connect data points obtained at different Co concentrations. [Ref. 30]

However, a paper published very shortly before that of Toyosaki *et al.*,[30] involving two of the same authors, reveals that a significant fraction of the Co in these films is present as Co(0).[57] Co K-shell XANES, EXAFS and Co $2p$ x-ray photoemission spectroscopy (XPS) reveal a mix of Co(0) and Co(II) in films grown at 10^{-6} torr. These authors rationalize that the Co(0) present in the films does not influence the magnetic properties because the measured optical magnetic circular dichroism (MCD) does not look like that of bulk Co metal, and is independent of temperature and magnetic field.[57] They argue that it is the Co(II), which they presume substitutes for Ti(IV) in the rutile lattice, that determines the magnetic properties. Co(0), they argue, must be present as very small nanoparticles, which are not expected to be FM at room temperature. Although the presence of Co(0) raises questions about the validity of these results, the scaling seen in the AHE data is consistent with true ferromagnetic semiconductor behavior, despite the presence of both Co(0) and Co(II). In contrast, Higgins *et al.*[51, 55] argue that the presence of Co metal particles near the superparamagnetic size limit (8-10 nm) which are embedded in a reduced TiO_2 rutile matrix can give rise to an AHE. These studies, perhaps as well as any, illustrate the need for thorough materials characterization in addition to conventional transport measurements, and the need to find a magnetic dopant that does not segregate in TiO_2. To this end, we now turn our attention to Fe and Cr as candidate magnetic dopants.

7.4 OTHER MAGNETIC DOPANTS IN TiO₂

7.4.1 Fe

Fe is an attractive candidate magnetic dopant for TiO_2 in that it exhibits a large magnetic moment per atom (2.22 μ_B) and a high Curie point (1043K) in its elemental form. Moreover, Fe(III) has an ionic radius (0.64Å) very close to that of Ti(IV) (0.68Å). One of its stable oxides (Fe_3O_4, or magnetite) is ferrimagnetic at room temperature. However, it has been shown that at least in polycrystalline films grown by reactive sputtering on unetched Si and glass substrates, the solid solubility of Fe in crystalline phases is fairly low, ~1.3 at. %.[58] Additionally, Fe induces a structural change from anatase to rutile for doping levels in excess of ~0.3 at % in polycrystalline films, and Fe goes from being a donor to an acceptor at ~0.1 at %.[58] The surprisingly low solid solubility of Fe in TiO_2 may result from the rich phase space of stable Fe-Ti-O compounds that form in the bulk. In any event, it is clear that in investigations of Fe-doped TiO_2 epitaxy, one must be exceedingly careful to determine the phase purity of these films and ensure that secondary phase formation has not occurred, particularly in the regime of higher dopant levels.

Wang *et al.*[59, 60] grew epitaxial reduced rutile doped with Fe ($Fe_xTi_{1-x}O_{2-\delta}$, where x = 0.02, 0.06 and 0.08) by PLD on α-Al_2O_3(012). The materials were FM at room temperature, exhibited an AHE, and it was concluded that $Fe_xTi_{1-x}O_{2-\delta}$ constitutes a new high-T_c DMS for x as high as 0.08. Yet, there were no spectroscopic measurements carried out that can *directly* detect secondary magnetic phases which have a high probability of being present, particularly at the higher Fe concentrations. On the basis of magnetic and magnetotransport measurements, these authors claim to be able to eliminate the presence of magnetic secondary phases for all Fe concentrations, despite having far exceeded the solid solubility of Fe in TiO_2 in at least two of the three concentrations investigated. The observed shape anisotropy suggests that *some* of the Fe was incorporated into a FM epitaxial film phase. However, the presence of oriented Fe_3O_4 particles would also give rise to magnetic anisotropy via the magnetocrystalline anisotropy effect. We have already shown that nonmagnetic host films containing magnetic particles can give rise not only to hysteretic magnetization curves, but also to magnetotransport such as AHE. AHE is thus a necessary but insufficient condition for intrinsic ferromagnetism. Moreover, inaccurate determination of the magnetic dopant concentration can lead to significant errors in the moment per dopant atom. The moments per Fe calculated from target compositions were used by Wang *et al.* to rule out the presence of secondary magnetic phases. Indeed, most practitioners of PLD tacitly assume that the dopant concentration in the film equals that in the target. Yet, it has been shown in the case of Fe-doped SnO_2 that films contained ~3x more Fe than the target.[61] A convincing case for true DMS behavior over a wide dopant concentration range cannot be made without spectroscopic measurements that can decisively eliminate the presence of magnetic secondary phases.

Hong *et al.*[62] attempted to grow Fe- and Ni-doped anatase TiO_2 on $SrTiO_3(001)$ and $LAlO_3(001)$ by PLD with Fe concentrations far in excess of the solid solubility limit. Preferential nucleation of rutile rather than anatase was observed for certain Ni doping levels, as was a wide range of magnetic properties. These results are not surprising in light of the absence of careful materials characterization, marked by the use of simple visual inspection to determine surface smoothness and XRD alone to determine the presence of impurity phases. We have learned from Co-doped anatase research that extreme care must be taken to adequately determine composition, magnetic dopant oxidation state and uniformity, and structural/morphological film homogeneity before drawing conclusions about the origin of any observed FM behavior. Depending upon the naked eye to ascertain film surface smoothness and conventional XRD with a laboratory x-ray source to eliminate the possibility of secondary phase formation is naïve at best.

In our investigations of epitaxial $Fe_xTi_{1-x}O_{2-\delta}$ (x = 0.07) rutile grown by OPAMBE on rutile $TiO_2(110)$ substrates, we also noted that the as-grown films were smooth to the eye, exhibited excellent, streaky reflection high energy electron diffraction (RHEED) patterns, and were FM at room temperature.[63] All Fe was in the +2 or +3 oxidation state, as judged by both Fe $2p$ XPS and the more bulk-sensitive Fe L-edge XANES, eliminating the possibility of Fe metal. However, AFM images revealed micron sized features, and cross sectional transmission electron microscopy (TEM) showed that that these features consist of Fe_3O_4 and TiO_2 crystallites. The TEM data are summarized in Fig. 5. The selected area diffraction pattern and the high resolution lattice image in the body of the figure were taken from an Fe-rich surface cluster. These are shown in the inset in the upper right. These images reveal the presence of (111) oriented twins of the inverse spinel structure exhibited by Fe_3O_4. The Fe_3O_4 presumably phase separated from TiO_2 during epitaxal growth as a result of exceeding the solid solubility of Fe in TiO_2. Thus, Fe_3O_4 is the source of the rather strong FM response exhibited by these films. This conclusion is supported by the striking resemblance between x-ray MCD spectra taken from these films with those measured for Fe_3O_4 standards.[63] Without the combination of detailed TEM and x-ray spectroscopies, the presence of Fe_3O_4 would have been missed, and we too might have erroneously concluded that another high-T_c DMS had been discovered.

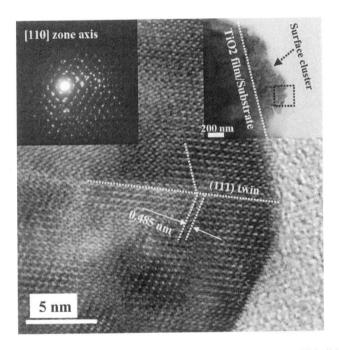

Figure 5: TEM cross sectional image for $Fe_{0.07}Ti_{0.93}O_2$ rutile grown on TiO_2(110) rutile by OPAMBE. The selected area diffraction pattern (left inset) and the high resolution image (body of panel) associated with the surface feature seen in the lower resolution image (right inset), together with EDX spectra, reveal the presence of Fe_3O_4 crystallites. [Ref. 63]

7.4.2 Cr

Cr is also of interest as a magnetic dopant in TiO_2 because of its successful use in III-V and II-VI DMSs.[64-66] Unlike Co and Fe, Cr metal is not FM, but rather antiferromagnetic. Therefore, one need not worry about the possibility of spurious ferromagnetism resulting from dopant metal phase separation. However, CrO_2 is a half-metallic ferromagnet,[67] so its formation during epitaxial growth must be closely monitored. Matsumoto *et al.*[19] stated that neither Cr-doped anatase nor rutile grown by PLD were FM. However, we have found that OPAMBE $Cr_xTi_{1-x}O_{2-\delta}$ is FM up to 690K for epitaxial anatase grown on $LaAlO_3(001)$, which is highly resistive as grown. However, OPAMBE rutile grown on $TiO_2(110)$, which is also insulating as grown, is not FM at 300K.{Droubay, 2005 #85} Phase-pure films of both anatase and rutile can be grown for x up to at least 0.10. Reducing both Cr-doped anatase and Cr-doped rutile by annealing in UHV results in room-temperature ferromagnetism. These results are corroborated by recent PLD growth of reduced Cr-doped rutile on α-$Al_2O_3(012)$.[69]

We show in Fig. 6 RHEED intensity oscillations for an OPAMBE growth of $Cr_xTi_{1-x}O_2$ on $LaAlO_3(001)$ in which the Cr to Ti flux ratio was ~1:9.

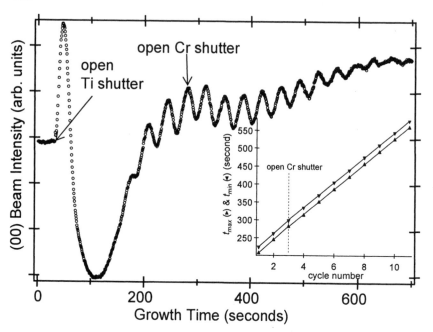

Figure 6: RHEED intensity oscillations for $Cr_xTi_{1-x}O_2$ anatase on $LaAlO_3(001)$. Inset - times at which intensities maximize (▲) and minimize (▼) vs. cycle number, measured from the first maximum.

Also shown as an inset in Fig. 6 is a plot of the times at which the (00) beam intensity maximizes and minimizes as a function of cycle number. The slopes of the line segments fit to these data are inversely proportional to the associated growth rate. Upon opening the Cr shutter, there is a ~6% decrease in both t_{max} and t_{min}, which signals the same percentage increase in growth rate. This result suggests that Cr substitutes for Ti in the lattice in that the abrupt increase in total metal flux is accompanied by a concomitant increase in growth rate. Such a correspondence is expected if the rate at which complete layers of anatase form directly follows the total metal flux under oxygen rich conditions. In addition, the RHEED beam intensity oscillation amplitude begins to decay after opening the Cr shutter, revealing a departure from layer-by-layer growth.Cr substitution for Ti is confirmed by scanned-angle x-ray photoelectron diffraction and EXAFS (not shown).[70] However, the question arises whether Cr has uniformly distributed throughout the film, or formed a secondary dopant-rich anatase phase similar to Co. This question is best answered by examining TEM lattice images and the associated energy dispersive x-ray (EDX) spectra. These results are shown in Fig. 7.

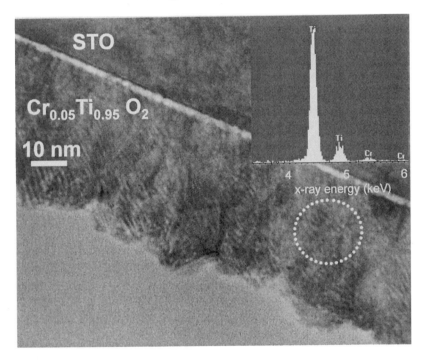

Figure 7: TEM cross section and EDX spectrum for $Cr_{0.05}Ti_{0.95}O_2$ anatase on $SrTiO_3(001)$. The spectrum was obtained for the beam location shown by the circle, and is typical of several such areas. [Ref. 68]

Here we show data for a film grown by OPAMBE on $SrTiO_3(001)$. Although the in-plane lattice mismatch is larger (-3.1%) than that associated with $LaAlO_3(001)$ (-0.26%), we chose STO for this measurement to eliminate interference from La L_β fluorescence. Inclusion of Cr into the film results in what we presume is strain-induced roughening, although the large lattice mismatch induces some roughening anatase on STO even in the absence of Cr. In spite of the film surface roughness, the material grows epitaxially, as judged by RHEED, XRD and high resolution TEM (not shown). EDX spectra measured in various cross sectional volumes consistently show that the Cr is uniformly distributed throughout the film (Fig. 7).

The film roughness is significantly enhanced by the addition of Cr. After opening the Cr shutter, the time required for the RHEED intensity oscillations to decay to zero (signaling the cessation of layer-by-layer growth) decreases with increasing Cr flux. The roughening is presumably a response to strain energy accumulation associated with Cr doping throughout the film.

We now address the issue of the formal oxidation state of the Cr. For this determination we use Cr- K-edge XANES, and these data are shown in Fig. 8. K-shell XANES probes the entire film depth because of the large x-ray attenuation lengths. Moreover, there are large chemical shifts for the various oxidation states of Cr, resulting in a high degree of sensitivity. The best match in the vicinity of threshold is with the Cr_2O_3 standard, revealing that Cr is in the +3 oxidation state in this material. This conclusion is corroborated by high resolution Cr $2p$ XPS (not shown).[70] As we shall see, the oxidation state of the magnetic dopant is critical to the magnetic properties of this material.

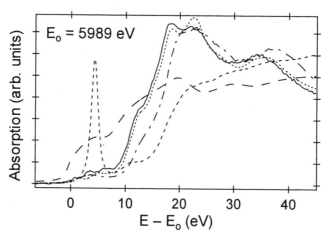

Figure 8: Cr- Kedge XANES spectra for 750Å of OPAMBE grown $Cr_{0.04}Ti_{0.96}O_2$ anatase on $LaAlO_3$(001) (solid), along with standards – Cr metal (long dash) Cr_2O_3 (dot), CrO_2 (dot dash), and Na_2CrO_4 (short dash).

We show in Fig. 9 in-plane VSM loops measured at room temperature for 300Å of $Cr_{0.11}Ti_{0.89}O_2$ on $LaAlO_3$(001) as grown by OPAMBE, and after 400°C anneals designed to reduce a few parts per thousand of Ti to the +3 formal oxidation state and thus introduce free carriers. The material is FM and insulating ($\rho > 4$ kΩ-cm) as grown, with a saturation moment of ~0.6 μ_B per Cr, a remanance of ~17%, and a coercivity of ~110 Oe. The saturation magnetization rises with the free carrier (electron) concentration, as seen in the figure. The enhanced magnetization with increasing conductivity is consistent with carrier mediated exchange. The question arises what gives rise to ferromagnetism in the highly resistive state?

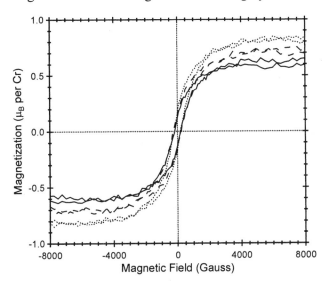

Figure 9: In-plane VSM loop measured at 25°C for 300Å of $Cr_{0.11}Ti_{0.89}O_2$ as grown by OPAMBE on $LaAlO_3(001)$ (solid curve), and after annealing at 400°C for two (dashed curve) and four hours (dotted curve).

One candidate mechanism for FM behavior in as-grown, highly resistive Cr-doped TiO_2 anatase is F-center mediated exchange interaction, as suggested by Coey *et al.*[61] for Fe-doped SnO_2. One way to maintain local charge neutrality when Cr(III) substitutes for Ti(IV) in the lattice is for one O vacancy to form for every two substitutional Cr(III) ions. Another way involves the formation of Ti cation interstitials. The O vacancy model is strongly supported by detailed EXAFS analysis for Co-doped anatase,[33] and the analysis for Cr-doped anatase is in progress. In the O vacancy model, the correct empirical formula is $Cr_xTi_{1-x}O_{2-x/2}$. Each such O atom vacancy contains two electrons which would effectively compensate two Cr(III) ions, even though these electrons are centered on an O vacancy and thus constitute an F center. In this model, these electrons mediate FM exchange between the Cr(III) ions. Significantly, no itinerant carriers are required for FM coupling. For a doubly-occupied F center, the two electrons must couple to a triplet spin state in order to induce FM coupling between Cr(III) ions within the F-center volume. For $Cr_{0.05}Ti_{0.95}O_{1.975}$, the Cr-Cr linear separation is ~13Å assuming uniform dispersion of the dopant atoms. Furthermore, the F-center radius estimated from the Bohr model is ~16Å, based on the DC dielectric constant of anatase films (31)[71] and the electron effective mass $(\sim 1 m_o)$.[72] Therefore, the F-center volume is sufficiently large to include multiple Cr(III) ions, giving rise to short-range FM coupling, provided the F-center electrons are in a triplet state. A triplet ground state is feasible in light of the large F-center volume. Cr-O hybridization would also permit a triplet ground state in the electrons formally associated with the F center. Overlapping F centers across the volume of the material would then permit long-range FM coupling of the Cr spins. A physically equivalent model is that of bound magnetic polarons, similar to what has been suggested to explain the ferromagnetism in highly resistive Mn-doped II-VI and III-V semiconductors.[73, 74] The observed enhancement in the saturation magnetization of a few tenths of a μ_B per Cr upon film reduction by UHV annealing to a resistivity of ~ 1 Ω-cm may result from additional FM coupling by the itinerant carriers introduced by excess O vacancies.

7.5 Co-DOPED TiO_2 ANATASE ON Si(001)

Successful integration with high-carrier-mobility nonmagnetic semiconductors creates a range of spin-based device possibilities for ferromagnetic oxide semiconductors. As mentioned in section 1, spin injection into Si represents an intriguing possibility that magnetically doped TiO_2 anatase may enable. However, efficient spin injection also requires that several demanding materials properties be simultaneously achieved. Some of these requirements are outlined in sections 1 and 2. Here, we focus on the physical and electronic structural requirements of the interface between the FM and nonmagnetic semiconductors. Efficient spin injection requires an interface with a high degree of crystallographic coherence, low edge dislocation density, and no undesired interface reaction products. A structurally and compositionally heterogeneous and/or defect-ridden interface is expected to result in spin decoherence to a significant extent. In addition, the band discontinuity for the majority carrier type must be sufficiently small that spin polarized carriers are not impeded during transport across the interface. Although the in-plane lattice mismatch between anatase $TiO_2(001)$ and Si(001) is reasonably small, TiO_2 is thermodynamically unstable on Si, leading to the expected formation of $TiSi_x$ and SiO_2.[75]

One way of circumventing this problem is to grow a thin buffer layer of some other material that is interface compatible with Si and TiO_2 anatase. $SrTiO_3$ (STO) represents such a material. By nucleating the first monolayer in a way that provides some level of oxidation resistance for the substrate, it has been shown that metastable, epitaxial STO(001) can be grown on Si(001) without any detectable SiO_2 or $TiSi_x$.[76-78] The idea is that the STO buffer would physically separate the anatase from the substrate and act a barrier to interfacial reaction. The STO layer would also nucleate anatase preferentially over rutile via epitaxial stabilization. Fig. 10 shows a structural drawing of the idealized interface. Here, we depict the STO/Si interface as including one monolayer (ML) of SrSi$_2$, which we have found to result in no measurable oxidation of the Si surface upon growth of 10 ML STO.[79] This interface structure has been predicted to be energetically stable against SiO$_2$ formation via first-principles calculations.[80] In addition, the conduction band offset predicted for this structure is in reasonable agreement with that measured by XPS, as discussed below. The anatase/STO interface we suggest in Fig. 10 is reasonable in light of the stabilizing effect of maintaining cation-anion spacings across the interface that are close to those found in the constituent materials.[81]

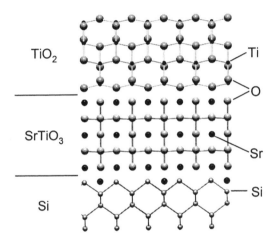

Figure 10: Idealized structural diagram for the TiO$_2$/SrTiO$_3$/Si(001) interface.

We have achieved successful heteroepitaxy of the TiO$_2$/SrTiO$_3$/Si(001) double heterojunction by MBE.[79, 82] Following growth of a one-ML template layer of SrSi$_2$, we co-evaporate stoichiometric quantities of Sr and Ti in an O$_2$ ambient at ~350°C, followed by full recrystallization of the STO layer by annealing at ~550°C in UHV. We then grow the magnetically doped anatase layer using activated (atomic) oxygen from the plasma source. It is not sufficient to transition from STO to anatase by simply closing the Sr shutter, as Sr metal aids in dissociating O$_2$ and enhancing the O flux at the growth front to the point that the Ti can be fully oxidized. We show in Fig. 11 RHEED patterns for 9 ML STO/Si(001) and 60 ML TiO$_2$/9 ML STO/Si(001) in two zone axes.

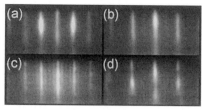

Figure 11: RHEED for 9 ML STO/Si(001) (a & b), and 60 ML TiO$_2$/9 ML STO/Si(001) (c & d) in the [100] (a & c) and [110] (b & d) zone axes.

The diffraction pattern for anatase in the [100] zone axis shows the (4x1) surface reconstruction characteristic of a well-ordered surface.[83, 84] However, TEM reveals the formation of a considerable amount of interfacial SiO$_2$. We show in Fig. 12 a representative cross sectional image for a specimen in which a 100Å anatase layer was grown on a 20Å STO buffer layer. The fact that the layered structure obeys the expected epitaxial relationship reveals that the amorphous SiO$_2$ layer, which is thicker than the STO layer, forms after nucleation of the epitaxial stack had occurred. In light of the rather slow growth rate required to nucleate 100% anatase and no rutile (~0.09 ML/sec),[34] and the relatively high substrate temperature required for good epitaxial growth (550°C), the most likely scenario for substrate oxidation is that during anatase deposition, O diffuses from the growth front, through the anatase and STO layers, to react with Si at the interface.[79] The thickness of the SiO$_2$ layer increases with time during anatase growth.

The problem of substrate oxidation is exacerbated by doping the anatase layer with Co because a higher O flux is required to oxidize Co compared with Ti. Analogous experiments in which Co$_x$Ti$_{1-x}$O$_{2-x}$/STO/Si(001) is grown reveal that a significant fraction of Co(0) is present in the anatase film, in addition to a thick SiO$_2$ layer at the interface.[79] It is clearly not thermodynamically favorable to grow Co-doped TiO$_2$/STO/Si without some SiO$_2$ formation at the interface, at least for STO buffer layer thicknesses up to 200Å.

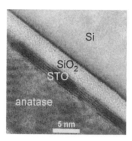

Figure 12: TEM image for 100 Å TiO$_2$/20 Å STO/Si(001). [Ref. 79]

Despite the difficulty of substrate oxidation, we have found that the conduction band offsets are favorable for spin polarized electron injection from magnetically doped TiO₂/STO into Si. We have used a combination of core and valence band photoemission spectroscopies[85, 86] to determine valence and conduction band offsets (VBOs & CBOs) at the pure TiO₂/STO/Si double heterojunction (DHJ).[82] The anatase film thickness was limited to 5 ML (~12Å) in order to permit substrate photoemission to be measured. The amount of SiO₂ at the interface is of order 1 ML for this anatase film thickness. BO determination is complicated by the fact that anatase and STO do not each possess a unique element not found in the other material. Unique elements on opposite sides of the interface allow the standard analytical expression of Kraut et al.[87, 88] to be used to find the VBO. Rather, a new graphical method was developed in which reference spectra for the pure materials were shifted in energy and weighted in intensity so their sum simulated the valence band spectrum of the DHJ.[82] The VBOs were then determined from these shifts, and the CBOs were in turn determined from the VBOs and the band gaps. The result of this exercise is shown in Fig. 13. The weak peak at ~102.3 eV in the Si $2p$ spectrum for the DHJ is due to those Si atoms at the interface which are part of a suboxide SiO$_x$ phase. This phase is the precursor of the more extensive SiO₂ phase seen Fig. 12. The CBOs extracted from this graphical method are within experimental error of zero, which means there is no electrostatic barrier to electron transport across the interface. The same is true for STO/Si(001) without anatase[89, 90].

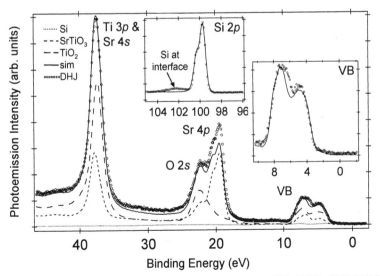

Figure 13: Simulated photoemission spectra for the 5 ML TiO₂/5 ML STO/Si(001) double heterojunction (DHJ) from which valence band offsets are determined. [Ref. 82]

7.6 SUMMARY

The materials science of magnetically doped TiO_2 is intriguing and potentially relevant to the rapidly emerging field of spin electronics. This review is of necessity a snapshot in time of a very fast moving field. The last three years have witnessed an explosion in research in magnetically doped oxide semiconductors. Some interesting results have emerged which show that in certain cases, robust ferromagnetism persists to room temperature and above. Some of the many observations of room temperature ferromagnetism in these materials can be correlated to substitution of the magnetic dopant at a host cation site, as expected for a DMS. In other cases, inadequate materials characterization was done to determine if the material is uniformly doped, or if secondary phases had formed. It is clear that care must be taken to provide adequate oxygen flux at the growth front in order to fully oxidize the magnetic dopant, and that adequate materials characterization be done to determine the oxidation state of the magnetic dopant, and the presence of secondary phases.

Investigators must also beware of magnetically contaminated substrates! We have discovered that several commonly available substrates, such as TiO_2, $SrTiO_3$ and $LaAlO_3$, frequently come from the various vendors with enough magnetic contamination that their hysteresis loops reveal saturation moments comparable to or greater than those of magnetically doped TiO_2 grown on clean substrates. We typically find that the contamination is present on the edges, presumably from the cutting tool, and that it can be removed by physical abrasion or etching in HNO_3. However, if it goes undetected and is not eliminated, this magnetic contamination can obviously lead one astray in a big way. It possible that some of the reports of "giant magnetic moments" are due to magnetic contamination in the substrate, particularly those that cannot be reproduced in other laboratories.

In the case of OPAMBE, atomic oxygen has been found to oxidize Co to Co(II), Fe to Fe(II) and Fe(III), and Cr to Cr(III) in $M_xTi_{1-x}O_{2-x/n}$ when grown on oxide substrates. All three exhibit ferromagnetism at room temperature. The dopant substitutes for Ti(IV) in the cases of Co and Cr, and no secondary phases are formed. However, Fe_3O_4 forms in the case of Fe, at least for a Fe mole fraction of 0.07, and the strong ferromagnetism seen for these specimens is attributed to the presence of Fe_3O_4. Although free carriers enhance the room temperature magnetization Cr- and Co-doped TiO_2, a FM interaction also occurs in the absence of free carriers. The most reasonable mechanism suggested by the experimental data for the Co and Cr dopants is that of F-center mediated or bound magnetic polaron exchange, in which electrons trapped in the O vacancies associated with dopant substitution for Ti(IV) mediate ferromagnetic coupling. This mechanism provides an explanation for the observation of ferromagnetism in insulating magnetically doped films. In the case of Co-doped TiO_2, the Co typically does *not* distribute in a uniform fashion, but rather tends to segregate into Co-enriched anatase surface particles. The Co is in the +2 oxidation state, and it substitutes for Ti(IV) in these particles. The Co mole fraction is locally high (x = ~0.2-0.4). The $Co_xTi_{1-x}O_{1-x}$ particles magnetically couple in a weak fashion in the absence of free carriers, but the coupling is strengthened by free electrons introduced via oxygen vacancies in excess of those required to compensate substitutional Co(II). Of the three, Cr appears to be the best behaved in terms of uniform dopant distribution and the absence of secondary phases.

In order for magnetically doped TiO_2, or any other FM oxide to be useful in spintronics, the ferromagnetism must be accompanied by a large spin polarization in the valence and/or conduction bands. Large spin polarization in turn requires exchange splitting in excess of kT. Future experiments on candidate materials should be oriented toward determining these important parameters.

Integration of magnetically doped TiO_2 anatase with Si is a daunting task and our work to date with Co as the dopant suggests that it may not be possible to simultaneously oxidize the host and magnetic dopant cations without also oxidizing Si at the interface. Although $SrTiO_3$ is an effective buffer layer for nucleating anatase, it is not an adequate oxygen diffusion barrier. We will investigate other magnetic dopants (such as Cr) that are more easily oxidized in the near future. The band offsets for the $TiO_2/SrTiO_3/Si(001)$ system are favorable for spin polarized electron injection.

The authors are indebted to S.M. Heald, J. Jaffe, K. Krishnan, K.A. Griffin, J.D. Bryan, D.R. Gamelin and J. Osterwalder for many stimulating and helpful discussions, and to C.H. Li and B.T. Jonker for the Hall measurements. This work was supported the US DOE Basic Energy Sciences Materials Science and Engineering Physics Division and the DARPA SpinS Initiative. The OPAMBE film growth and characterization described in this chapter were performed in the Environmental Molecular Sciences Laboratory, a national scientific user facility sponsored by the Department of Energy's Office of Biological and Environmental Research and located at Pacific Northwest National Laboratory. The XANES and EXAFS measurements were performed at the PNC-CAT facilities at the Advanced Photon Source, located at Argonne National Laboratory. This facility is supported by the US DOE Office of Science Grant No. DEFG03-97ER45628, the University of Washington, a major facilities access grant from NSERC, and Simon Fraser University. Use of the Advanced Photon Source was supported by the U.S. Department of Energy, Office of Science, Office of Basic Energy Sciences, under Contract No. W-31-109-ENG-38.

244

REFERENCES

[1] G. A. Prinz, *Science* **282**, 1660 (1998).

[2] S. A. Wolf, D. D. Awschalom, R. A. Buhrman, R. A. Daughton, S. von Molnar, M. L. Roukes, A. Y. Chtchelkanova and D. M. Treger, *Science* **294**, 1488 (2001).

[3] S. DasSarma, *Amer. Sci.* **89**, 516 (2001).

[4] D. D. Awschalom, M. E. Flatte and N. Samarth, *Sci. Amer. June issue*, 67 (2002).

[5] S. Datta and B. Das, *Appl. Phys. Lett.* **56**, 665 (1990).

[6] S. Bandyopadhyay and M. Cahay, *Appl. Phys. Lett.* **85**, 1433 (2004).

[7] Y. A. Bychkov and E. I. Rashba, *J. Phys. C* **17**, 6039 (1984).

[8] Y. Q. Jia, R. C. Shi and S. Y. Chou, *IEEE Transactions: Magnetics* **32**, 4707 (1996).

[9] P. R. Hammar, B. R. Bennett, M. J. Yang and M. Johnson, *Phys. Rev. Lett.* **83**, 203 (1999).

[10] S. Gardelis, C. G. Smith, C. H. W. Barnes, E. H. Linfield and D. A. Ritchie, *Phys. Rev. B* **60**, 7764 (1999).

[11] D. R. Loraine, D. I. Pugh, H. Jenniches, R. Kirschman, S. M. Thompson, W. Allen, C. Sirisathikul and J. F. Gregg, *J. Appl. Phys.* **87**, 5161 (2000).

[12] G. Schmidt, D. Ferrand, L. W. Molenkamp, A. T. Filip and B. J. van Wees, *Phys. Rev. B* **62**, R4790 (2000).

[13] A. T. Hanbicki, B. T. Jonker, G. Itskos, G. Kioseoglou and A. Petrou, *Appl. Phys. Lett.* **80**, 1240 (2002).

[14] F. Matsukura, H. Ohno, A. Shen and Y. Sugawara, *Phys. Rev. B* **57**, R2037 (1998).

[15] A. M. Nazmul, S. Sugahara and M. Tanaka, *cond-mat/0208299* (2002).

[16] T. Dietl, H. Ohno, F. Matsukura, J. Cibert and D. Ferrand, *Science* **287**, 1019 (2000).

[17] S. J. Pearton, C. R. Abernathy, D. P. Norton, A. F. Hebard, Y. D. Park, L. A. Boatner and J. D. Budai, *Mat. Sci. Eng. R* **40**, 137 (2003).

[18] Y. Matsumoto, M. Murakami, T. Shono, T. Hasegawa, T. Fukumura, M. Kawasaki, P. Ahmet, T. Chikyow, S.-Y. Koshihara and H. Koinuma, *Science* **291**, 854 (2001).

[19] Y. Matsumoto, M. Murakami, K. Hasegawa, T. Fukumura, M. Kawasaki, P. Ahmet, K. Nakajima, T. Chikyow and H. Koinuma, *Appl. Surf. Sci.* **189**, 344 (2002).

[20] S. A. Chambers, S. Thevuthasan, R. F. C. Farrow, R. F. Marks, J. U. Thiele, L. Folks, M. G. Samant, A. J. Kellock, N. Ruzycki, D. L. Ederer and U. Diebold, *Appl. Phys. Lett.* **79**, 3467 (2001).

[21] D. H. Kim, J. S. Yang, K. W. Lee, S. D. Bu, T. W. Noh, S.-J. Oh, Y.-W. Kim, J.-S. Chung, H. Tanaka, H. Y. Lee and T. Kawai, *Appl. Phys. Lett.* **81**, 2421 (2002).

[22] C. B. Murray, S. Sun, H. Doyle and T. Betley, *MRS Bull.* **26**, 985 (2001).

[23] A coherent state of a spin ensemble is one that can be expressed in terms of a single spinor. For example, a coherent spin state in which the spins are all oriented along the +x axis can be expressed as a linear combination of spin up (+z or ↑) and spin down (-z or ↓) basis vectors (e.g. $|S_x> = c_1|↑> + c_2|↓>$, with $c_1 = c_2 = 1/\sqrt{2}$). Coherence requires that c_1 and c_2 be well defined over time. Spin decoherence is the process by which the overall spin state of the particle collapses into a *statistical mixture* of spin up and down (with equal probability, in this example.) Such an incoherent or "mixed" state cannot be described by a single spinor. However, the entire process can be described in terms of the spin density matrix; see E. Merzbacher, *Quantum Mechanics* (2nd Ed.) (Wiley, New York, 1970) p. 278-289.

[24] H. Shimizu, T. Hayashi, T. Nishinaga and M. Tanaka, *Appl. Phys. Lett.* **74**, 398 (1999).

[25] T. Hayashi, Y. Hashimoto, S. Katsumoto and Y. Iye, *Appl. Phys. Lett.* **78**, 1691 (2001).

[26] S. J. Potashnik, K. C. Ku, S. H. Chun, J. J. Berry, N. Samarth and P. Schiffer, *Appl. Phys. Lett.* **79**, 1495 (2001).

[27] R. M. Stroud, A. T. Hanbicki, Y. D. Park, G. Kioseoglou, A. G. Petukhov and B. T. Jonker, *Phys. Rev. Lett.* **89**, 166602 (2002).

28   J. Zhang and H. Banfield, *J. Phys. Chem. B* **104**, 3481 (2000).

29   L. Forro, O. Chauvet, D. Emin, L. Zuppiroli, H. Berger and F. Levy, *J. Appl. Phys.* **75**, 633 (1994).

30   H. Toyosaki, T. Fukumura, Y. Yamada, K. Nakajima, T. Chikyow, T. Hasegawa, H. Koinuma and M. Kawasaki, *Nat. Mat.* **3**, 221 (2004).

31   B. G. Hyde and S. Andersson, *Inorganic Crystal Structures* (John Wiley & Sons, 1989).

32   S. A. Chambers, *Materials Today* **5**, 34 (2002).

33   S. A. Chambers, T. Droubay and S. M. Heald, *Phys. Rev. B* **67**, 100401(R) (2003).

34   S. A. Chambers, C. M. Wang, S. Thevuthasan, T. Droubay, D. E. McCready, A. S. Lea, V. Shutthanandan, and C. F. Windisch Jr., *Thin Solid Films* **418**, 197 (2002).

35   *CRC Handbook of Physics and Chemistry* (CRC Press, 2003).

36   S. R. Shinde, S. B. Ogale, S. Das Sarma, J. R. Simpson, H. D. Drew, S. E. Lofland, C. Lanci, J. P. Buban and N. D. Browning, *Phys. Rev. B* **67**, 115211 (2003).

37   B.-S. Jeong, Y. W. Heo, D. P. Norton, J. G. Kelly, R. Rairigh, A. F. Hebard, J. D. Bidai and Y. D. Park, *Appl. Phys. Lett.* **84**, 2608 (2004).

38   S. A. Chambers, T. Droubay, C. M. Wang, A. S. Lea, R. F. C. Farrow, L. Folks, V. Deline and S. Anders, *Appl. Phys. Lett.* **82**, 1257 (2003).

39   S. A. Chambers and R. F. C. Farrow, *MRS Bull.* **28**, 729 (2003).

40   P. A. Stampe, R. J. Kennedy, Y. Xin and J. S. Parker, *J. Appl. Phys.* **92**, 7714 (2002).

41   J.-Y. Kim, J.-H. Park, B.-G. Park, H.-J. Noh, S.-J. Oh, J. S. Yang, D.-H. Kim, S. D. Bu, T.-W. Noh, H.-J. Lin, H.-H. Hseih and C. T. Chen, *Phys. Rev. Lett.* **90**, 017401 (2003).

42   T. Droubay and S. A. Chambers, unpublished.

43   The O vacancies introduced with reduced O atom flux are distinct from those accompanying Co(II) subsitition for Ti(IV) in the lattice. The latter are required to maintain local charge neutrality in the crystal. Although each O atom vacancy in TiO_2 leaves behind two electrons, those associated with vacancies accompanying substitutional Co(II) are effectively localized near Co(II) sites and are thus not available for conduction. To obtain *n*-type semiconducting behavior, O vacancies in excess of those needed to compensate Co(II) must be incorporated, and this is accomplished by reducing the O flux during growth. With these additional O vacancies, the correct empirical formula is $Co_xTi_{1-x}O_{2-x-y}$, where y is the number of shallow donor O vacancies..

44   Z. Yang, G. Liu and R. Wu, *Phys. Rev. B* **67**, 060402(R) (2003).

45   J. E. Jaffe, T. C. Droubay and S. A. Chambers, *J. Appl. Phys.*, in press (2005).

46   An expansion in *c* is at best indirect evidence for Co substitution at lattice sites and the resulting strain. Simultaneous measurement of the in-plane lattice parameter, *a*, along with *c* is required to specify the true strain state of the film. This measurement apparently was not done. However, a simultaneous determination of *a* and *c*, even if done, does not necessarily establish Co cation site substitution. Varying strain states of the film may result for other reasons. These include choice of substrate (STO vs. LAO) and the associated lattice mismatch, differing film morphology, and variable twin densities in the LAO substrate. Due to its atom specific nature, Co K-edge EXAFS is a much more direct probe of Co lattice site substitution.

47   E. A. Stern, *Phys. Rev. Lett.* **15**, 62 (1965).

48   L. Berger, *Phys. Rev. B* **2**, 4559 (1970).

49   J. C. Denardin, M. Knobel, X. X. Zhang and A. B. Pakhamov, *J. Magn. Magn. Mat.* **262**, 15 (2003).

50   P. Xiong, G. Xiao, J. Q. Wang, J. Q. Xiao, J. S. Jiang and C. L. Chien, *Phys. Rev. Lett.* **69**, 3220 (1992).

51   J. S. Higgins, S. R. Shinde, S. B. Ogale, T. Venkatesan and R. L. Greene, *Phys. Rev. B* **69**, 073201 (2004).

246

52 Y. Matsumoto, R. Takahashi, M. Murakami, T. Koida, X.-J. Fa, T. Hasegawa, T. Fukumura, M. Kawasaki, S.-Y. Yoshihara and H. Koinuma, *Jap. J. of Appl. Phys.* **40**, L1204 (2001).

53 W. K. Park, R. J. Ortega-Hertogs, J. S. Moodera, A. Punnoose and M. S. Seehra, *J. Appl. Phys.* **91**, 8093 (2002).

54 N. H. Hong, J. Sakai, W. Prellier and A. Hassini, *Appl. Phys. Lett.* **83**, 3129 (2003).

55 S. R. Shinde, S. B. Ogale, J. S. Higgins, H. Zheng, A. J. Millis, V. N. Kulkarni, R. Ramesh, R. L. Greene and T. Venkatesan, *Phys. Rev. Lett.* **92**, 166601 (2004).

56 V. Shutthanandan, S. Thevuthasan, S. M. Heald, T. Droubay, M. H. Engelhard, T. C. Kaspar, D. E. McCready, L. Saraf, S. A. Chambers, B. S. Mun, N. Hamdan, P. Nachimuthu, B. Taylor, R. P. Sears and B. Sinkovic, *Appl. Phys. Lett.* **84**, 4466 (2004).

57 M. Murakami, Y. Matsumoto, T. Hasegawa, P. Ahmet, K. Nakajima, T. Chikyow, H. Ofuchi, I. Nakai and H. Koinuma, *J. Appl. Phys.* **95**, 5330 (2004).

58 A. R. Bally, E. N. Korobeinikova, P. E. Schmid, F. Levy, and F. Bussy, J. Phys. D **31**, 1149 (1998).

59 Z. Wang, J. Tang, T. LeDuc, W. Zhou and L. Spinu, *J. Appl. Phys.* **93**, 7870 (2003).

60 Z. Wang, W. Wang, J. Tang, T. LeDuc, L. Spinu and W. Zhou, *Appl. Phys. Lett.* **83**, 518 (2003).

61 J. M. D. Coey, A. P. Douvalis, C. B. Fitzgerald and M. Venkatesan, *Appl. Phys. Lett.* **84**, 1332 (2004).

62 N. H. Hong, W. Prellier, J. Sakai and A. Hassini, *Appl. Phys. Lett.* **84**, 2850 (2004).

63 Y. J. Kim, T. Droubay, A. S. Lea, C. M. Wang, V. Shutthanandan, S. A. Chambers, R. P. Sears, B. Taylor and B. Sinkovic, *Appl. Phys. Lett.* **84**, 3531 (2004).

64 S. E. Park, H.-J. Lee, Y. C. Cho, S.-Y. Jeong, C. R. Cho and S. Cho, *Appl. Phys. Lett.* **80**, 4187 (2002).

65 S. G. Yang, A. B. Pakhamov, S. T. Hung and C. Y. Wong, *Appl. Phys. Lett.* **81**, 2418 (2002).

66 H. Saito, V. Zayets, S. Yamagata and K. Ando, *Phys. Rev. Lett.* **90**, 207202 (2003).

67 R. J. Soulen Jr, J. M. Byers, M. S. Osofsky, B. Nadgorny, T. Ambrose, S. F. Cheng, P. R. Broussard, C. T. Tanaka, J. Nowak, J. S. Moodera, A. Barry and J. M. D. Coey, *Science* **282**, 85 (1998).

68 T. Droubay, S. M. Heald, V. Shutthanandan, S. Thevuthasan, S. A. Chambers and J. Osterwalder, *J. Appl. Phys.*, in press (2005).

69 Z. Wang, J. Tang, H. Zhang, V. Golub, L. Spinu and T. LeDuc, *J. Appl. Phys.* **95**, 7381 (2004).

70 J. Osterwalder, T. Droubay, T. C. Kaspar, J. R. Williams and S. A. Chambers, *Thin Solid Films*, submitted (2004).

71 S. Roberts, *Phys. Rev.* **76**, 1215 (1949).

72 H. Tang, K. Prasad, R. Sanjines, P. E. Schmid and F. Levy, *J. Appl. Phys.* **75**, 2042 (1994).

73 A. C. Durst, R. N. Bhatt and P. A. Wolff, *Phys. Rev. B* **65**, 235205 (2002).

74 A. Kaminski and S. Das Sarma, *Phys. Rev. Lett.* **88**, 247202 (2002).

75 D. G. Schlom and J. H. Haeni, *MRS Bull.* **27**, 198 (2002).

76 R. A. McKee, F. J. Walker and M. F. Chisholm, *Phys. Rev. Lett.* **81**, 3014 (1998).

77 J. Lettieri, J. H. Haeni and D. G. Schlom, *J. Vac. Sci. Technol. A* **20**, 1332 (2002).

78 H. Li, X. Hu, Y. Wei, Z. Yu, X. X. Zhang, R. Droopad, A. A. Demkov and J. Edwards Jr, *J. Appl. Phys.* **93**, 4521 (2003).

79 T. C. Kaspar, T. Droubay, C. M. Wang, S. M. Heald, A. S. Lea and S. A. Chambers, *J. Appl. Phys.*, submitted (2004).

80 C. J. Forst, C. R. Ashman, K. Schwarz and P. E. Blochl, *Nat. Mat.* **427**, 53 (2003).

81 R. A. McKee, F. J. Walker, E. D. Specht, G. E. Jellison Jr, and L. A. Boatner, *Phys. Rev. Lett.* **72**, 2741 (1994).

[82] A. C. Tuan, T. C. Kaspar, T. Droubay, J. W. Rogers Jr, and S. A. Chambers, *Appl. Phys. Lett.* **83**, 3734 (2003).

[83] Y. Liang, S. Gan, S. A. Chambers and E. I. Altman, *Phys. Rev. B* **63**, 235402 (2001).

[84] M. Lazzeri and A. Selloni, *Phys. Rev. Lett.* **87**, 266105 (2001).

[85] S. A. Chambers, T. Droubay, T. C. Kaspar, M. Gutowski and M. van Schilfgaarde, *Surf. Sci.* **554**, 81 (2004).

[86] S. A. Chambers, T. Droubay, T. C. Kaspar and M. Gutowski, *J. Vac. Sci. Technol. B* **22**, 2205 (2004).

[87] E. A. Kraut, R. W. Grant, J. W. Waldrop and S. P. Kowalczyk, *Phys. Rev. Lett.* **44**, 1620 (1980).

[88] E. A. Kraut, R. W. Grant, J. W. Waldrop and S. P. Kowalczyk, *Phys. Rev. B* **28**, 1965 (1983).

[89] S. A. Chambers, Y. Liang, Z. Yu, R. Droopad, J. Ramdani and K. Eisenbeiser, *Appl. Phys. Lett.* **77**, 1662 (2000).

[90] S. A. Chambers, Y. Liang, Z. Yu, R. Droopad and J. Ramdani, *J. Vac. Sci. Technol. A* **19**, 934 (2001).

Interfaces and Surfaces: Correlated Electron Systems

CHAPTER 8

INTERFACES IN MATERIALS WITH CORRELATED ELECTRON SYSTEMS

J. Mannhart

Center for Electronic Correlations and Magnetism, Augsburg University, Exp. Phys. VI, Institute of Physics, D-86135 Augsburg, Germany

8.1 INTRODUCTION

Interfaces and surfaces play an important role in the application of advanced materials, as well as for fundamental research in solid state physics. In the realization of spintronics, for example, the injection of spin-polarized electrons across ferromagnet - semiconductors interfaces is a major challenge.[1] Most Josephson junctions used in sensors and other electronic devices also rely on interfaces. In high-T_C superconductivity, the reduction of the critical current density of grain boundaries with boundary angle poses strong demands on the architecture of high-T_C cables (see Fig. 1).[275] More than 70 million dollars are annually invested in research programs pertaining to the development of coated conductors using optimized grain boundaries, revealing clearly the relevance of such interfaces for applications.

Most experimental studies aimed at elucidating high-temperature superconductivity utilize interfaces or surfaces and rely on specific properties of the interfaces, which in many cases are only barely known. In fact, these studies predominantly provide information on interfaces rather than on the bulk. Examples are photospectroscopic measurements of the band structure and phase sensitive measurements of the order parameter symmetry. Obviously, for applications as well as for the interpretation of such measurements, comprehensive knowledge of the involved interfaces is required.

Interfaces in conventional superconductors are well understood and have been accordingly optimized. The interface physics of conventional superconductors, however, captures only a few aspects of the far more complex behavior of interfaces in high-T_C superconductors. This is due to fundamental differences of the electronic systems of the two superconductor families. These arise from the strong electronic correlations, the resulting unconventional order parameter symmetry, the small carrier density and coherence lengths and the large electric susceptibilities, which are characteristic for the cuprates only.[3]

The goal of this contribution is to provide an overview of the fundamental physics underlying the properties of interfaces in materials with highly correlated electronic systems. I will demonstrate the direct connection between these basic questions and large scale applications of these materials. Because of their outstanding importance for applications, and since a large body of experimental data is available, special attention will hereby be given to interfaces in the high-T_C cuprates.

Due to the complexity and size of the field, this article, which is based on conference presentations and on previous overviews of interfaces in

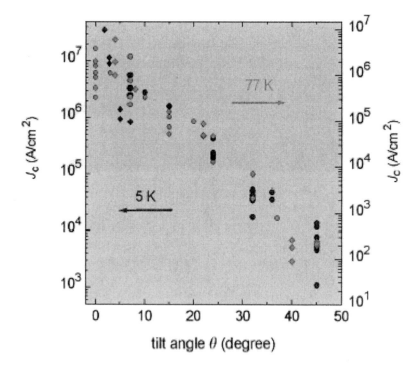

Figure 1: Critical current density J_c of grain boundaries in YBa$_2$Cu$_3$O$_{7-\delta}$ bicrystalline films as a function of boundary angle [Refs. 4, 5, 6, 7]. Figure replotted from Mannhart and Chaudhari, 2001.

complex superconductors,[3] can not be comprehensive. It is noted that several comprehensive review articles on Josephson junctions in high-T_C superconductors have already been published, such as overviews on superconductor-normal metal-superconductor junctions,[8] grain boundary junctions[9] and the current phase relation of Josephson junctions.[10]

8.2 INTERFACES IN COMPOUNDS WITH CORRELATED ELECTRON SYSTEMS

The electronic systems of the cuprate superconductors are strongly correlated. That is, the behavior of the charge carriers is governed by specific, local interactions or interactions with a small number of neighbors, and therefore can not be described by interactions with a uniform mean field generated by a background of charges. These correlations give rise to the high transition temperatures. Interfaces and surfaces obviously modify the correlations.[9,11-16]

8.2.1 Changes of Conventional Electronic Properties

At the interface, the bulk electronic system is altered, due to many reasons, and sometimes with dramatic consequences. Interfaces obviously break the translational, and where present, the rotational symmetry of the bulk. Important effects result such as reductions of bandwidths and dispersion, changes of the crystal fields, or shifts of the Madelung energies. Stress or strain is induced by the interfaces, which alter the distances and bonds between the ions. The lattice may even be structurally or electronically reconstructed. At the interface, point defects, dislocations and stacking faults tend to be incorporated into the lattice. These defects have their own electronic or magnetic properties and may act as scattering centers. The chemical composition may change close to the interface, as the mechanical strain and built-in potentials can cause ions to drift (see, *e.g.,* Maier, 2004).[17] It is evident that these phenomena depend in size and magnitude, sometimes even in sign, on the specific microscopic properties of the respective surface or interface.

Due to these and other phenomena, interfaces shift and distort the electronic states and energy levels at the ions and modify the bands. These effects, some of which can be directly probed by resonant photoelectron spectroscopy,[18] change the electronic and, where applicable, the magnetic susceptibility and modify the polarizability and screening. Interfaces further affect orbital and spin order, and, where present, spin-orbit coupling.

8.2.2 Modifications of Electronic Correlations

Since the electronic states of the individual ions and the screening are altered at interfaces, their correlations are strongly modified as well. Because the correlations control the electronic behavior of the material, their modification can induce dramatic changes of the collective electronic and magnetic properties, to the extent that phase transitions, for example from the superconducting to the insulating state, are induced.

Take, for example, the Hubbard model. An interface-driven reduction of the electronic screening will enhance the on-site Coulomb energy U. Conversely, U is readily lowered by several eV should the screening be enhanced by image charges, as is the case for a transition metal oxide and conventional metal interface.[13,19] The same argument applies with minor changes to the charge transfer energy Δ, which at such interfaces is reduced by the polarizability and the Madelung energies of the transfer's initial and final states. Besides U and Δ, the hopping matrix element t and the energies of direct and indirect exchanges J_i may also be altered. Obviously, it would be delightful if a set of simple rules of thumb could be established to describe the influence of the interfaces on the correlation energies. Since the correlations are depending on the detailed aspects of a specific interface, these rules, unfortunately, may not exist.

Since U or Δ may decrease to values smaller than the bandwidth W, a phase transition to a metallic phase may be initiated at the interfaces or surfaces of Mott- or band-insulators. For the (110)-surface of the Mott-insulator NiO this has been predicted by G. Sawatzky and coworkers.[13] Such an interface-driven phase transition is schematically illustrated in Fig. 2.

Intriguingly, interfaces may also strengthen the correlation effects. In $La_xCa_{1-x}VO_3$, for example, the bands are expected to be narrowed at the surface, due to the reduced atomic coordination. On these simple grounds, the ratio of U/W is expected to be enhanced.[20] Indeed, for a range of dopant concentrations x, these compounds, while being metallic in the bulk, establish insulating surfaces.[12] A detailed account of this phenomenon has been provided by calculations on $(Sr,Ca)VO_3$ using local density approximation merged with dynamical mean-field theory (LDA+DMFT). These point towards the emergence of a quasi-one dimensional surface band in such compounds.[15,21-22] Such a band with reduced dimensionality is very susceptible to correlation effects and could thus provoke the metal-insulator transition at the surface.

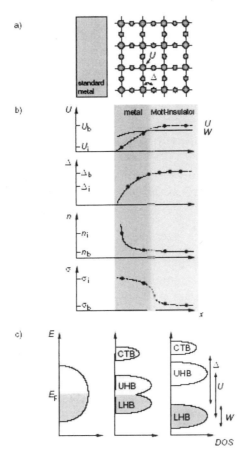

Figure 2: Schematic illustration of a possible electronic behavior of a Mott-Hubbard insulator brought into contact with a normal metal. a) Sketch of the lattice, the electronic system of the Mott-insulator is characterized by the on-site repulsion energy U the charge transfer energy Δ and the hopping matrix element t (not drawn). The interface modifies the correlation parameters, therefore in b) possible spatial dependencies of U, the bandwidth W, Δ, carrier density n and conductivity σ are sketched. The indices i and b refer to the interface and the bulk, respectively. c) Schematic band structures of the Mott-insulator (right), the metal (left) and the interface region (middle), showing the upper and lower Hubbard bands (UHB, LHB), a charge transfer band (CTB) and the Fermi energy E_F.

8.2.3 State Occupancy, Charging

At interfaces there is not only a change of the energy spectrum and of the correlations, but due to charge transfer, the state occupancy also varies. Note that polar, charged surfaces are readily obtained in transition metal oxides, a prominent example being the (111)-surface of NiO. Due to ionic imbalances, most interfaces modify the electronic potential and thereby produce space charge layers, altering the density n of mobile charge carriers. This effect is well known from conventional systems, such as heterojunctions in standard semiconductors, like Si and GaAs, and grain boundaries in oxides like $SrTiO_3$ and ZnO. Interface charging is the active mechanism of ceramic capacitors or of varistors fabricated from these materials.

The role of interface charging has been crisply demonstrated by Ohtomo and Hwang, 2004.[23] Stacking two insulators, $LaTiO_3$ and $SrTiO_3$, they obtained a conducting superlattice. It is the interface between the two insulators that is conducting. This is due to charge transfer, but interestingly the interface is only conducting if doped with electrons. For a discussion of these effects see also Okamoto and Millis, 2004.[16]

8.2.4 Phase Separation

Electronic phases may separate in correlated electron systems to minimize the total energy, for example into charged metallic and insulating phases. Their spatial distribution is usually irregular and may be fluctuating. The various phases interact in different ways with interfaces, such that the phase structure may be pinned at the interface. For example, depletion layers found at grain boundaries in high-T_c cuprates may be described as pinned stripe phase. Phase separation within the interface plane, generated either by the interface itself or resulting from interaction with the phase-separated bulk, seems reasonable to expect, but has not been identified yet. It would obviously results in inhomogeneous transport across the interface.

8.2.5 Spatial Profiles

Evidently, due to these interdependent phenomena, any parameter of the electronic system may be altered by interfaces. It is obvious that at the interfaces, electronic phases or phase-like electronic systems with special properties may be obtained.

Intriguingly, the profiles and length scales with which the electronic parameters relax from the interface to the bulk behavior differ. As an example, the carrier density relaxes on the scale of the Thomas-Fermi screening length, whereas coupling strengths may be tied to the strain relaxation, measured in unit cells. In superconductors, the Cooper pair density changes on the coherence length scale, and correlation parameters have again their own lengths. The spectrum of these scales, exemplified also in Fig. 2, may provoke exotic and complex electronic properties in the inhomogeneous transition region between bulk and interface.

8.2.6 Transport across Interfaces in Correlated Electron Systems

While interfaces modify electronic correlations in the bulk in an intricate manner, transport across interfaces is even more complex. As pointed out by J. Freericks, correlation effects within the barriers can alter transport in surprising ways.[24] Correlations can, for example, strongly influence the temperature dependence of the critical current of a Josephson junction,[25] or the sensitivity of a junction to barrier inhomogeneities, giving rise to an intrinsic pin-hole effect.[26] This pin-hole effect could explain the pertinacious parameter spread, typically of the order of 10%, which is found for all high-T_c Josephson junctions, blocking large scale integration of high-T_c electronics.

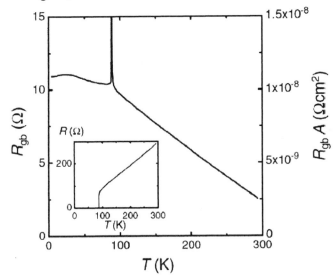

Figure 3: Temperature dependence of the resistance R_{gb} of a (100)(110), [001] tilt grain boundary in a $YBa_2Cu_3O_{7-\delta}$ bicrystalline film. The insets show dR_{gb}/dT of the boundary and the corresponding $R(T)$ curve of one grain next to the boundary. [Ref. 27]

Fig. 3 illustrate how exotic the transport properties of interfaces in correlated electronic systems may be. The figure shows the resistance of a 45° grain boundary in a $YBa_2Cu_3O_{7-\delta}$ film as a function of temperature. While the resistance of the grains increases linearly as a function of temperature, the resistance of the boundary linearly decreases, a completely unexpected effect,[27] which reflects the special properties of such 45° grain boundaries even at temperatures far above T_c.

8.2.7 Tailoring and Use of Interfaces in Correlated Electron Systems

Having such outstanding properties, is it possible to benefit from the interfaces in correlated electron systems, putting them to work in applications? This is indeed the case. Grain boundary Josephson junctions, for example, are used in highly sensitive gradiometers searching for mineral deposits[28] (see below) or as sensors in biomagnetic applications.[29,30]

For the practical use of the interfaces, it is of course desirable to optimize them for the requirements specific to an application. Although our understanding of the interfaces is still very limited, in first cases, they are indeed being optimized. By selectively modifying the electronic parameters of the interfaces, these can be tailored without degrading the bulk. Grain boundaries in the cuprates are selectively doped to enhance the grain boundary coupling (see below),[31] and compositionally graded interfaces have been designed to strengthen interface magnetism in $La_{0.6}Sr_{0.4}MnO_3$ - $SrTiO_3$ multilayers.[32] Freericks et al.[24] propose optimized barriers consisting of correlated materials to enhance the speed of high-T_c Josephson junctions.

In some cases it has already been possible to even control the interface phases, creating, for example, ferromagnetic interfaces in antiferro- or paramagnetic host lattices.[33] In systems in which the properties of the interfaces are superior to those of the bulk, optimized architectures may be used to exploit the interface contributions. Nano-grained compounds, for example, offer huge grain boundary areas. Alternatively, large interface areas are provided by superlattices. Clearly, phase control of interface phases is a great challenge, but one that may yield ample opportunities for the design of materials with unusual properties, useful, e.g., for the fabrication of novel devices.

In 1996, for example, Hwang et al.[34] and Gupta et al.[35] discovered that grain boundaries in the colossal-magnetoresistance (CMR) materials strongly enhance the magnetoresistance in the small magnetic field range, which is the range that is important for most applications. Despite extensive work on this intriguing topic,[36-38] there is still considerable speculation regarding the nature of these grain boundaries and their conduction mechanisms. As summarized by Blamire et al.,[38] the reasons for this large magnetoresistance are complex. Models include spin polarized tunneling, inelastic tunneling, hopping conduction through a nonmagnetic interface, and tunneling through an interfacial depletion layer. Similarly to the grain boundaries in the cuprates it has been found possible to tune the grain boundaries in the manganites by selective doping[38] or by incorporating foreign phases.[39]

The option to utilize the interface phases in correlated materials shall be illustrated with two more examples: an intriguing proposal and a speculative thought.

The proposal has been presented by Altieri et al.[13] Studying the example of MgO films grown on Ag (100), this group proposes to use the screening by metals to tune the correlation parameters of thin oxide films grown on metallic substrates to obtain novel material properties in the film. Comparably, Körting et al.,[40] showed that the critical temperatures of thin $YBa_2Cu_3O_{7-\delta}$ films used as drain-source channels in FETs can be tuned by electronic interactions with the gate insulator, thereby enhancing the performance of the device.

For the speculative example, consider the interface between a high-T_c superconductor with a large number of CuO_2 planes per unit cell and a second phase, forming for example a precipitate. The superconductor can be doped by charge transfer from this particle, such that along its c-direction a uniform doping profile in the set of CuO_2 planes is obtained. Because the nonuniformity of the doping is a main hindrance to the enhancement of T_c in high-T_c cuprates with more than two CuO_2 planes per unit cell, the interface phase may have a very high T_c, supposing the pairing interaction is not suppressed by detrimental interface effects. Such interface effects may be underlying some of the irreproducible reports of cuprates with a super-high T_c.

8.2.8 Controlled Phase Transitions

Phase transitions induced by interfaces are of particular scientific appeal and key to many applications. At grain boundaries, contacts to normal metals, pn-junctions and at surfaces the cuprates may undergo phase transitions into the insulating state. The control of such phase transitions by doping, with electric fields, by irradiation with light, quasiparticle injection and pressure is an exciting field of research (see Fig. 4), summarized by Ahn et al.[41] Electric field- and quasiparticle injection-controlled phase transitions in drain-source channels offer an intrinsic amplification for transistors.[42] Such phase transition transistors have been proposed to break the scaling limits experienced by conventional semiconducting FETs.[43] Phase transitions at interfaces have also been demonstrated in structures fabricated from non-superconducting materials. Charge transfer, for example, causes magnetic interface phases in $CaMnO_3/CaRuO_3$ heterostructures[33] and in $LaMnO_3/SrMnO_3$ superlattices.[44]

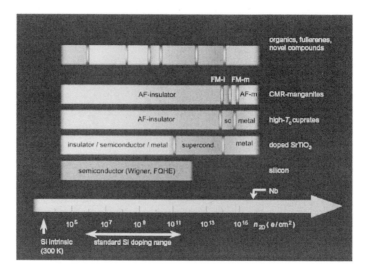

Figure 4: Illustration of the zero-temperature behavior of various correlated materials as a function of sheet charge density (n_{2D}). Silicon is shown as reference. The examples for high-T_c superconductors and for colossal magnetoresistive manganites reflect $YBa_2Cu_3O_{7-\delta}$ and $(La,Sr)MnO_3$, respectively. The top bar has been drawn to illustrate the richness of the phases of organic materials. AF: antiferromagnetic, FM: ferromagnetic, I: insulator, M: metal, SC: superconductor, FQHE: fractional quantum Hall effect, Wigner: Wigner crystal. [Ref. 41]

8.3 FUNDAMENTALS OF INTERFACES IN THE HIGH-T_C CUPRATES

Now the focus will be directed to the two main mechanisms by which electronic correlations modify the charge transport across interfaces in the cuprates. These are the unconventional order parameter symmetry and the charge-transfer-induced metal-insulator phase transition. Before doing so, it should be noted that the cuprates react sensitively to interfaces because their superconducting coherence lengths are extremely short and because their complex crystal structure makes them susceptible to defects. The coherence lengths of the cuprates are in the Angstrom range and are therefore comparable to the size of the unit cells. Typically in such materials the boundary conditions imposed by the interface on the pair potential induce a drastic reduction of the order parameter.[45]

In addition, due to the large mismatch in carrier density, the order parameter is abruptly reduced at interfaces between high-T_C superconductors and conventional metals such as silver or gold.[8,46] Due to this sensitivity of the order parameter to interfaces, the fabrication of useful Josephson junctions from the cuprates was considered to be close to impossible.

A similar conclusion was drawn from the sensitivity of the cuprates to chemical degradation. Compared to Nb or Al, for example, the solid state chemistry of the oxide superconductors is far more complex. A brief exposure of the surface to air may result in the formation of carbonate or hydroxide phases which due to the short coherence lengths involved, strongly affect the electronic transport across a contact formed at this surface.

For several years, the chemical and structural sensitivity of the cuprates and their short coherence lengths were seen as main factors in controlling the interface properties. While these effects are indeed important, the electronic correlations which cause phase transitions at the interface and the d-wave symmetry are equally consequential.[47]

8.3.1 Influence of the Unconventional Order Parameter Symmetry on Interface Properties

The symmetry of the superconducting order parameter in the high-T_C cuprates is dominated by a robust d_{x2-y2}-component.[48,49] This is a direct consequence of the electronic correlations described in chapter 2. Because most of the experiments elucidating the order parameter symmetry have been done by studying either surfaces of high-T_C superconductors, interfaces between high-T_C superconductors and normal superconductors or grain boundaries within the cuprates, the d-wave symmetry is primarily known for the superconducting layers next to the interface. Far less information is available on the order parameter symmetry of the bulk, for which no phase-sensitive measurements are available. In fact, K.A. Müller points out that it may only be the interface which is dominated by the d_{x2-y2}-wave symmetry.[50] The unconventional order parameter symmetry, revealed by studies of interfaces, strongly influences their properties, as will be illustrated by a few examples:

First, at grain boundary junctions with a finite boundary angle, the lobes of the superconducting order parameter face each other with this misorientation (see Fig. 5). Such boundaries are conveniently fabricated using the bicrystal technique illustrated in Fig. 6. It seems natural that the superconducting coupling across the interface is reduced with increasing boundary angle, as modeled in Sigrist and Rice[51] and Barash et al.[52] But, as illustrated by Fig. 7, there are more startling effects to be noted. This figure shows an $I_C(H)$-characteristic typical for (100)/(110) grain boundary Josephson junctions built from high-T_C superconductors.[9] Here, I_C is the grain boundary critical current, and H a magnetic field applied in the grain boundary plane. Although symmetric, the characteristic is highly anomalous. A finite magnetic field of a few Gauss enhances the critical current above the zero field value by an order of magnitude. Such characteristics have never been reported for Josephson junctions built from conventional superconductors and directly prove the existence of areas in the junctions with a negative critical current density J_C. In these regions,

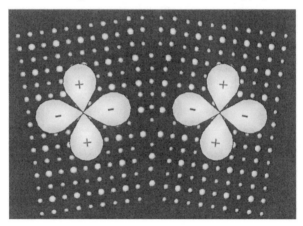

Figure 5: Illustration of the configuration of the $d_{x^2-y^2}$-wave order parameter lobes at a grain boundary with a symmetric [001]-tilt misorientation of 15° (sketch courtesy of G. Hammerl).

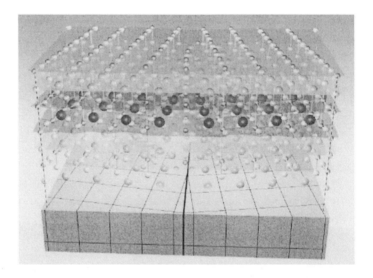

Figure 6: Illustration of the bicrystal principle. The sample material, here $YBa_2Cu_3O_7$, is epitaxially grown on a bicrystalline substrate. By this, the grain boundary of the substrate is replicated in the film. [Ref. 53]. Figure courtesy of G. Hammerl).

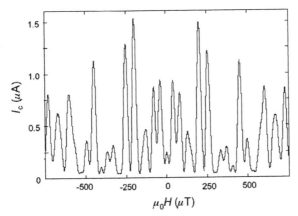

Figure 7: Critical current I_c as a function of magnetic field H applied in the grain boundary plane parallel to the c-axis direction for a 6 μm wide and 120 nm thick bridge across an asymmetric 45° [001] tilt $YBa_2Cu_3O_{7-\delta}$-grain boundary at 4.2 K. [Figure courtesy of C.W. Schneider, reprinted from Ref. 9]

even for $H=0$, the Josephson current flows backward across the interface. The only viable explanation of this effect known today, is based on the $d_{x^2-y^2}$-wave symmetry component and grain boundary faceting. This mechanism is sketched in Fig. 8. The sign-difference of the phases of neighboring lobes of the $d_{x^2-y^2}$-wave order parameter may shift the local phase differences across the junction by π, which for the Josephson current $J=J_c\ sin(\varphi+\pi)$ is equivalent to a negative J_c. Grain boundary junctions with such π-phase shifts were grown in the tricrystal experiments of C.C. Tsuei and J.R. Kirtley.[49] The observation of supercurrents which were induced in such samples and generated half flux quanta (see Fig. 9) gave evidence for the $d_{x^2-y^2}$ wave symmetry of the cuprates.

The facets with negative critical current densities give rise to a range of surprising effects. Induced by the facet structure, these asymmetric 45° grain boundaries spontaneously generate magnetic flux. Due to the small value of the inductances involved, this flux is unquantized and flips its polarity stochastically along the length of the grain boundary. This spontaneous flux, which locally breaks the time reversal symmetry, has been imaged by scanning SQUID microscopy.[54] Such a micrograph is shown in Fig. 10. As predicted by R. Mints *et al.*,[55] Josephson vortices are unstable at such boundaries and may splinter into pieces. This behavior, too, could be imaged by scanning SQUID microscopy.[56]

At junctions between misoriented superconductors, the order parameter is suppressed, even without any artificial barrier, due to the finite size of the coherence length 'ξ'.[52,57] Up to distances of the order of ξ from the

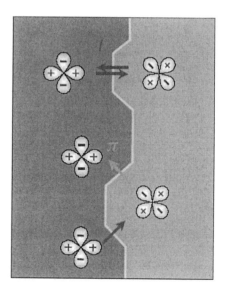

Figure 8: Sketch of a faceted (100)(110), [001] tilt grain boundary. As shown in the middle part of the drawing the faceting may cause an additional phase shift of π, leading to areas in which the Josephson current flows backward across the junction.

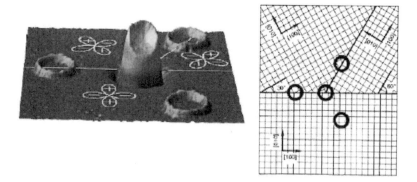

Figure 9: Scanning SQUID microscopy image of four superconducting rings with inner diameters of 48 μm and widths of 10 μm patterned into a tricrystalline $YBa_2Cu_3O_{7-\delta}$ film. As shown by in the lower half of the figure, the middle ring is centered on the tricrystal point of the substrate. In this ring, a supercurrent generating half of a magnetic flux quantum h/4e is generated due to the occurrence of a π-phase shift of the superconducting order parameter across at least one of the grain boundaries. The sample was cooled in a field < 2 mG and measured at 4.2 K. [Ref. 63, figure courtesy of J.R. Kirtley and C.C. Tsuei].

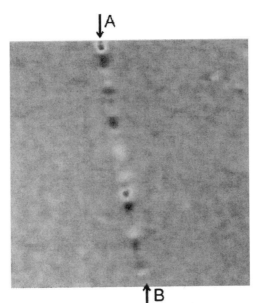

Figure 10: Scanning SQUID microscope image of an asymmetric 45° [001]-tilt YBa$_2$Cu$_3$O$_{7-\delta}$ bicrystal grain boundary (film thickness ≈ 180 nm, scanning area 400 μmx400 μm). The boundary is marked by the arrows *A* and *B*. Self-generated magnetic flux is apparent along the grain boundary. The bottom section shows a cross section through the data along the grain boundary measured in units of flux quanta penetrating the SQUID's pickup loop. The measurement was done at 4.2 K. [Ref.54]

boundary, the order parameter is affected by the orientation of the adjacent grain and thus reduced due to frustration. For such interfaces the complex admixture of additional symmetry components is expected.[58,59] Remarkably, in several experiments designed to detect the admixtures, these additional symmetry components have not been found.[60] The intrinsic suppression of the order parameter is one of the mechanisms limiting the critical current of grain boundaries in the high-T_c cuprates and therefore the performance of superconducting high power cables.

A property of Josephson junctions especially interesting for device applications is the relation $I(\varphi)$ between the Josephson current and the difference φ of the phases of the superconducting wave functions on both sides of the junction. For various configurations of Josephson junctions incorporating superconductors with unconventional order parameter symmetry, deviations from the standard sinusoidal current-phase dependence have been predicted. Extreme deviations from the harmonic current-phase relations were indeed found for large angle boundaries.[10] For asymmetric 45° grain boundaries, Il'ichev and coworkers[10] reported the occurrence of a strong second harmonic component to the current-phase relation. Further consequences of the unconventional order parameter symmetry include the occurrence of bound states in the barrier[57,61] and of midgap states at zero energy caused by the sign change of the pair potential, which may be experienced by quasiparticles reflected at such interfaces.[62]

As described, for interfaces between $d_{x^2-y^2}$-wave superconductors, the orientation of the superconductors with respect to the junction and to each other control the junction properties. Likewise, the roughness of the barrier is much more important than in conventional junctions. Most principles and relations known from conventional s-type junctions do not hold for Josephson junctions involving unconventional superconductors; this must be taken into account for junction design, fabrication and analysis.

8.3.2 Interface Charging and Band Bending

Charge transfer processes play a decisive role in the high-T_c cuprates - again in contrast to many conventional superconductors. A famous example is given by holes transferred from the block-layers to the CuO_2-planes.[64] As has been analyzed for the first time using the example of superconducting pn-junctions,[65] in the cuprates charge transfer processes are usually associated with bending of the electronic band structure of the superconductor. In the high-T_c cuprates, band bending, induced for example by local variations of the work function, electrostatic charging or surface states, is an effect strong enough to control superconductivity. In these materials, due to their large dielectric constant and the small carrier density n, the Thomas-Fermi screening lengths λ_{TF} are in the range of 0.5 – 1 nm, and thus are comparable to the superconducting coherence lengths. In addition, the transport properties of the cuprates depend much more sensitively on the carrier concentration than the transport properties of conventional superconductors, for which it is virtually impossible to induce a phase transition into an insulating state.

Although for exemplary grain boundaries the positions of the cations are already known from microscopy studie,[66] a generally applicable and exact quantitative description of the influence of charge transfer processes on the electronic properties of the interfaces is hard to develop. The typical length-scales involved, such as the electrostatic screening length, the inverse of the Fermi wave number $1/k_F$, and the coherence lengths are of the same order and are comparable to the lattice-spacings, which invalidate the average medium approaches typically used to assess the distribution of charge and potential. A correct description of the charge transfer requires detailed knowledge of the microscopic electronic properties including the pairing mechanism, which is not available. To make things worse, there is no reason why the generic phase diagrams established for neutral correlated electrons systems should be directly applicable to charged layers.

The space charge layers present at the grain boundaries in $Bi_2Sr_2CaCu_2O_{8+x}$ and in $YBa_2Cu_3O_{7-\delta}$ have recently been imaged by electron-beam holography.[67,68] Strong charging of the dislocations was found, the space charge cylinders having a radius of ≈ 1.7 nm for $YBa_2Cu_3O_{7-\delta}$. Surprisingly, the dislocation cores were measured to be negatively biased (-2.18 eV for $Bi_2Sr_2CaCu_2O_{8+x}$ and -2.4 eV for $YBa_2Cu_3O_{7-\delta}$), which lead to the suggestion that the boundaries are insulating because strong overdoping of the boundaries empties the lower Hubbard band. The widths and heights of these potential barriers well exceed the effective values estimated from the critical current densities, normal state resistivities or boundary capacities, which typically yield a few hundred meV for the height and $\approx 1-2$ nm for the total width of the barrier (see also Ref. 69).

With the uncertainties given, charge transfer and band bending are expected to form an insulating layer at the grain boundary with a width of very few nanometers as proposed by Mannhart and Hilgenkamp,[11] 1998 (see Fig. 11). Transport of quasiparticles and Cooper-pairs through such an insulating layer occurs by tunneling. The fraction of multi-step tunneling versus direct tunneling processes is controlled by the density and nature of interface states, expected to be associated with the dislocation cores or other structural features of the boundary.

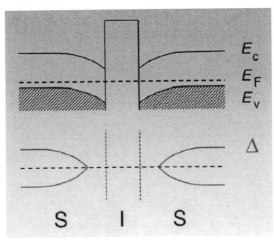

Figure 11: Illustration of a possible scenario based on a semiconductor band-picture for band-bending at a grain boundary in an oxide superconductor leading to the formation of depletion-layers at the interface. For clarity, the full band structure of a cuprate is modeled in this sketch as the one of a strongly p-doped semiconductor.

Charge transfer occurring at interfaces in oxide superconductors implies that doping of the superconductor and of the interface to reduce the internal grain boundary potential and the width of the space charge layers enhances the critical currents of the boundaries.[70] Therefore, in contrast to conventional superconductors, more as in semiconductors, doping provides a valuable tool to optimize the superconductors for application. Why should the carrier concentration that maximizes the T_C of a single crystal provide the best coupling across grain boundaries? Indeed, for $YBa_2Cu_3O_{7-\delta}$ it was found that the grain boundary critical currents are greatly enhanced if the superconductor is doped by substituting Y with Ca (see Fig. 12). Since overdoping lowers the T_C, homogeneous Ca-doping enhances the critical currents only to temperatures of about 70 K. As was shown in studies on bicrystalline samples[31] and on coated conductors[72], the

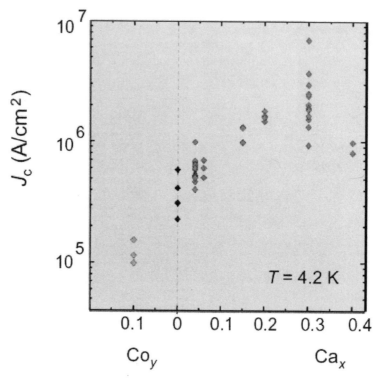

Figure 12: Dependence of the critical current density of 24° grain boundaries in $Y_{1-x}Ca_xBa_2Cu_3O_{7-\delta}$ and in $YBa_2Cu_{3-y}Co_yO_{7-\delta}$ films on the Ca and Co concentrations x and y at 4.2 K. [Ref. 70]

T_C reduction can be avoided by using doping multilayers to selectively overdope the grain boundaries (Fig. 13). In such doping heterostructures the grain boundary critical currents are enhanced up to T_C at all temperatures (see Fig. 14). Indeed, the electron holography measurements give direct evidence that Ca-doping greatly reduces the boundary potential in width and height.

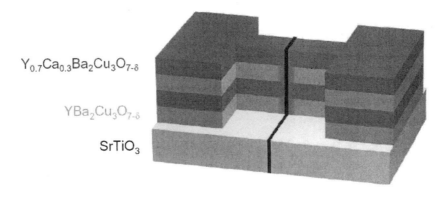

$Y_{0.7}Ca_{0.3}Ba_2Cu_3O_{7-\delta}$

$YBa_2Cu_3O_{7-\delta}$

$SrTiO_3$

Figure 13: Illustration of the selective doping of grain boundaries induced by grain boundary diffusion in doping heterostructures. [Ref. 71]

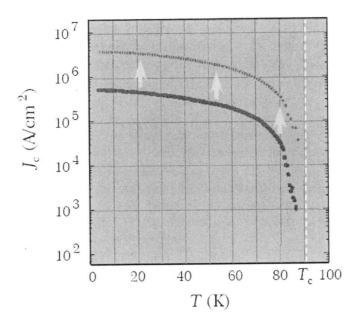

Figure 14: Temperature dependence of the critical current densities of 24° boundaries in a standard $YBa_2Cu_3O_{7-\delta}$ film (lower curve) and a $Y_{0.7}Ca_xBa_2Cu_3O_{7-\delta}$ / $YBa_2Cu_3O_{7-\delta}$ $Y_{0.7}Ca_xBa_2Cu_3O_{7-\delta}$ trilayer (upper curve). [Ref. 31]

8.4 INTERFACES IN HIGH-T_c CUPRATES – APPLICATIONS

The physics of interfaces and surfaces involving materials with correlated electron systems is highly complex, as described in chapter 3. Their properties are difficult to understand and even more difficult to optimize and to control. Yet, such interfaces are critical for many applications. The optimization of interfaces is a burning issue in high-T_C superconductivity. In the following, two important examples shall be discussed.

8.4.1 Josephson Junctions

In the first years of high-T_C superconductivity it was impossible to fabricate useful Josephson junctions. The underlying problems were ascribed to the short coherence length effects and to the cuprates' complex structure and chemistry. A good Josephson junction requires a superconductor to abut the junction's interface with a large superconducting energy gap. This gap is reduced if the coherence length of the superconductor is short, even more so if close to the interface the superconductor is chemically or structurally disturbed. On top of this, it seemed technologically impossible to grow multilayers of the oxide superconductors in high quality, as required for planar tunnel junctions, the successful junction technology for the classical low-T_C superconductors. As insurmountable as these problems seem, they were partially solved or circumvented by taking advantage of the special fundamental properties of the cuprates.

First, although the coherence length argument is obviously correct, it turned out that in all technologically relevant cuprates the value of the gap is so large that for applications based on superconducting quantum interference devices (SQUIDs) a substantial gap reduction can be tolerated. Second, the epitaxy of complex oxide films progressed vigorously. Now, epitaxial cuprate heterostructures are grown by sputtering, thermal evaporation, and by pulsed laser deposition, a technology that itself advanced significantly by meeting the challenges of high-T_C superconductivity. Third, novel Josephson junctions such as bicrystal and step-edge junctions were invented which utilize the strong correlations, large anisotropy, and tunable carrier density of the cuprates and do not use as active elements c-axis oriented tunnel barriers grown by deposition processes (see Fig. 15).

Combining the advanced epitaxial techniques and the novel Josephson junctions with optimized noise rejection schemes and tailored device designs, robust and low noise SQUIDs are now fabricated routinely.[9,73,74] These SQUIDs achieve at 77 K white noise levels of $\approx$10-50 fT/$\sqrt{\text{Hz}}$, exceeding only marginally the noise of commercial Nb-SQUIDs (a few fT/√Hz). As high-T_c SQUIDs can be operated outside shielded rooms, first commercial applications have been realized to provide functions that cannot be accomplished otherwise. Products include scanning SQUID microscopes to image magnetic fields and electric currents; for example they are used to localize wiring defects in chip packages (summarized in ref. 9), and systems for geological prospecting.[28]

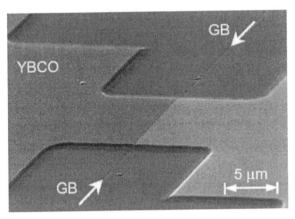

Figure 15: SEM micrograph of a bicrystalline Josephson junction. A YBa$_2$Cu$_3$O$_{7-\delta}$ film patterned by ion-beam etching into a superconducting bridge, crosses a 8° grain boundary (GB)

Further, SQUID-based sensors offer excellent performance for a wide spectrum of applications in biomagnetism, in non-destructive evaluation, and in magnetometry. Which of these devices will evolve into profitable products remains to be seen. Cardiography is an example where SQUID systems may prove to be especially beneficial, as SQUIDs can measure spatially resolved the magnetic fields generated by the heart. Here again, advancements in the device technology enable operation in standard clinical environments outside shielded rooms (see Fig. 16). For overviews see refs. 9, 53, 75 and more extended in refs. 30, 73.

8.4.2 Superconducting Cables

At first, the discovery of superconductors with transition temperatures well exceeding the boiling temperature of liquid nitrogen seemed to solve all problems on the way to energy efficient cables. It turned out, however, that the critical current density of the standard polycrystalline high-T_c superconductors is much too small for applications. For cost-effective

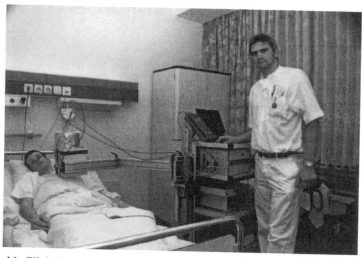

Figure 16: Clinical magnetocardiography test performed by the University of Jena. The magnetometer close at the patient's chest uses a bicrystalline SQUID sensor operated at 77 K to measure the time dependent magnetic fields generated by the heart currents. Figure courtesy of Prof. P. Seidel, University of Jena, Germany.

transmission cables, for motors, generators and transformers current densities of 10^4- 10^5 A/cm^2 are required, in some cases in fields up to 5 T, at the operation temperature of preferably 77 K. These current densities are engineering current densities, i.e. they refer to the complete crossection of the conductor. However, the critical current density of standard, polycrystalline $YBa_2Cu_3O_{7-\delta}$ pellets in self field equals only a few hundred A/cm^2 at 77 K. The bicrystal experiments described revealed that the critical current density of conventional polycrystalline high-T_c superconductors is limited by the grain boundaries.[2,76] For all high-T_c superconductors for which such measurements have been performed, including $ReBa_2Cu_3O_{7-\delta}$ (Re = rare earth) and the BSCCO compounds $Bi_2Sr_2CaCu_2O_{8+x}$ and $Bi_2Sr_2Ca_2Cu_3O_{10+x}$, the critical current density of grain boundaries decreases exponentially as a function of grain boundary angle (for an overview see ref. 9). How can cables with useful critical currents be built? The first option is to engineer grains with large aspect ratios, so that large effective grain boundary areas are obtained.[77] This is indeed possible for BSCCO cables,[78] since BSCCO forms platelet-like grains, which naturally yield boundaries with huge areas (see Fig. 17). This effect is used in all BSCCO wires fabricated today to achieve critical current densities of currently up to $\approx$ 4-5*10^4 A/cm^2, corresponding to engineering critical current density of $\approx$ 2*10^4 A/cm^2 ($Bi_{1.8}Pb_{0.3}Sr_2Ca_2Cu_3O_{10+x}$, 77 K).[79] Unfortunately, in these wires the superconductor has to be embedded in silver sheets. Due to the silver, these cables seem to be useful only for applications for which costs are a secondary issue, such as large-scale ship motors. A 5 MW BSCCO motor operated at 30 K is being tested by the US Navy, and a 30 K motor with 36.5 MW is being built by American Superconductor. To set the scale: the Queen Mary 2 is powered by four 21.5 MW pods. The use of superconductors can cut the losses by two and the size and weight (several hundred tons) of such large motors by factors of 5 to 8. If used in power lines, superconductors are estimated to halve the net losses.

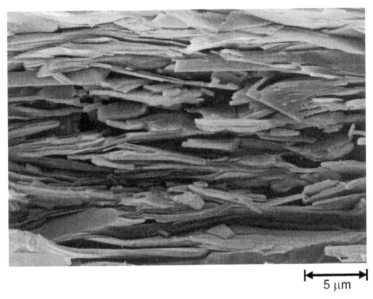

5 μm

Figure 17: SEM micrograph of a $Bi_2Sr_2Ca_2Cu_3O_{10+x}$ a wire made (longitudinal section) showing the platelet-like grains of the BSCCO compound.

The costly silver can be avoided by using $ReBa_2Cu_3O_{7-\delta}$ superconductors. A second advantage of these compounds is their relatively small anisotropy. Therefore thermally activated flux motion is much smaller than in the BSCCO compounds, such that the $ReBa_2Cu_3O_{7-\delta}$ superconductors can be used at 77 K in magnetic fields of several Tesla. Also for $ReBa_2Cu_3O_{7-\delta}$ the grain architecture is preferably optimized and, as described, the interfaces can be selectively doped to achieve large current densities (Hammerl *et al.*, 2002)[71]. Of course, the most obvious route to large critical currents is to align the grains along all axes, and indeed, the critical current density of polycrystalline $ReBa_2Cu_3O_{7-\delta}$ superconductors has been enhanced by orienting the grains, at present to FWHM angular spreads of 5° or better. The alignment raises the critical

274

current density because it ameliorates the detrimental correlation effects of band bending and the $d_{x^2-y^2}$-wave symmetry. It is achieved by growing the superconductor as films on buffered metallic carrier tapes, orienting the grains in the metallic tape or in the buffer layer. The first technique is known under the name of RABiTS (rolling assisted biepitaxially textured substrate), the second as IBAD (ion beam assisted deposition), or in a variant as ISD (inclined substrate deposition). The fabrication of aligned metallic tapes, currently Ni alloys, is a low-cost, mass-scale process. Since high-T_C films with 77 K current densities beyond $2*10^6$ A/cm^2 are now efficiently deposited on such tapes by non-vacuum processes such as dip- and spray-coating, this process will likely provide cost-effective superconducting cables (see Fig. 18). In the present summer of 2004

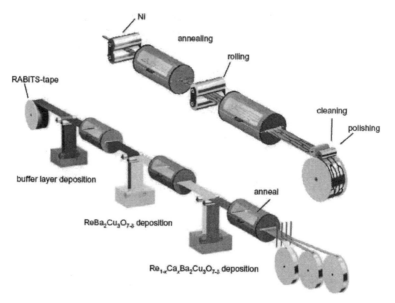

Figure 18: Sketch illustrating large scale production of ReBa2Cu3O7-δ -based RABiTS tapes using dip-coating. [Ref. 71]

coated conductor tapes with critical current densities exceeding $2*10^6$ A/cm^2 (J_e> 10^5 A/cm^2, corresponding to 250 A /cm of tape width) have been fabricated routinely, and thick IBAD layers have been produced that carry more than 1000 A/cm.. Conductors are fabricated with a length of about one hundred meter. Since fast and continuous deposition and scalable processes are being established, it will be possible to fabricate the desired length of kilometer in the next years, in the long run possibly at costs below the equivalent costs of a copper-wire, which ranges at $\approx$ 20 €/kAm! It is amazing to see how intimately these large scale applications are connected to the fundamental, at best barely understood, problems of the physics of interfaces in correlated electron described in chapter 2 systems (see also ref. 53). Ted Geballe summarized it crisply: "Who, besides a cable company should care about the symmetry of a macroscopic order parameter?". For an overview see Malozemoff et al., 2005.[75]

8.5 SUMMARY

Interfaces in compounds with correlated electronic systems are characterized by highly unusual electronic and magnetic properties. While such interfaces and surfaces offer great potential for the growth of materials or devices with outstanding properties, their fundamentals are only vaguely understood. In high-temperature superconductivity, these interfaces are having a direct impact on the realization and costs of large scale applications, such as high power cables, generators or superconducting ship motors. The design and optimization of interfaces in correlated electron systems is an enticing and rewarding challenge.

ACKNOWLEDGEMENTS

Parts of this contribution are based on previous articles written together with H. Hilgenkamp, A. Malozemoff and D. Scalapino. The author thanks them for the collaborations and V. Eyert, D. Khomskii, T. Kopp, A. Malozemoff, K. Ohmatsu, C. W. Schneider and D. Vollhardt for helpful comments and extensive discussions. I thank G. Hammerl, A. Herrnberger , and K. Wiedenmann for fabricating several of the figures, as well as E. Saladie and J. Askin for their help in writing and editing. This work was supported by the BMBF (project No. 13N6918), by the DFG via the SFB484 and by the ESF via the THIOX programme.

REFERENCES

1 I. Zutic, J. Fabian, and S. Das Sarma, *Rev. Mod. Phys.* **76**, 323 (2004).
2 D. Dimos, P. Chaudhari, J. Mannhart and F.K. LeGoues, *Phys. Rev. Lett.* **61**, 219 (1988).
3 J. Mannhart and H. Hilgenkamp, *Physica C* 317-318, 383 (1999).
4 Z.G. Ivanov, P.A. Nilsson, D. Winkler, A.J. Alarco, T. Claeson, E.A. Stepantsov and A. Ya Tzalenchuk, *Appl. Phys. Lett.* 59, 3030 (1991).
5 H. Hilgenkamp and J. Mannhart, *Appl. Phys. Lett.* **73**, 265 (1998).
6 N.F. Heinig, R.D. Redwing, J.E. Nordman and D.C. Larbalestier, *Phys. Rev. B* **60**, 1409 (1999).
7 D.T. Verebelyi, C. Cantoni, J.D. Budai, D.K. Christen, H.J. Kim and J.R. Thompson, *Appl. Phys. Lett.* **78**, 2031 (2001).
8 K.A. Delin and A.W. Kleinsasser, *Supercond. Sci. Technol.* **9**, 227 (1996).

[9] H. Hilgenkamp and J. Mannhart, *Rev. Mod. Phys.* **74**, 485 (2002).

[10] A.A. Golubov, M.Yu Kupriyanov and E. Il'ichev, *Rev. Mod. Phys.* **76**, 411 (2004).

[11] J. Mannhart and H. Hilgenkamp, *Mat. Science and Eng. B* **56**, 77 (1998).

[12] K. Maiti and D.D. Sarma, *Phys. Rev. B.* **61**, 2525 (2000).

[13] S. Altieri, L.H. Tjeng and G.A. Sawatzky, *Thin Sol. Films* **400**, 9 (2001).

[14] A. Liebsch, *Phys. Rev. Lett.* **90**, 096401-1 (2003).

[15] A. Sekiyama, H. Fujiwara, S. Imada, S. Suga, H. Eisaki, S.I. Uccida, K. Takegahara, H. Harima, Y. Saitoh, I.A. Nekrasov, G. Keller, D.E. Kondakov, A.V. Kozhevnikov, Th. Pruschke, K. Held, D. Vollhardt and V.I. Anisimov, *Phys. Rev. Lett.* **93**, 156402 (2004)

[16] S. Okamoto and J. Millis, *Nature* **428**, 630 (2004).

[17] J. Maier, *Physical Chemistry of Ionic Materials. Ions and Electrons in Solids.* John Wiley and Sons (2004).

[18] M. Lippmaa, T. Ohnishi, K. Shibuya, N. Nakagawa, H. Kunigashira, D. Kobayashi, R. Hashimoto, A. Chikamatsu, M. Oshima, H. Wadati, A. Fujimori, M. Kawasaki and H. Koinuma. *Presentation at the CERC/ERATO-SSS Workshop*, Maui Island, October 1-4 (2003).

[19] D.M. Duffy and A.M. Stoneham, *J. Phys. C: Solid State Phys.* **16**, 4087 (1983).

[20] A. Liebsch, *cond-mat. 0301537:1* (2004).

[21] D. Vollhardt, *private communications* (2004).

[22] I.A. Nekrasov, G. Keller, D.E. Kondakov, A.V. Kozhevnikov, Th. Pruschke, K. Held, D. Vollhardt and V.I. Anisimov. *cond-mat. 0211508:1* (2004).

[23] A. Ohtomo and H.Y. Hwang. *Nature* **427**, 423 (2004).

[24] J.K. Freericks, B.K. Nikolic and P. Miller, *IEEE Trans. Appl. Supercond.* **13**, 1089 (2003).

[25] J.K. Freericks and B.K. Nikolic, *Appl. Phys. Lett.* **82**, 970 (2003).

[26] J.K. Freericks, B.K. Nikolic and P. Miller, *Phys. Rev. B* **64**, 054511 (2001)

[27] C.W. Schneider, S. Hembacher, G. Hammerl, R. Held, A. Schmehl, A. Weber, T. Kopp and J. Mannhart, *Phys. Rev. Lett.* **92**, 257003-1 (2004).

[28] C.P. Foley, *Presentation at the International Superconducting Electronics Conference*, Sidney, 7-11 July, (2003). Available at: http://www.tip.csiro.au/ISEC2003/talks/ITh4.pdf

[29] V. Pizzella, S. Della Penna, C. Del Gratta and G.L. Romani, *Supercond. Sci. Technol.* **14**, 79 (2001).

[30] H. Itozaki, K. Sakuta, T. Kobayashi, K. Enpuku, N. Kasai, Y. Fujinawa, H. Iitaka, K. Nikawa and M. Hidaka, "Applications of HTSC SQUIDs." In *Vortex Electronics and SQUIDs*, T. Kobayashi, H. Hayakawa, M. Tonouchi (Eds.). ed. Berlin Heidelberg: Springer-Verlag, (2003).

[31] G. Hammerl, A. Schmehl, R.R. Schulz, B. Goetz, H. Bielefeldt, C.W. Schneider, H. Hilgenkamp and J. Mannhart, *Nature* **407**, 162 (2000).

[32] H. Yamada, Y. Ogawa, Y. Isii, H. Sato, M. Kawasaki, H. Akoh and Y. Tokura, *Science* **305**, 646 (2004).

[33] K.S. Takahashi, M. Kawasaki and Y. Tokura, *Appl. Phys. Lett.* **79**, 1324 (2001).

[34] H.Y. Hwang, S.-W. Cheong, N.P. Ong and B. Batlogg, *Phys. Rev. Lett.* **77**, 2041 (1996).

[35] A. Gupta, G.Q. Gong, X. Gang, R.R. Duncombe, P. Lecoeur, P. Trouilloud, Y.Y. Wang, V.P. Dravid and J.Z. Sun, *Phys. Rev. B.* **54**, R15629 (1996).

[36] N.D. Mathur, G. Burnell, S.P. Isaac, T.J. Jackson, B.S. Teo, J.L. MacManus Driscoll, L.F. Cohen, J.E. Evetts and M.G. Blamire, *Nature* **387**, 266 (1997).

[37] R. Gunnarsson, A. Kadigrobov, and Z. Ivanov, *Phys. Rev. B.* **66**, 024404-1 (2002).

[38] M.G. Blamire, C.W. Schneider, G. Hammerl and J. Mannhart, *Appl. Phys. Lett.* **82**,

2670 (2003).

39 S.A. Köster, V. Moshnyaga, K. Samwer, O.I. Lebedev, G. van Tendeloo, O. Sahpoval and A. Belenchuk, *Appl. Phys. Lett.* **81**, 1648 (2002).

40 V. Körting, Q. Yuan, P.J. Hirschfeld, T. Kopp and J. Mannhart, *Phys. Rev. B* **71**, 104510 (2005).

41 C.H. Ahn, J.M. Triscone and J. Mannhart, *Nature* **424**, 1015 (2003).

42 J. Mannhart, *Supercond. Sci. Technol.* **9**, 49 (1996).

43 D.M. Newns, J.A. Misewich, C.C. Tsuei, A. Gupta, B.A. Scott and A. Schrott, *Appl. Phys. Lett.* **73**, 780 (1998).

44 T. Koida, M. Lippmaa, T. Fukumura, K. Itaka, Y. Matsumoto, M. Kawasaki and H. Koinuma, *Phys. Rev. B.* **66**, 144418-1 (2002).

45 G. Deutscher and K.A. Müller, *Phys. Rev. Lett.* **59**, 1745 (1987).

46 G. Deutscher and R.W. Simon, *J. Appl. Phys.* **69**, 4137 (1991).

47 J. Mannhart, H. Bielefeldt, B. Goetz, H. Hilgenkamp, A. Schmehl, C.W. Schneider and R. Schulz, *Phil. Mag. B.* **80**, 837 (2000).

48 D.J. Van Harlingen, *Rev. Mod. Phys.* **67**, 515 (1995).

49 C.C. Tsuei and J.R. Kirtley, *Rev. Mod. Phys.* **72**, 969 (2000).

50 K.A. Müller, *Phil. Mag. Lett.* **82**, 279 (2002).

51 M. Sigrist and T.M. Rice, *Rev. Mod. Phys.* **67**, 503 (1995).

52 S. Yu Barash, A.V. Galaktionov and A.D. Zaikin, *Phys. Rev. B.* 52, 665 (1995).

53 J. Mannhart and P. Chaudhari, *Physics Today* Nov.48-53 (2001).

54 J. Mannhart, H. Hilgenkamp, B. Mayer, Ch. Gerber, J.R. Kirtley, K.A. Moler and M. Sigrist, *Phys. Rev. Lett.* **77**, 2782 (1996).

55 R.G. Mints, *Phys. Rev. B* **57**, 3221 (1998).

56 R.G. Mints, I. Papiashvili, J.R. Kirtley, H. Hilgenkamp, G. Hammerl and J. Mannhart, *Phys. Rev. Lett.* 89, 067004-1 (2002).

57 H. Hilgenkamp and J. Mannhart, *Appl. Phys. A* **64**, 553 (1997).

58 S.R. Bahcall, *Phys. Rev. Lett.* **76**, 3634 (1996).

59 M. Fogelström, D. Rainer and J.A. Sauls, *Phys. Rev. Lett.* **79**, 281 (1997).

60 C.W. Schneider, W.K. Neils, H. Bielefeldt, G. Hammerl, A. Schmehl, H. Raffy, Z.Z. Li, S. Oh, J.N. Eckstein, J. Mannhart and D.J. van Harlingen, *Eur. Phys. Lett.* **64**, 489 (2003).

61 S. Yu Barash, *Phys. Rev. B* 61, 678 (2000).

62 C.-R. Hu, *Phys. Rev. Lett.* **72**, 1526 (1994).

63 J. R. Kirtley, C.C. Tsuei, J.Z. Sun, C.C. Chi, L.S. Yu-Jahnes, A. Gupta, M. Rupp and M.B. Ketchen, *Nature* 373, **225** (1995).

64 Y. Tokura and A. Takahisa, *Jap. Journal Appl. Phys.* **29**, 2388 (1990).

65 J. Mannhart, A. Kleinsasser, J. Ströbel and A. Baratoff, *Physica C* **216**, 401 (1993).

66 N.D. Browning, J.P. Buban, P.D. Nellist, D.P. Norton, M.F. Chisholm and S.J. Penny-cook, *Physica C* **294**, 183 (1998).

67 M.A. Schofield, L. Wu and Y. Zhu, *Phys. Rev. B* **67**, 224512-1 (2003).

68 M.A. Schofield, M. Beleggia, Y. Zhu, K. Guth and C. Jooss, *Phys. Rev. Lett.* **92**, 195502-1 (2004).

69 J.H. Ransley, P.F. McBrien, G. Burnell, E.J. Tarte, J.E. Evetts, R.R. Schulz, C.W. Schneider, A. Schmehl, H. Bielefeldt, H. Hilgenkamp and J. Mannhart, *Phys. Rev. B*, in press (2004).

70 A. Schmehl, B. Goetz, R.R. Schulz, C.W. Schneider, H. Bielefeldt, H. Hilgenkamp and J. Mannhart, *Eur. Phys. Lett.* **47**, 110 (1999).

71 G. Hammerl, A. Herrnberger, A. Schmehl, A. Weber, K. Wiedenmann, C.W. Schneider and J. Mannhart, *Appl. Phys. Lett.* **81**, 3209 (2002).

72 A. Weber, G. Hammerl, A. Schmehl, C.W. Schneider, Mannhart J., B. Schey, M. Kuhn, R. Nies, B. Utz and H.-W. Neumueller, *Appl. Phys. Lett.* **82**, 772 (2003).

[73] D. Koelle, R. Kleiner, F. Ludwig, E. Dantsker and J. Clarke, *Rev. Mod. Phys.* **71**, 631 (1999).

[74] K. Enpuku, S. Kuriki and S. Tanaka, "High-T_c SQUIDs." In *Vortex Electronics and SQUIDs*, T. Kobayashi, H. Hayakawa, M. Tonouchi (Eds.). ed. Berlin Heidelberg: Springer-Verlag (2003).

[75] A.P. Malozemoff, J. Mannhart and D. Scalapino, *Physics Today* (to be published)

[76] D. Dimos, P. Chaudhari and J. Mannhart, *Phys. Rev.* B **41**, 4038 (1990).

[77] J. Mannhart and C.C. Tsuei, *Z. Phys. B* **77**, 53 (1989).

[78] K. Heine, J. Tenbrink and M. Thöner, *Appl. Phys. Lett.* **55**, 2441 (1989).

[79] R.M. Scanlan, A.P. Malozemoff and D.C. Larbalestier, Submitted to *IEEE Proceedings* (2004).

CHAPTER 9

ELECTRONIC RECONSTRUCTION AT SURFACES AND INTERFACES OF CORRELATED ELECTRON MATERIALS

A.J. Millis

Department of Physics, Columbia University
538 West 120th Street, NY NY 10027 USA

9.1 INTRODUCTION

The intellectual challenges posed by "correlated electron" materials (those classes of compounds which exhibit interesting and unusual electronic behavior such as "colossal" magnetoresistance, high temperature superconductivity, magnetic, charge and orbital order, "heavy fermion" behavior, quantum criticality, etc) are a central focus of condensed matter physics. It is generally accepted that the unusual behavior is due in large part to strong electron-electron interactions, whose effects cannot be adequately represented by density functional band theory techniques. Over the past decade our understanding of the bulk properties of correlated electron materials has dramatically improved: for reviews, see e.g. ref. 1 and 2. A fundamental observation has been that in bulk, 'correlated electron' materials exhibit a wide range of novel phases, involving various permutations of metallic and insulating behavior, and of superconducting, magnetic, charge and orbital order. It is argued here that the time is now ripe for a systematic investigation *correlated electron surface/interface science.* , i.e. of the changes in correlated electron behavior which may occur near the surface of a correlated electron compound or in the vicinity of an interface between a correlated electron material and another material (either strongly correlated or not). It is further argued that just as the fundamental question in traditional surface science is of surface reconstruction", in other words, of how atomic positions differ on the surface relative to bulk, so the fundamental question of correlated electron surface/interface science is *electronic reconstruction:* how the rich and interesting electronic phase behavior characteristic of the correlated electron materials changes near a surface or interface.[3]

The issue of surface and interface induced changes in correlated electron behavior is of fundamental interest as a basic question in condensed matter physics and materials science, but is also important from other points of view. The possibility of exploiting the unusual properties of correlated electron systems to make novel devices has intrigued workers for many years; for example, an important motivation for the study of 'colossal' magnetoresistance manganites has been the possibility of exploiting the very high degree of spin polarization in some kind of "spin valve" device.[4] However, almost any prospective device involves an interface through which electrons pass; an understanding of the factors controlling the near-interface behavior of correlated electron compounds is therefore essential to rational design and optimization of possible devices. Similarly, there has been much discussion recently of "correlated electron nanostructures" and "correlated electron nanoparticles". But by definition, nanostructures and nanparticles possess a very large surface to volume ratio, so their electronic properties are likely to be determined by surface effects.

There are three main classes of 'correlated electron' materials: transition metal oxides, where the key physics is a strong repulsion leading (in appropriate circumstances) to Mott or charge transfer insulating behavior and/or to magnetism; heavy fermion compounds, where the key physics is a carrier spin interaction which leads via the Kondo effect to the formation of heavy-mass quasiparticles, and the organic compounds, where low dimensionality leads to the enhancement of relatively weak interactions. In this review the focus will be mainly on transition metal oxides, because an impressive body of controlled experimental data is accumulating and interesting effects have been revealed. Much less is reliably known about surfaces and interfaces of Becchgaard salts and other organic compounds displaying correlated electron behavior. The available data suggest that surface effects on the fundamental aspects of heavy fermion physics are not strong,[5] while the exquisite sensitivity of the Kondo temperature to changes in parameters implies that effects of chemical nonstoichiometry and of other forms of disorder will cause considerable practical complications.

Surface science is a very well developed subject, and much important work on surfaces of transition metal oxides has been done.[6] The field of correlated electron surface science is however still in its infancy. Only recently has systematic experimental attention been focussed on the surface or interface-induced changes in the novel electronic properties which make the correlated electron materials so interesting,[7-10] and only recently have appropriate theoretical techniques[11-13] become available. In this article we summarize what has been accomplished so far, and outline a program for future progress. The rest of the article is organized as follows. We first list the factors which control the correlated electron phenomena of interest and provide (where available) experimentally realized examples. We then outline the theoretical tools available and describe some of the recent results, and give a brief conclusion.

9.2 CONTROLLING FACTORS

The central novel physics of transition metal oxide compounds is the 'Mott' metal insulator transition occurring at commensurate band fillings and associated with magnetic, charge and orbital order. The basic questions, therefore, are how the Mott insulating behavior differs at the surface from the bulk, and whether the associated magnetic, charge and orbital ordering patterns (if any) are the same at a surface or interface as they are in the bulk, or different.

Important factors controlling the Mott physics include carrier concentration, relative strength of interaction and band-width, orbital degeneracy and disorder. These are changed by proximity to a surface or interface. Some key issues include:

• *Strain fields,* induced by lattice mismatch at an interface or by reconstruction at a surface: these may extend a long distance from a surface or an interface, and act to lift orbital degeneracy and also (although this may be a less significant effect) to change hopping amplitudes.

• *Interdiffusion* of atoms across an interface may be an important source of disorder: expecially in heterostructures composed of different types of transition metal oxides, the chemical similarity makes it difficult to prevent interdiffusion, while the different physical properties associated with different transition metals implies that the associated 'disorder' will be very large.

• *Atomic reconstructions*: electronic band-width is controlled by the interatomic distance and by the geometrical arrangement of atoms, both of which may change because the structure of a surface (or sometimes an interface) may be reconstructed relative to bulk, while interaction strength is strongly affected by the polarizability of surrounding ions which may also change near a surface or an interface.

• *Surface charge layers,* due for example to the need to compensate a polar surface will lead to substantial band-bending and thus to near-surface charge variation.

• *Leakage of carriers* across an interface from one material to another also implies local variation of charge density.

In the following section we summarize recent experimental work relating to each of these phenomena. The identification of well-controlled experimental systems which isolate (to the extent possible) one factor or the other is important, because many factors may contribute to a difference in behavior between surface and bulk, and what is most important in practice may not always be clear *a priori*.

9.2.1 Strain

Strain Growth of thin (10–100 *nm*) films and heterostructures of 'interesting' oxides on (relatively) inert substrates is an important activity pursued by many groups world-wide. Most suitable substrates have a signficant (~ 1–2%) lattice mismatch with material of interest. This mismatch strains the material of interest, and research over the last five years has made it clear that this strain can strongly affect physical properties, especially of materials, such as the "colossal" magnetoresistance manganites, in which orbital degeneracy plays an important role. Early evidence of the importance of anisotropic strain was provided by the author in collaboration with A.Migliori and T. M. Darling,[14] who showed via theoretical calculations and ultrasound experiments that an infinitesimal anisotropic strain strongly suppressed the ferromagnetic transition temperature of manganite materials. Subsequently, the University of Maryland group[15,16] and, independently, the University of Tokyo group[8] showed that 1–2% substrate-induced *compressive* strain could actually change the ground state of $La_{0.7}Ca_{0.3}MnO_3$ from metallic to insulating, while the University of Tokyo group[8] also showed that a similar level of *tensile* strain had no such effect. The length scales relevant for strain effects in manganite films appear to be of the order of 2–400Å; beyond this length one form of structural defect or another appears, allowing the substrate-induced strain to heal.

The theoretical understanding of these results is not complete. Fang and colleagues[17] presented local spin density approximation band calculations which indicated that compressive strain of this order of magnitude could change the magnetic state from ferromagnetic to antiferromagnetic. However, it is not clear how antiferromagnetism in a partly filled band could produce the observed insulating behavior. In bulk manganite materials, insulating behavior is now known to be due to charge and orbital order (which often, although not always, leads to antiferromagnetism as well). Biswas and co-workers[15,16] showed that their compressively strained films exhibited a field driven insulator-metal transition with transport properties very similar to those observed in bulk materials with charge and orbital ordering. These results suggest that strain induces a charge and orbitally ordered insulating state, but this state has not yet been directly detected and theoretical explorations of the possibility are lacking.

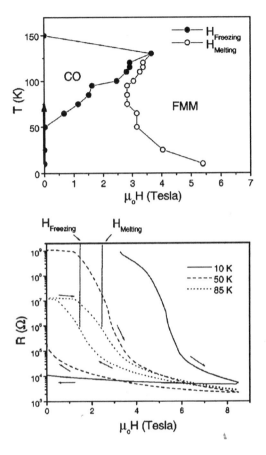

Figure 1: **Upper panel**: Phase diagram, obtained on the basis of magnetic field and temperaturate dependent resistivity measurements such as those shown in the lower panel, for strained thin film of $La_{0.7}Ca_{0.3}MnO_3$ which in bulk does not exhibit a charge ordered phase. [Ref. 16]

9.2.2 Interdiffusion

Interdiffusion of ions across an interface is a fundamental difficulty in the preparation of heterostructures. The situation is exacerbated in the case of transition metal oxide compounds by the relative similarities in size and chemical bonding tendencies of different transition metals. For example, fabrication of high quality bulk 'double-perovskite' samples of materials such as Sr_2FeMoO_6 (a high Curie-temperature 'half-metal' which could be very promising for 'spintronic applications'[18,19,20]) is severely complicated by the tendency towards mis-site disorder[21] (*Fe* sitting where *Mo* should and vice-versa). This dramatically degrades the properties and is very difficult to prevent even though *Fe* is in the 3*d* series and *Mo* is in the 4*d* series and in a different column of the periodic table.

In recent years the Kawaski group at the University of Tokyo, for example, has undertaken a systematic study of many transition metal oxide superlattices (see, e.g. ref. 22) and has presented evidence that interdiffusion of transition metal ions across interfaces fundamentally affects the properties when layers are thinner than about 10Å. Experimentalists however are making continuing impressive progress on the difficult problem of controlling growth on the atomic scale, and it seems likely that interdiffusion will become less of an issue in the future.

9.3 ATOMIC RECONSTRUCTION AND OTHER CHANGES NEAR A SURFACE

Atoms near a surface or interface experience a different local environment than those in the bulk of a material, with evident consequences for near-surface electronic behavior. One obvious effect is that the coordination of atoms near the surface is lower than in bulk. The concomitant reduction of net carrier delocalization will tend to increase the effects of correlations. Photoemission studies of $Li/CaVO_3$[23] and V_2O_3 [24,25] have been interpreted in this way.[12,26,27] The effects may be expected to be relatively subtle because the relative change in coordination is often small (1/6 for the (001) surface of a cubic lattice, for example). For this reason some of the theoretical works[27] have argued that a 'fine tuning' of a interaction parameter to close proximity to the Mott critical value is necessary to understand the data.

Other effects however also may occur. As noted by Duffy and Stoneham[28] and verified, extended and emphasized by G. A. Sawatzky and collaborators[29], one important consequence of the change in local environment is a change in screening, which leads to a change in the effective 'Hubbard U' or 'charge transfer gap', which is the crucial interaction parameter[30] controlling the nature of the electronic phase. The change may become larger, for a transition-metal-oxide-normal metal interface (the case studied by Ref. 29) or smaller, for a transition metal oxide-vacuum interface.

The difference in local environment between a surface or interface and the bulk may lead to changes in positions of near-surface atoms. These changes may lead to a different lattice symmetry at the surface or simply to changes in interatomic distances and bond angles. In either case, changes in near surface electronic structure will result, changing the interplay between bandwidth and interaction, and possibly changing the electronic phase. For example, Ref. 31 reports photoemission evidence for a correlation-driven *charge disproportionation* at the surface of $CalSrVO_3$ which, it is suggested, may be due to a rotation of the VO_6 octahedra at the surface.

An important set of very recent experiments adds substantial insight on this point. A group at Oak Ridge National Labs[10] has studied surfaces of $Ca_{1.9}Sr_{0.1}RuO_4$, a material which in bulk undergoes 'Mott' metal to insulator transition when the temperature is reduced below $T_{MI,bulk} \approx 150K$, and have found that the surface remains metallic down to the lower temperature of $T_{MI,surface} \approx 125K$. The group also shows that at the surface the RuO_6 octahedra tilt less than in bulk; they suggest that this different tilt increases the electronic overlap and therefore the electronic bandwidth. An additional issue is that the distances between the Ru and the apical oxygens decrease, perhaps changing level energies and shifting the occupancy of electrons between different orbitals. Some of the experimental results of this group are summarized in Figure 2. A complete theoretical understanding of these results has not been achieved, but recent work[32] by S. Okamoto and the author has presented evidence that the physics is driven by changes in the $Ru–O$ breathing phonon frequency at the surface.

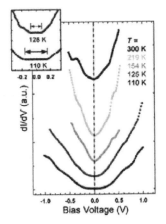

J. Zhang, E. W. Plummer et. al
Surface Metal-Insulator
Transition in $Ca_{0.19}Sr_{0.1}RuO_3$

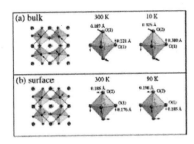

Figure 2: **Left panel**: STM tunnelling data demonstrating onset of surface insulating behavior (gap) only below T ≈ 125K in $Ca_{1.9}Sr_{0.1}RuO_3$ (bulk metal insulator transition at T ≈ 150K. **Right panel**: results of structural measurements showing rotation of RuO_6 octahedra at surface differs from that in bulk. Data courtesy of J. Zhang and E. W. Plummer, Oak Ridge National Laboratory.

9.3.1 Surface charge layers

One important class of surface (or interface) is the polar surface. Issues associated with polar surfaces are particularly important for strongly correlated transition metal oxides, many of which form in some variant of the simple ABO_3 perovskite structure. For many materials in this structure most of the obvious surface planes (for example, the (001) plane of $LaTiO_3$) are polar in the absence of significant reconstruction relative to bulk. While the conventional expectation is that the charge imbalance caused by a polar surface is compensated by atomic surface reconstruction (either a high degree of vacancies or adatoms, or by faceting so the surface is locally nonpolar), it has been proposed theoretically that instead an *electronic reconstruction* may occur, leading for example to a metallic layer at the surface of insulating ZnO.[33,34] This effect would certainly change correlated electron properties; for example, insulating phases are favored at commensurate densities, and metallic phases at incommensurate densities. G. Sawatzky and collaborators[7] have presented evidence that this behavior occurs at polar surfaces a correlated system (namely K_3C_{60}).

9.3.2 Charge leakage

At an interface between two different materials, leakage of charge density from one material to the other means that the electronic density near the interface is different from the bulk density in either material. This effect leads, of course, to the Schottky barrier physics essential to semiconductor junctions. It may also have important implications for correlated electron physics, because the behavior of correlated electron systems depends strongly on the carrier density. Kawasaki and collaborators have presented experimental evidence of the importance of this effect in the context of heterostructures grown of different magnetic transition metal oxides.[22]

Recent experimental work provides a dramatic illustration of the effects of charge leakage. Ohtomo, Muller, Gredul and Hwang have succeeded in fabricating digital heterostructures of $LaTiO_3/SrTiO_3$. $LaTiO_3$ is a 'Mott' insulator in which the formal valence is such that there is one d-electron per Ti and La is in the +3 state, whereas $SrTiO_3$ is a band insulator, with (in the formal valence language) no d-electrons per Ti and with Sr in the +2 state. These two materials have essentially identical crystal structures and lattice constants, and in a tour de force of film growth, Ohtomo et al succeeded in growing a wide range of heterostructures in which an arbitrary (small) number n of $LaTiO_3$ layers alternated with a different, arbitrary number m of $SrTiO_3$ layers.[9] An dark-field image of one of the heterostructures is shown in the left panel of Fig 3. Ohtomo et al were further able to measure the longitudinal and Hall electrical conductivities (some Hall results are shown in the right hand panel of Fig 3). For all heterostructures studied, metallic conduction in the plane of the heterostructure was observed, whereas the bulk material is insulating.

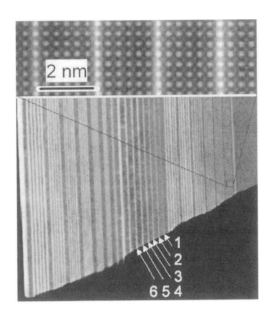

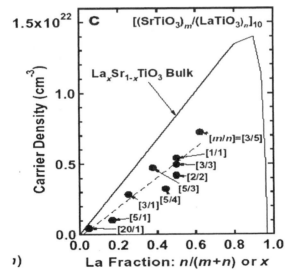

Figure 3: **Upper panel:** Dark field TEM image of *LaTiO₃-SrTiO₃* heterostructure grown by Ohtomo et. al. [Ref. 9] **Lower panel:** hall resistivity for different heterostructures, indicating metallic behavior; for all structures fabricated.

Finally, Ohtomo et al were able to use the electron energy loss mode of a transmission electron microscope to map out the distribution of *Ti* d-electrons. Signficant 'leakage' of the *Ti* d-electron density from the *LaTiO₃* region to the *SrTiO₃* region was observed; sample results are shown in Fig

288

4. It seems likely that this "leakage" is responsible for the observed metallic behavior.

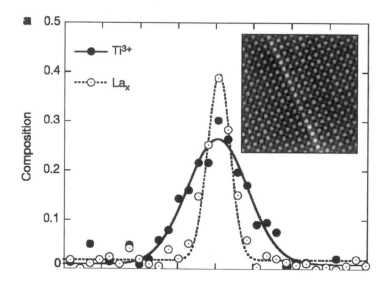

Figure 4: **Main panel:** charge distribution (filled circles) and *La*-ion distribution (open circles) measured by [Ref. 9] for lone $LaTiO_3$ layer in infinite $SrTiO_3$ system. **Inset:** transmission electron micrograph of heterostructure, showing plane of *La* ions as light dots.

9.4 THEORETICAL APPROACHES

In this section we summarize recent theoretical work relevant to the issue of surface and interface physics of transition metal oxides. The problem is challenging: strong correlations are difficult to treat, and the difficulties increase in the presence of a surface or interface. For bulk systems, two basic approaches exist: density functional based band theory calculations, and model-system based treatments. Very recently, hybrid approaches including 'LDA+U'[35] and LDA+'DMFT'[36,37] have been devised. Very recent work has begun to apply both techniques to surface and interface problems.

Density functional based approaches[38] are believed to treat very well the basic energetics of solids, and the effects of charge screening. For example, as is noted below we have obtained evidence from model system calculations that the basic structure of the 'charge leakage' part of the problem is insensitive to the 'strong correlations' aspect of the problem. Band theory methods are also very useful for relating atomic reconstruction and strain to changes in electron itineracy. However, band theory methods tend to associate insulating behavior with long ranged order, so are not as helpful for the Mott/charge-transfer aspects of the problem; also by construction they are optimized for ground state properties and are not as useful for many aspects of experimental interest such as excitation spectra, transport properties or $T > 0$ properties.

Local density approximation band theory methods have been used to study the effect of strain on the phase diagram of 'colossal' magentoresistance manganite materials.[17] It is found that compressive strain favors antiferromagnetic states and tensile strain ferromagnetic ones, in qualitative accord with experiment. However, the staggered charge and orbital order which Biswas et. al. argue[15,16] occurs in compressively strained films was not studied. Fang and co-workers also used LDA and LDA+U methods to study the surface of the $La_{1-x}Sr_xMnO_3$ system [40], finding that lattice distortions at the surface can change the electronic phase, leading to antiferromagnetism in a system in which the bulk is ferromagnetic. While this prediction has not been confirmed experimentally (perhaps in part because it is not easy to obtain a high quality surface of a pseudocubic perovskite manganite) the result is conceptually important as a demonstration that a different electronic phase may occur on a surface than in bulk. These researchers applied similar methods to the surface of $Sr_{2-x}Ca_xRuO_4$,[40] predicting that a surface-induced distortion of the RuO_6 octahedron would lead to a ferromagnetic phase. This prediction seems not to be borne out experimentally, A possible insight into the difficulty is provided by the Hartree-Fock calculations,[32] which suggest that alternative states (too expensive to investigate in a full band theory calculation) may be favored.

Very recently, generalized gradient approximation (GGA)[41] and local density (LDA) and LDA+U methods[42] have been applied to calculate the charge profile of the 'Hwang' $La/SrTiO_3$ heterostructure. The LDA results appear to agree well with the charge profiles previously calculated by applying the Hartree-Fock approximation to a tight binding model of the band theory;[39,46] the GGA methods indicate a very strong 'overscreening' of the La charges, leading to an unbinding of the charges from the La layer. One may expect much more work along these lines in the near future.

Model system based approaches amount to isolating a subset of relevant orbitals, treating this part of the band theory via a tight binding model, and adding various essentially phenomenologically chosen interactions. The simplicity of model systems relative to full band calculations allows the use of more sophisticated methods to treat the many body physics. For transition metal oxides the procedure seems likely to be reasonable because the relevant near fermi-surface bands are relatively narrow and relatively isolated from other bands.

A key issue in the model-system based approach is the value assigned to interaction parameters such as the 'Hubbard U' repulsion. It is difficult to determine the interactions *a' priori* , so calculations should scan a range of interaction parameters and should determine experimentally measureable quantities which will allow these parameters to be reliably estimated.

Analysis of model systems may be performed by numerical diagonalization (or quantum Monte Carlo analysis) of small clusters, by various static mean field methods, and by the 'dynamical mean field' approximation. In the special case of one dimensional systems a wide variety of techniques are available and have been elegantly exploited to study end effects in spin chains.[43] It seems very likely that the techniques can be applied to end effects in conducting chains. We suspect that the system sizes accessible via numerical diagonalization or Monte Carlo are in the main too small to be useful (although we note that in the particular case of carriers coupled only to classical spins, Calderon, Brey and Guinea have made an important study of the surface magnetism and its implications for spin transport across a barrier[44]). Therefore, it seems likely that the various static and dynamic mean field techniques hold the most promise for future work.

Hartree-Fock: The Hartree-Fock approximation, while far from a perfect representation of the many-body physics, provides a reasonable first look at the ground state phase diagram of a correlated system. From a technical point of view it amounts to minimizing an approximation to the ground state energy by solving a one-electron problem in the presence of a self consistently determined field, and in this sense has something in common with band theory calculations. It may be straightforwardly extended to spatially inhomogeneous situations (it was used, for example, to provide early evidence of 'stripes' in models of doped transition metal oxides[45]), and in particular the long-ranged Coulomb interaction may naturally be included.

The computational requirements of the Hartree-Fock approach are relatively modest; the needed calculations may be carried out on a PC. It is also flexible: in addition to a range of electronic interactions, the effects of lattice distortions may easily be included. Also, consideration of small fluctuations about the Hartree-Fock solution leads to collective modes. However that the Hartree-Fock method is known to overestimate the tendency to order.

Hartree-Fock calculations of the Hwang heterostructure[9] were undertaken by Okamoto and Millis.[39,46] These calculations revealed the existence of a roughly three unit cell wide 'transition regime over which the charge density changed from the value 1 per cell representative of bulk $LaTiO_3$ to the value 0 representative of bulk $SrTiO_3$. The precise form of the crossover and width of the crossover regime are controlled by the self-consistent screening of the long ranged Coulomb interaction, but the approximate 3 layer thickness is a robust result, occurring for a wide range of physically reasonable parameters. Representative results for the charge density profile are shown in Figure 5.

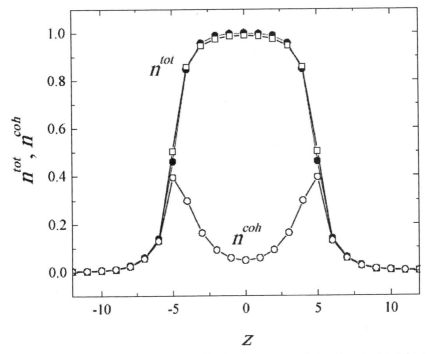

Figure 5: Layer dependent total charge density (upper curves) density calculated for ten-layer one-orbital "Hwang" heterostructure using Hartree-Fock (filled symbols) and dynamical mean field (open symbol) methods. Three unit cell transition region is visible. The two methods are seent o give essentially the same charge density. Lower curves: metallic quasiparticle density calculated using dynamical mean field method.

The key implication of this result is that in structures of the 'Hwang' type, there is an approximately three unit cell wide edge region in which the electronic behavior differs dramatically from the bulk behavior characteristic of either constituent. Thus only for systems in which the Mott insulator is more than six unit cells wide will the behavior begin to approach that of the bulk. This conclusions is reinforced by Hartree-Fock calculations of the ground state phase diagram,[39,46] which found that the Hartree-Fock phase diagram of a 'Hwang' heterostructure with n $LaTiO_3$ layers only reverted to the bulk (Hartree-Fock) behavior for $n > 10$. Very recent evidence[47] suggests that although the general finding (that the phase behavior changes in the vicinity of a surface or interface) is robust, the specific predictions of the Hartree-Fock approximation are likely to change substantially in a more accurate theory including the beyond-Hartree-Fock effects discussed below.

Beyond Hartree-Fock The Hartree-Fock approximation amounts to approximating an interacting problem by a self-consistently chosen noninteracting one. Quantum fluctuations about the self-consistently chosen state are not included, and the predictions for electron spectral functions and non-zero temperature are notoriously inaccurate. Mothods for including thermal and quantal fluctuations and for providing a more reasonable view of the excitation spectrum are needed.

A promising route to further progress is the 'dynamical mean field' method which stems from important work of Mueller-Hartmann and of Metzner and Vollhardt,[48] has been extended and applied in important ways by Jarrell[49] and co-workers, and has been developed to a high degree by Kotliar and co-workers.[50,11,13] The method is most simply presented by analogy to density functional band theory. Just as the various implementations of density functional theory may be regarded as approximations to an exact, but unknown, functional of the electron density, whose minimum at given external potential gives the ground state energy and charge density; the various[13,51,52] implementations of the dynamical mean field method may be regarded as approximations to an exact, but unknown, 'Luttinger-Ward' functional Φ of the electron self-energy, Σ whose functional derivatives at given external potential give all of the exact energies and response functions of the system. The key point, first enunciated in ref. 13 (we follow here the treatment in ref. 52) is that approximating the full electron self energy $\Sigma(r,r';\omega)$ as a sum of a finite number, N of functions of frequency:

$$\Sigma(r, r'; \omega) \approx \sum_{m=1...N} \phi_m(r, r')\Sigma_m(\omega) \tag{1}$$

corresponds to approximating the 'Luttinger-Ward functional' by a finite number N of coupled 'quantum impurity models': 0+1 dimensional field theories which are nontrivial but tractable. In Eq 1 the functions ϕ_m have a prescribed dependence on the space coordinate (see e.g. ref. 52 for details), the frequency dependent functions Σ_m have to be determined by solving the quantum impurity models, and the impurity models are fixed by a self-consistency condition obtained by relating the full (lattice) Green function $G(r,r';\omega)$, computed with the approximate self energy Eq 1, to the appropriate Green functions g_{imp} of the impurity models, using a relation specified by the functions ϕ_m.[52]

A variety of approximations to Σ have been introduced. The simplest and most widely used is to neglect the space dependence of Σ entirely, approximating $\Sigma \rightarrow \Sigma(\omega)$. This 'single-site' approximation becomes exact for a monatomic Bravais lattice in the limit of spatial dimensionality $d \rightarrow \infty$ and is in many cases believed to be quite accurate in $d=3$.[11] In this case there is only one term in the sum on the right hand side of Eq 1 and $\phi_m(r,r') \rightarrow 1$ so only a single quantum impurity model needs to be solved, and the self-consistency condition is $G_{lattice}(R=0,\omega)=\int[(d^3p)/((2\pi)^3)]\ [1/(\omega-\varepsilon_p-\Sigma(\omega))]=$ $g_{impurity}(\omega).^{11}$

The 'single-site dynamical mean field approximation' is widely studied, but many other choices are possible. These correspond to approximations which are uncontrolled (not exact in any limit), but seem likely to be reasonable (as with most mean-field theories). In particular, Potthoff and Nolting[12] have suggested treating surface (and, by extension, heterostructure) phenomena by giving each layer a different self energy:

$$\Sigma \longrightarrow \Sigma_n(\omega) \tag{2}$$

with n a layer index. Then each layer is described by a different quantum impurity model and the self consistency condition fixing the impurity model for layer n is

$$G_{lattice}(R_\| = 0, n; \omega) = \int \frac{d^2 p_\|}{(2\pi)^2} G(p_\|, n, n; \omega) = g_{impurity,n}(\omega)$$

Here $R_\|$ ($p_\|$) is the position (momentum) coordinate in the plane of the layer, n is the layer index and $g_{impurity,n}$ is the Green function corresponding to the impurity-model for layer n. $G(p_\|,n,n;\omega)$ is the projection onto layer n of the full Green function obtained by inverting the combination of the band Hamiltonian, the potential obtained from the (self-consistently screened) Coulomb interaction and the (layer-dependent) self-energy.

This procedure has been implemented by Potthoff and Nolting for the simple case of a one-band Hubbard model with a surface modelled by simply setting to zero the appropriate hopping matrix element,[12,25] and the approach was applied to a multiband model of the $CaVO_3$ system by Liebsch[27]: good agreement with photoemission measurements[23] was found, if the interaction was chosen to be near the critical value at which the Mott transition occurs. The correspondence between theory and experiment is at present incomplete because the other physical effects listed above have not been included in the theoretical calculations; but these effecs should be straightforward to add to the theory.

In another pioneering work, Freericks has used dynamical mean field methods to study electron transport through a simple model heterostructure.[54] These calculations directly relevant to any experimental system because the 'Falicov-Kimball' model is used, rather than a physically relevant interaction, and parameters are chosen so that the electron density does not change as one moves across the heterostructure (no charge leakage), but have produced interesting insights into transport.

294

The dynamical mean field procedure was applied by us to the single-orbital 'Hubbard model' version of the 'Hwang' heterostructure.[47] The key physical output is the layer resolved spectral function $A(z,z;\omega)=\int[(d^2k)/((2\pi)^2)] \, Im \, G(z,z;k,\omega)$. Sample results are shown in Fig. 6.

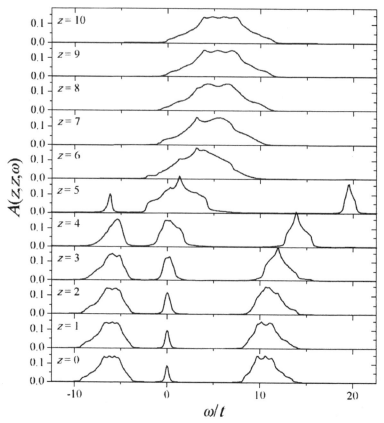

Figure 6: **Main panel:** Layer-resolved spectral function of ten-layer one-orbital 'Hwang' heterostructure, showing spatial evolution of many body density of states from the noninteracting form characteristic of the empty band (top of figure) to the upper and lower Hubbard band form characteristic of bulk Mott insulator (bottom of figure). The small near fermi surface density of states at the center of the heterostructure (bottom of figure) arises from the quantum mechanical tail of the quasiparticle states in the transition region.

One sees that far from the heterostructure that material has the spectral function of a band insulator (density of states takes the noninteracting form with chemical potential below the bottom of the band) whereas in the center of the heterostructure the density of states takes essentially the Mott insulating form (upper and lower Hubbard bands, with a very weak fermi surface feature due to quantum mechanical tailing of wave functions from the interface region, where reasonably robust metallic behavior (high fermi surface density of states) is seen to exist. Results have not yet been considered for the phase diagram, but preliminary indications are that the phase diagaram found using the dynamical mean field method is substantially different from that found previously using the Hartree-Fock.[39,46]

9.4 SUMMARY

The surface and interface behavior of correlated electron systems is emerging as an important topic in condensed matter physics and materials science. In this review we have outlined the main factors which may be expected to control the physics, have listed some apparently promising model systems which isolate one or another of these control parameters, and have summarized the theoretical 'state of the art'. We have identified the main issue as *electronic reconstruction*: the way in which the electronic phase at the surface or interface differs from that in bulk. We expect that the next few years will see rapid progress.

ACKNOWLEDGEMENTS

Much of the work reported here was performed in collaboration with Dr. Satoshi Okamoto. The author's work in this area has been supported by the US National Science Foundation through grants DMR-033876 and the University of Maryland-Rutgers MRSEC and by the US Department of Energy under grant ER-46169.

REFERENCES

1. M. Imada, A. Fujimori, and Y. Tokura Rev. Mod. Phys. **70**, 1039-1263 (1998).
2. See the articles in the special issue, vol **288** of *Science* magazine.
3. The phrase was apparently coined by G. Sawatzky, who applied it to the specific case of charge rearrangement at a polar surface. We suggest that this useful phrase should be applied more generally to cover all changes to electron phase behavior occuring near a surface or an interface.
4. See, e.g. A. Gupta and J. Z. Sun, J.M.M.M. **200** 24-43 (1999).
5. A. D. Chinchure, E. Munoz-Sandoval and J. Mydosh, Phys. Rev. **B64**. 020404/1-4 (2001).
6. C. Noguera, *Physics and chemistry at oxide surfaces*, (Cambridge University Press: Cambridge, 1996)
7. R. Hesper, L. H. Tjeng, A. Heeres, and G. A. Sawatzky Phys. Rev. **B62**, 16046-16055 (2000).
8. M. Izumi, Y. Ogimoto, Y. Konishi, T. Makano, M. Kawasaki and Y. Tokura, Mat. Sci. Eng. **B84** 53-7 (2001).
9. M. Ohtomo, D. A. B. Muller, D. Grezul and H. Y. Hwang, Nature **419** 378 (2002).
10. R. G. Moore, J. Zhang, S. V. Kalinin, A. P. Ismail, R. Baddorf, D. Jin, D. G. Mandrus, and E. W. Plummer, Phys. Stat. Sol. in press (2004).

11. A.Georges, B. G. Kotliar, W. Krauth and M. J. Rozenberg, Rev. Mod. Phys., **68**, 13 (1996).

12. M. Potthoff and W. Nolting Phys. Rev. B60, 7834-7849 (1999).

13. S. Savrasov and B. G. Kotliar, pps 259-301 in *New Theoretical Approaches to Strongly Correlated Systems*, A.M. Tsvelik Ed., (Kluwer Academic Publishers: 2001).

14. A. J. Millis, A. Migliori and T. Darling, Journal of Applied Physics 83 1588 (1998).

15. A. Biswas, M. Rajeswari, R. C. Srivastava, Y. H. Li, T. Venkatesan, R. L. Greene and A. J. Millis, Phys. Rev. **B61** 9665-8 (2000).

16. A. Biswas, M. Rajeswari, R. C. Srivastava, T. Venkatesan, R. L. Greene, Q. Lu, A. L. DeLozanne, and A. J. Millis, Phys. Rev. **B63** 184424/1-7 (2001).

17. Z. Fang, I. V. Solovyev, and K. Terakura Phys. Rev. Lett. **84**, 3169-3172 (2000).

18. A W Sleight, J. M. Longo and R. Ward, Inorg. Chem 1 245 (1962).

19. K. I. Kobayashi, T. Kimura, Y. Tomioka, H. Sawada, K. Terakura and Y. Tokura, Phys. Rev. **B59** 11159-62 (1998).

20. J. Gopalakrishnan, A. Chattopadhyay, S, B, Ogale, T. Venkatesan, R. L. Greene, A. J. Millis, K. Ramesha, B. Hannoyer and G. Marest, Phys. Rev. **B62** 9538-42 (2000).

21. Y. Tomioka, T. Okuda, Y. Okimoto, R. Kumai, K.-I. Kobayashi, and Y. Tokura, Phys. Rev. **B 61**, 422 (2000).

22. T. Koida M. Lippmaa, T. Fukumura, K. Itaka, Y. Matsumoto, M. Kawasaki and H. Koinuma Phys. Rev. **B 66**, 144418 (2002).

23. K. Maiti, P. Mahadevan, and D. D. Sarma, Phys. Rev. Lett. **80**, 2885 (1998).

24. Hyeong-Do Kim, J. H. Park, J. W. Allen, A. Sekiyama, A. Yamasaki, K. Kadono, S. Suga, Y. Saitoh, T. Muro, and P. Metcalf, cond-mat/0212110 (unpublished).

25. S. Schwieger, M. Potthoff, and W. Nolting Phys. Rev. **B67**, 165408 (2003).

26. A. Liebsch, cond-mat/0301418.

27. A. Liebsch, cond-mat/0301537

28. D. M. Duffy and A. M. Stoneham, J. Phys. **C16** 4087 (1983).

29. S. Altieri, C. J. Tjeng and G. Sawatzky, Thin Solid Films **400** 9-15 (2001) and Phys. Rev. **B61** 16948 (2001).

30. J. Zaanen, G. Sawatzky and P. W. Allen, Phys. Rev. Lett. **55** 418-21 (1985).

31. K. Maiti, D. D. Sarma, M. J. Rozenberg, I. H. Inoue, H. Makino, O. Goto, M. Pedio, and R. Cimino, Europhys. Lett. 55, 246 (2001).

32. S. Okamoto and A. J. Millis, cond-mat/ 0402267.

33. See, for example, section 3.3 of C. Noguera, *Physics and chemistry at oxide surfaces*, op. cit. (Cambridge University Press: Cambridge, 1996) and references therein.

34. A. Wander, F. Schedin, P. Steadman, A. Norris, R. McGrath, T. S. Turner, G. Thornton, and N. M. Harrison Phys. Rev. Lett. **86**, 3811-3814 (2001).

35. Vladimir I. Anisimov, Jan Zaanen, and Ole K. Andersen, Phys. Rev. **B 44**, 943-954 (1991).

36. S Savrasov, B. G. Kotliar and Elihu Abrahams, Nature **410** 793 (2001).

37. K. Held, G. Keller, V. Eyert, D. Vollhardt, and V. I. Anisimov Phys. Rev. Lett. **86**, 5345-5348 (2001).

38. See, e.g. R. M. Martin, *Electronic Structure: Basic Theory and Practical Methods*, Cambridge University Press 2004.

39. S. Okamoto and A. J. Millis, Nature **428**, 630 (2004).

40. Zhong Fang and K. Terakura, cond-mat/0103226.

41. D. R. Hamann (private communication) finds that at $T=0$ with a standard DFT-supercell calculation the ferroelectric tendency of $SrTiO_3$ is so strong that the Coulomb interaction becomes *overscreened* by the lattice, so that the electrons move away from the $LaTiO_3$ layers and exist in the bulk of the $SrTiO_3$! This finding suggests that a deeper analysis of the ferroelectric properties of $SrTiO_3$ is essential.

42. Z. S.. Popovic and S. Satpathy, cond-mat0409281.

43. S. Eggert and I. Affleck, Phys. Rev. Lett. **75** 934 (1995).

44. M. J. Calderón, L. Brey, and F. Guinea Phys. Rev. B 60, 6698-6704 (1999).
45. J. Zaanen and P. B. Littlewood Phys. Rev. **B 50**, 7222-7225 (1994).
46. Okamoto and A. J. Millis, cond-mat/0404275 (Phys. Rev. B (in press)).
47. S. Okamoto and A. J. Millis, cond-mat/0407592.
48. W. Metzner and D. Vollhardt, Phys. Rev. Lett. **62** 324 (1989); E. Mueller-Hartmann, Z. Phys. **74** 507 (1989) and U. Brandt and C. Mielsch, Z. Phys. **75** 365 (1989).
49. M. Jarrell Phys. Rev. Lett. **69**, 168-171 (1992); Th. Pruschke, D. L. Cox, and M. Jarrell Phys. Rev. **B 47**, 3553-3565 (1993).
50. A. Georges and G. Kotliar Phys. Rev. **B 45**, 6479-6483 (1992).
51. M. Potthoff, cond-mat/0306278.
52. S. Okamoto, A. J. Millis, H. Monien and A. Fuhrmann, Phys. Rev. B in press (2003) (cond-mat/0306178).
53. M. Potthoff Phys. Rev. B 64, 165114 (2001).
54. J. K. Freericks, cond-mat/0408225.

Wide Band Gap Semiconductors

CHAPTER 10

WIDE BAND GAP ZnO AND ZnMgO HETEROSTRUCTURES FOR FUTURE OPTOELECTRONIC DEVICES

R.D. Vispute*, S. S. Hullavarad, D.E. Pugel**, V. N. Kulkarni. S. Dhar, I. Takeuchi, and T. Venkatesan.

Center for Superconductivity Research, Department of Physics, University of Maryland, College Park, MD 21045, USA.

Also at Blue Wave Semiconductors, Inc, Columbia, MD 21045.

**Also at NASA Goddard Space Flight Center, Greenbelt, MD 20771, USA.*

10.1 INTRODUCTION

Oxide-based novel thin-films of homo- and hetero-structures are technologically attractive for future optoelectronic devices because of their exciting fundamental intrinsic and extrinsic optical, electrical, magneto-optical, and piezoelectric properties[1,2]. In the class of optoelectronic materials, ZnO is emerging as a potential candidate due to its direct and wide band gap and its ability to tailor electronic, magnetic, and optical properties through doping, alloying, quantum wells and heterostructures, and nano-engineering[3-5]. Tailored materials and heterostructures are necessary to build many technologically important devices like laser diodes, visible and solar blind detectors, transparent electronics based on thin film field effect transistors, and spintronic devices.

In this chapter, we focus on various novel features, fundamental issues and recent advances on zinc oxide and its alloys. Particularly, the emphasis is given towards optimization of thin film growth process for the fabricating heterostructures of hexagonal and cubic alloys, and homo- and hetero-epitaxy. Applications of these heterostructures advanced in our

302

laboratories for fabrication of visible blind and solar blind UV detectors are highlighted along with enumeration of the commercial advantages. Finally, issues related to doping of *n* as well as *p* type dopants in the ZnO system and the future development of electronics are highlighted.

10.1.1 Properties of ZnO

ZnO is a wide band gap, optoelectronic material belonging to II-VI family of semiconductors. It crystallizes stably in wurtzitic hexagonal crystal structure, as shown in figure 1, in which Zn and O planes are alternately stacked along the *c*-axis direction. The coordinations of Zn^{+2} and O^{-2} are both four-fold. Its allowing with MgO with Mg> 62% leads to cubic phase system.

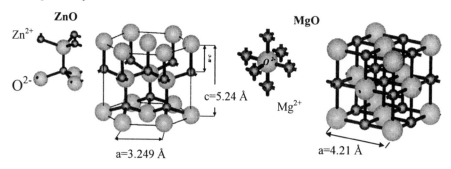

Figure 1: Schematics of ZnO wurtzite (left) and cubic MgZnO NaCl (right) crystal structures.

Table 1 describes structural, electrical, and optical properties of ZnO. ZnO has a direct energy bandgap of 3.3 eV at room temperature. The energy band parameters, such as electron and hole effective masses, as well as optical and electrical properties are quite comparable to that of GaN, a material that is well established for blue LED and semiconductor laser diode. In fact, ZnO is a much more intense and efficient UV emitter (at least 4 to 5 times more intense) at its band edge than GaN, owing to the fact that it has a higher free exciton binding energy (60 meV) that is more than twice than that of GaN (25meV). It is worth mentioning here that the radiative recombination in ZnO is mediated by excitons. This should lead to narrow emission linewidths, whereas it is band-to-band in GaN system yielding broad bandwidths. Based on these properties, it is envisioned that ZnO system has a niche for a wide range of devices such as blue and UV light emitting diodes, heterojunction diode lasers[6-7], visible and solar blind ultra-violet light detectors[8,9], optical waveguides, spintronic devices based on dilute magnetic semiconductors[10-13], piezoelectric devices, surface acoustic wave-based photodetector devices[14], and resonant tunneling devices[15], as well as transparent thin film transistors for displays[16].

Table 1. Properties of ZnO (Note: Unless indicated otherwise, all properties are at 300K).

Basic Properties		Crystal Lattice Properties	
Density [g/cm³]	5.67	Crystal Structure Space Group	Wurtzite $P6_3mc - C^4_{6v}$
Molecular weight [atomic unit]	81.38	Lattice constants [Å]	$a = 3.24$ $c = 5.20$
Ion radii [Å]	$r_{Zn}^{2+} = 0.60$ $r_O^{2-} = 1.40$	Shear Modulus [GPa]	50.0
Electronic configuration	Zn: [Ar] $3d^{10}4s^2$ O: [He] $2s^2 2p^4$	Stacking Fault Energy [mJ]	100
Thermal Properties		**Electrical Properties**	
Melting point [K]	2242	Energy bandgap [eV]	$E_g = 3.27$ (300K) $E_g = 3.44$ (6K)
Coefficient of Linear Expansion [10^{-6}/K]	$\alpha_{//c} = 25.6$ at 260 K $\alpha_{\perp c} = 45.0$ at 260 K	Electron Hall mobility [cm²/V·s] Thin film Bulk single crystal	$\mu_{n\,\perp c} = 70$ $\mu_{n\,//c} = 170$ $\mu_{n\,\perp c} = 150$
Thermal conductivity [W/m·K]	$\kappa = 54$	Effective Mass of Electrons Effective Mass of Holes [m_e]	$m_n^* (\Gamma \to A) = 0.28$ $m_h^* (\Gamma \to A) = 0.59$
		Dielectric Constant	$\varepsilon_{(0)\,\perp c} = 7.80$ $\varepsilon_{(0)\,//c} = 8.75$ $\varepsilon_{(\infty)\,\perp c} = 3.70$ $\varepsilon_{(\infty)\,//c} = 3.75$

10.1.2 Novel semiconducting ZnMgO alloy heterostructures

In this section, we discuss novel features of ZnO that have recently emerged while studying alloying of thin film heterostructures. Interestingly, band gap alloying is possible and it can be only achieved in thin film heterostructures. Three major substitutional elements have emerged as candidates for bandgap tailoring: Mg[17-19], V[20] and Cd[21]. Mg and V are known to broaden the bandgap, whereas Cd is known to narrow it. Out of these dopants Mg has following advantages: First, the ionic radius of Mg^{2+} is comparable to that of Zn^{2+}: 0.57 Å and 0.60 Å, respectively. Second, as shown in Fig. 2, it is possible to widen band gap of the alloy in the same way that band gap widening has been shown in III-V compounds. In principal, the band gap of ZnMgO alloy can be tuned

304

from 3.3 up to 7.8 eV by tuning the Mg content from 0 – 100 %[19]. The corresponding cutoff wavelength range is ~ 400-157 nm. This variation of band gap as a function of alloy composition is the largest one than that found in well established semiconductor systems such as GaAs, AlGaN, and Si[22]. However, structural transition from hexagonal wurtzite type to cubic structure could pose limitations near mixed phase region. Details of the fabrication and characterization of ZnMgO alloys along with their device applications are discussed in the latter sections.

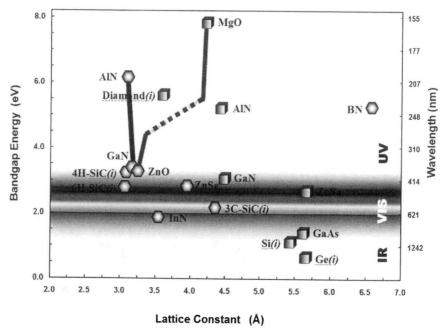

Figure 2: Band gap and crystallography of various electronic materials.

10.1.3 Advantages of ZnO based heterostructures for processing

Before discussing thin film growth and applications, we note some commercial advantages of the ZnO system. Firstly, high-quality ZnO single crystals are readily available. Wafers are currently at 50 mm in diameter. Secondly, it is believed that ZnO wafer production process will scale at a faster rate than that of GaN, or other common non-oxide wafer candidates for ultraviolet applications because of the differences in the wafer growth processes[23]. Sapphire is also known to be a most common substrate for heteroepitaxy despite the large lattice mismatch (~17%) with ZnO. Additionally, ZnO lattice matches with GaN (within 2%), making epi-GaN/sapphire as an ideal substrate for heterostructure junction with minimal bandgap distortion from lattice strain that could facilitate hybrid device technologies. Moreover, ZnO can be deposited at low- deposition temperatures by low-cost deposition techniques, such as RF sputtering, pulsed laser deposition (PLD), and chemical vapor deposition (CVD). Another interesting feature of ZnO is its ionic nature that lends itself to ease in the microfabrication process using both wet[24] and dry etch[25] techniques.

10.2 PROCESSING AND CHARACTERIZATION OF HETEROSTRUCTURES

Optimization of deposition parameters of thin film heterostructures and the post-deposition annealing conditions are the most important steps for synthesis of high-quality heteroepitaxial films. Particularly, crystallinity of the heterostructures and multilayers is of utmost importance for the investigation of fundamental properties and the development of any device technology. Growth of competitive orientations, especially in anisotropic materials such as ZnO, can lead to variable resistance in devices and variable reliability. Control of defects and undesirable impurities is equally important, as these defects and impurities can serve as scattering or recombination centers, affecting the overall response of optoelectronic devices. For these reasons, we have devoted an extensive effort to optimize deposition conditions. Controlling composition is also equally important as the bandgaps in these materials determine the characteristics of device operation.

We have used the PLD technique for the fabrication of thin films and multilayers as well as for metallization of devices. A KrF Excimer laser ($\lambda = 248$ nm) delivering nanosecond scale pulses with $1 - 30$ Hz repetition rate was used as a laser ablation source. A plano-convex lens and other optics are external to the growth chamber and directing UV beam through a quartz window into a chamber maintained at the base pressures of $\sim 10^{-6}$ to 10^{-7} Torr. The ablated material is ejected off of the target, generating a plasma plume composed of excited target constituents. Thin film heterostructures of ~ 5000 Å thick were grown on various substrates including c-plane (0001) sapphire, $SrTiO_3$ (001), MgO(001) with an average deposition rate of ~ 0.9 Å/pulse. The deposition temperature was varied from room temperature to 800°C, under an oxygen (O_2) pressure ranging from $10^{-5} - 10^{-1}$ Torr. For further introduction to PLD, we refer the reader to the Chapter of this book by Rjinders and Blank. We have used a diverse range of analytical tools to determine *crystallinity, composition and optical properties*. Specifically, crystallnity and composition of heterostructures were evaluated using four-circle x-ray diffraction (XRD), and Rutherford backscattering spectroscopy (RBS) / ion channeling using well-collimated 1.5-3.1 MeV He$^+$ ions using our in-house 1.7 MV accelerator facility. The surface morphology was studied by atomic force microscopy (AFM) and optical properties were explored with UV-visible spectroscopy, photoluminescence (PL) and cathodoluminesence (CL).

10.2.1 ZnO heterostructures on Sapphire (0001)

Here, we discuss the optimization of processing variables required for the synthesis of high quality heterostructures of ZnO as well as MgZnO. Dependence of the process variables on the surface and interfaces, structure and optoelectronic properties of the heterostructures is discussed in detail.

Figure 3 shows XRD θ–2θ, ω, and φ scans of the ZnO films grown at 750°C and at O$_2$ background pressures of 10^{-3} Torr. All ZnO layers grown in a wide range of O$_2$ pressure were found to be highly c-axis oriented. In

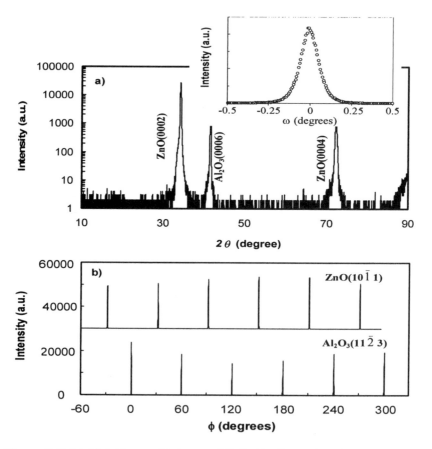

Figure 3. XRD (a) θ-2θ scan (b) φ-scan of ZnO thin films epitaxially grown on(0001) Al$_2$O$_3$. Inset in Figure (a) indicates rocking curve for ZnO (0002) peak.

the lower oxygen pressure regimes (10^{-5}-10^{-4} Torr), the films have a c-axis lattice parameter 0.25% larger than that of the bulk value. This is expected due to both oxygen deficiency and the compressive strain induced by the sapphire substrate (lattice mismatch between sapphire and ZnO is 17 %). However, in the higher oxygen pressure regimes (10^{-2}-10^{-1} Torr), the c-axis lattice constant was found to approach the bulk value. The XRD ω-rocking curve measurements were carried out to determine the epitaxial quality of the layers and the film alignment with respect to the substrate. A representative XRD rocking scan for (0002) peak is shown in the inset of Fig. 3 (a). We found that the FWHM of the rocking curve was moderately dependent of the O_2 pressure used during thin film growth. The lowest FWHM was found to be 0.07° (corresponds to ~ 4 arc-min) for the film grown at an O_2 pressure of 10^{-4} Torr. This value of Δω is among the lowest for ZnO epitaxial films on sapphire reported so far. *In-plane* epitaxy, as determined by XRD-φ scans of the ZnO($10\bar{1}1$) planes (see Fig.3.(b)), however, was strongly influenced by the oxygen pressure. The FWHM of the ZnO ($10\bar{1}1$) peak in the φ-scan for the ZnO films grown at 10^{-4} Torr and 10^{-1} Torr are 0.43° and 0.78°, respectively. The increase of FWHM in the Φ scan for the film grown at higher O_2 pressure clearly indicates a large misalignment of the ZnO grains and hence poor epitaxy. Also note that a 30° rotation of unit cell of ZnO with respect to the sapphire substrate[26]. Such rotation of lattice minimizes interface strain associated with the lattice mismatch and could result in reduced dislocation densities which is essential for efficient operation of electronic and photonic devices.

The effects of O_2 pressure on the degree of heteroepitaxy and defect density were quantitatively measured by using RBS and ion channeling techniques. In these measurements, χ (minimum yield), the ratio of the backscattered yield with the 1.5 MeV He^+ beam incident along [0001] (channeled) to that of a random direction, reflects the epitaxial quality (as well as defect density) of the film. Figure 4 shows the aligned and random backscattering spectra for 5000Å thick ZnO films grown at various oxygen pressures.

Fig 4(c) shows the systematic dependence of χ, near the surface and the interface (ZnO/sapphire), on the O_2 pressure. It can be seen that the χ near the surface region for the films grown in O_2 pressure regimes of 10^{-4} to 10^{-3} Torr is ~2-3 % and is comparable to that of high quality single crystals, while it is worsen (50-60%) for the film grown at an O_2 pressure of 10^{-1} Torr, indicating poor epitaxy. From Fig. 4(c), it should also be noted that the defect densities near the *interface* are strongly dependent upon the O_2 pressure indicating the influence of O_2 during the initial stages of ZnO film growth. The χ near the interface is about 10-12% for the film

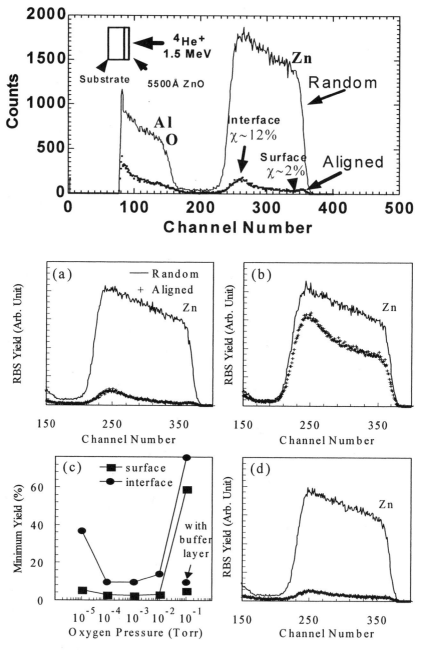

Figure 4. RBS and ion channeling results for the ZnO films. Top plot (unlabelled) full-scale RBS channeling for film grown at 10^{-4} Torr. Effect of the oxygen pressure on channeling is shown in plots (a),10^{-4} Torr, (b) 10^{-1} Torr, and (c) minimum yield vs O_2 pressure for growth. Plot (d) is for 10^{-1} Torr with a nucleation layer of 100 Å grown at 10^{-4} Torr.

grown at 10^{-4} Torr and increases substantially to 50-60% as the O_2 pressure increased to 10^{-1} Torr. The χ of 10-12% near the interface is due to unavoidable threading dislocations. In this case, an estimated threading dislocation density is $\sim 10^9/cm^2$ that is due to a large lattice mismatch between the ZnO films and the sapphire. The increase in χ for the film grown at higher O_2 pressure clearly indicates that the film contains additional defects such as low angle grain boundaries and interstitials[27].

In order to elucidate the origin of change in the crystalline quality and growth mechanism as a function of O_2 pressure, we have investigated the surface morphology of the ZnO films by AFM. Figure 5 shows the surface morphology of the ZnO films grown at various O_2 pressures. The

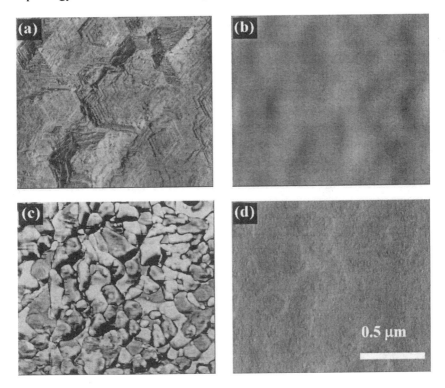

Figure 5: AFM images of ZnO films grown at various oxygen pressures: (a) 10^{-4} Torr, (b) 10^{-2} Torr, (c) 10^{-1} Torr, and (d) 10^{-1} Torr with nucleation layer of 100 Å grown at 10^{-4} Torr.

morphology of the films grown at 10^{-5}-10^{-4} Torr is dominated by a typical "honeycomb" like structure with three-dimensional (3D) growth features as evidenced by well faceted hexagons (Fig. 5(a)). These features could be due to high interfacial energy associated with ZnO film on sapphire, a high surface mobility of the Zn adatom, and a substantial desorption of Zn as a result of a reduction in the O/Zn ratio. The surface desorption mechanism

310

roughens the film surface. Indeed, we observed increased RMS roughness (150 Å) for the films grown at O_2 pressure of 10^{-5}-10^{-3} Torr. We also observed a high density of atomic steps on the facets with a step height of about 5.0Å which corresponds to a single unit cell height. These high densities of steps formed in a low oxygen pressure during the initial stages of the growth can significantly influence the growth of ZnO film subsequently deposited, and is an important issue which is discussed later. The transition towards the growth of a smooth film was found at O_2 pressure of 10^{-2} Torr. This change in growth mode results in a substantial reduction of RMS roughness to 10-20 Å for a flat surface. A further increase of O_2 pressure, to 10^{-1} Torr, showed an adverse effect on the surface morphology (Fig. 5(c)), with typical features of high nucleation densities, irregular grains with different sizes, and increased surface roughness to about 400 Å.

Now we discuss the influence of oxygen pressure on the optoelectronic properties of the ZnO films which are crucial for the development of ZnO based optoelectronic devices. In general, the electrical and optical properties of the conventional semiconductor heterostructures are governed by the nature of thin film growth process, growth mechanisms, lattice misfits, defects, surfaces, and interfaces. In the case of ZnO, we have also observed dependence of the optoelectronic properties on the epitaxy, defects, surfaces and interfaces.

Figure 6 shows the variation in electrical transport properties, namely the Hall mobility (μ) and carrier concentration (n), for the ZnO films grown under various oxygen pressures. The highest electron mobility (72 cm^2/V.s) was obtained for the film grown at 10^{-4}-10^{-3} Torr, but that film also showed the highest carrier concentration (7×10^{17}/cm^3). Further increase in O_2 pressure reduces the Hall mobility, possibly as a result of electron scattering from defects.

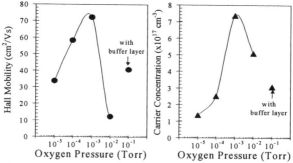

Figure 6: Hall mobility (μ) and carrier concentration (n) Vs oxygen pressure for ZnO growth. For comparison the data for the ZnO film grown at 10^{-1} Torr with nucleation layer grown at 10^{-4} Torr is also provided. Figures 7 (a) and (b) show 80K

Photoluminescence (PL) spectra for the ZnO films grown at 10^{-4} Torr and 10^{-1} Torr, respectively. In both cases, the spectra indicate distinct peaks due to the free and bound ($D°X$) excitons and features due to the donor-acceptor pair transitions at 3.32 eV with phonon replicas at 3.25 and 3.18 eV (see inset of Fig. 7). The free and bound exciton peaks have FWHMs of 27 meV and 7 meV, respectively. The optical quality of the ZnO epilayer grown at 10^{-4} Torr is such that the excitonic luminescence intensity produced is two orders of magnitude higher than that of the textured film grown at 10^{-1} Torr. This indicates that a high concentration of defects in ZnO film affects the radiative processes. In addition, no green band (often attributed to structural defects in ZnO) near 500 nm is observed for the high quality epilayers grown at 10^{-4} Torr, while it is clearly present for the textured film. We did not see any broadening in the exciton peaks probably due to the very small radius (~17Å) of excitons in the ZnO.

From the above results, we have clearly established that the growth of high quality epitaxial ZnO films with smooth surface morphology and desirable electrical and optical properties have different sets of optimum oxygen pressure regimes. The ZnO films in the lower oxygen pressure regime lead to high quality epitaxy but have rough surface morphology

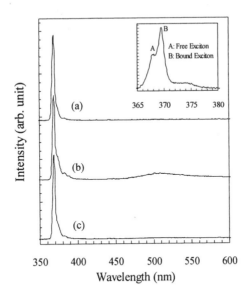

Figure 7: 80 K PL spectra for the ZnO films grown at various oxygen pressures: (a) 10^{-4} Torr, (b) 10^{-1} Torr, (c) 10^{-1} Torr with nucleation layer grown at 10^{-4} Torr. The inset shows a high-resolution spectrum indicating free and bound excitons.

which may pose a problem in the fabrication of quantum wells and superlattices. The increase in oxygen pressure to 10^{-2} Torr resulted in films with flat surfaces because of reduced Zn desorption; however, they suffer from poor electron mobility due to an increase in the defect density near the interface as seen in the ion channeling measurements. A further increase in oxygen pressure (10^{-1} Torr) results in high nucleation density and poor epitaxy. To circumvent these problems, a two-step growth process has been developed for the fabrication of high quality ZnO films. In this process, the nucleation layer is grown at low oxygen pressure (10^{-4} Torr) which produces a high quality crystallographic template for subsequent growth of ZnO at a high oxygen pressure (10^{-1} Torr). This two step process produces ZnO films with superior epitaxy resulting in ion channeling characteristics (Fig. 4(d)) similar to those in films grown at 10^{-4} Torr oxygen pressure. The surface morphology (Fig. 5(d)) and electrical properties are also improved as compared to those in films grown at 10^{-1} Torr. High optical luminescence quality without a green band (Fig. 7(c)) is also achieved in these films. These results are important in the context of fabrication of high quality ZnO based heterostructures, superlattices, and quantum wells for which smoother films with good electrical and optical properties are required. Additionally, growth of ZnO films under high oxygen pressure is desirable for p-type doping studies in order to avoid compensation due to oxygen vacancies.

10.2.2 Heteroepitaxy of ZnO on GaN/Sapphire

We also studied the ZnO epitaxy on high-quality MOCVD grown epi-GaN/sapphire (0001) substrates[28]. The ZnO films grown on epi-GaN/sapphire were found to be single crystalline due to the match of stacking order and a low lattice misfit (1.9%) between GaN and ZnO, as compared to those grown directly on sapphire. A comparison on structural, thermal, and optical properties of ZnO and GaN is given in Table 2.

Table 2: Structural, Optical and Thermal Properties of ZnO and GaN compared to Sapphire

Material	Crystal Structure	Lattice Constant (A)	Thermal Expn. Coeff. (/°K)	Thermal Conductivity (W/cm K)
ZnO	Wurtzite	a=b=3.249	6.51 X10^{-6}	0.60
		c=5.206	3.02 X10^{-6}	
GaN	Wurtzite	a=b=3.186	5.59 X10^{-6}	1.35
		c=5.178	7.75 X10^{-6}	
Sapphire (Al_2O_3)	Corundum/ Rhombohedral	a=b=4.758 c=12.99	7.50 X10^{-6} 8.50 X10^{-6}	0.50

ZnO/GaN/Sapphire heterostructures were investigated by a four circle x-ray diffraction technique. The x-ray diffraction measurements showed highly c-axis oriented films with a rocking curve FWHM of 0.05° (3 arc-min). The *in-plane* epitaxial relationship in these heterostructures was found to be ZnO[10$\bar{1}$0] || GaN[10$\bar{1}$0] || Al₂O₃ [1120]. The FWHM of the (1011) peak for both ZnO and GaN was about 0.25° (15 arc min), and no tilting was observed between the ZnO and GaN films. These measurements indicate that this misorientation is much smaller than that for the ZnO epilayers grown directly on sapphire (0001).

The ion channeling measurements on these films showed a minimum yield near the surface region of ~1 to 2%. The dechanneling analysis shows that the dislocation densities near the ZnO/GaN and ZnO/sapphire interfaces are $2x10^8/cm^2$ and $4x10^9/cm^2$, respectively. Figure 8 shows the high-resolution transmission electron microscopy (HRTEM) lattice image of the ZnO/GaN interface, illustrating that the lattice planes of ZnO are perfectly aligned with those of GaN, and the interface is fairly sharp. These results confirm the suitability of a GaN buffer layer for ZnO growth.

Figure 8: High Resolution Transmission Electron Microscopy image of ZnO on a GaN substrate. The interfacial region is marked with an "I".

Figure 9 shows a cathodoluminesence (CL) spectrum obtained at 8K for a 5000Å ZnO epilayer grown on epi-GaN/sapphire. The CL spectra for the ZnO film on GaN/sapphire demonstrate distinct peaks due to the free A-exciton and D°X bound exciton. In addition to this, the CL spectra show pronounced features due to the donor-acceptor pair transitions at 3.32 eV, with phonon replicas at 3.25 and 3.18 eV. From the CL studies, we also note the FWHM of a free A-exciton line width is about 20 meV, which is comparable to that of device quality GaN films. The intense CL peak and

314

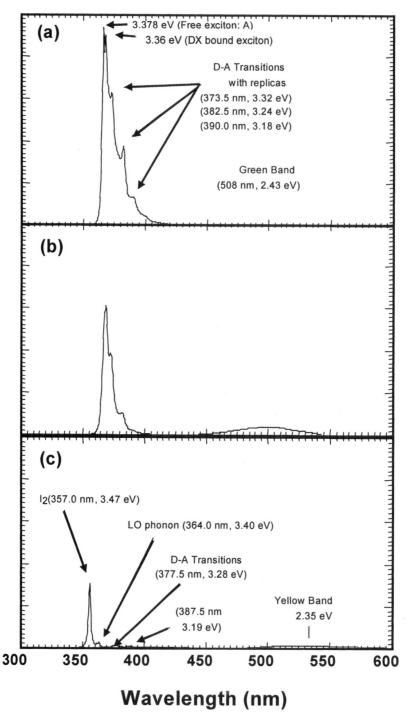

Figure 9: Cathodoluminesence spectra at 8K for epi-ZnO grown on epi-GaN/Sapphire.

the absence of a green band clearly indicate the high-quality epitaxy of ZnO on GaN/sapphire.

Due to their lattice matching epitaxy, thermal, and optical compatibility, the ZnO/GaN heterostructures on sapphire may be useful for the fabrication of hybrid optoelectronic devices exploiting advantages of both ZnO and GaN. Hamdani *et al*[29] demonstrated high optical quality epitaxial GaN films on ZnO crystals by molecular beam epitaxy (MBE). This means that under MBE and PLD growth conditions, high quality ZnO/GaN and GaN/ZnO heterostructures can be integrated on sapphire. Another interesting feature of a thick GaN buffer layer for ZnO is its high thermal conductivity which may be beneficial for ZnO thin film lasers. At this juncture, we feel optimistic about the possibility of being able to fabricate novel p-n junctions based on n-type ZnO (Al or Ga doped) and p-type GaN (Mg-doped) semiconductors for light emitting devices. The doped films can be grown by using doped targets. The high electrical conductivity and optical transparency achieved in doped ZnO epi-layers may also be used as the transparent electrodes for blue GaN optical devices.

10.3 BAND GAP TAILORING IN MgZnO

Band gap tailoring refers to the ability to tune the band structure of a semiconductor which is useful for tailoring the performance of optoelectronic devices. Band gap tailoring of semiconductors can be typically achieved by:

- alloying two or more semiconductors.
- use of heterostructures to cause quantum confinement or formation of superlattices.
- use of strained epitaxy.

Here we focus on alloying of ZnO with MgO. Before we discuss properties of thin film alloys, it is important to consider novelty of our approach in alloying oxide systems as realization of Mg-Zn-O alloy is challenged by the incompatible electronic configuration and crystal structure of MgO (NaCl type cubic structure with lattice constant a= 4.24 Å) and ZnO (wurtzite, hexagonal structure with lattice constants a= 3.24 Å and b= 5.20 Å). According to the phase diagram of ZnO-MgO binary system, solid solubility of MgO in ZnO is less than 4 mol% [30]. This situation is in contrast to the standard semiconductor alloying e.g. AlGaN and AlGaAs, where two (or more) components of the alloy have the same crystal structure so that the final alloy also has the same crystalline structure[31].

In spite of a large structural dissimilarity and limited solid solubility between ZnO and MgO, Ohtomo *et al*[17] have reported the solid solubility of MgO in ZnO to be 33 mol% for the thin film alloys grown under metastable conditions. Above 33%, MgO is reported to segregate from the wurtzite MgZnO lattice limiting its band gap maximum up to 3.9 eV. While studying the optimizing conditions, we found that the solubility of MgO in ZnO films is a function of growth temperature and this factor is critical for stabilizing the structure and chemical phase purity of the alloys. However, due to dissimilarity of crystal structures between ZnO and MgO, the change of a crystal structure will be the inherent drawback of the Mg-Zn-O system that can inevitably cause phase separation and mixed phases for certain composition. Thus, discontinuity is expected when tuning the bandgap from one end to the other. Minimizing the phase separation region is thus a necessary step toward realization of high quality $Mg_xZn_{1-x}O$ films and alloy heterostructures. Additionally, the success of alloying ZnO with MgO relies on the deposition technique. In our case, use of PLD technique in which the non-thermal equilibrium thin film deposition process enables a metastable $Mg_xZn_{1-x}O$ alloy with higher x value and shorter range of mixed phase region.

Shown in Figure 10 is the summary of the band gap alloying in thin films grown by pulsed laser deposition as a function of Mg content in ZnO

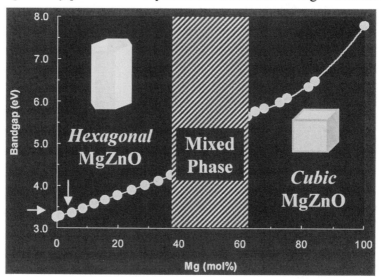

Figure 10: Compositional tuning of structure and bandgap with Mg concentration in ZnMgO alloy fabricated using pulsed laser deposition technique.

and its optical properties. The films grown at 550°C in an oxygen pressure of 10^{-4} Torr have a single phase wurtzite hexagonal structure. $Mg_xZn_{1-x}O$ thin film alloys with Mg content between 45 to 60 mol% are not single phased. For x > 0.6, all Mg-rich alloys are found to be cubic phase similar to that of MgO structure. We have also successfully extended alloying studies for cubic systems in which ZnO is alloyed with MgO for the formation of Mg rich cubic alloys. In the following sections, we discuss the processing aspects of MgZnO thin films and the fabrication of hexagonal and cubic alloy heterostructures.

10.3.1 Growth of thin film alloy heterostructures

For growth of $Mg_xZn_{1-x}O$ films, a series of sintered $Mg_xZn_{1-x}O$ ($x=0\sim1$) targets were fabricated and used for laser ablation experiments. Due to the higher vapor pressure of Zn than Mg, the composition of the $Mg_xZn_{1-x}O$ film deviated significantly from that of the target[18]. The change of the film composition in turn changes $Mg_xZn_{1-x}O$ energy bandgap. Thus, the optimal deposition temperature is determined by both the crystalline quality and the Mg solid solubility with the consideration of the film composition. All alloy films discussed here were grown at 700-750 °C and oxygen partial pressure of 10^{-4} Torr. To determine the effect of composition on the band gap, we used UV-Visible Spectrophotometry to extract a given alloy's bandgap. Figure 11 is the ultraviolet-visible transmission spectra of $Mg_xZn_{1-x}O$ thin films deposited on double-side polished c-plane sapphire substrates. With an increase of the Mg mole fraction, the absorption edge shifts continuously to the short wavelength side, clearly indicating the capability of tuning the $Mg_xZn_{1-x}O$ bandgap by composition. Initially, the absorption edge shifts at a rate of ~2.5 nm for every mole percent increase of Mg. This number drops to ~2.1 nm around the wavelength of 300 nm. Between the wavelengths of 230 nm to 290 nm, there is no transmittance line reaching zero indicating that the bandgap of $Mg_xZn_{1-x}O$ can not be tuned between 4.2 eV and 5.3 eV. A double-edge transmission curve, which is the fingerprint of the hexagonal-cubic mixed phase is also visible around this regime. Beyond the phase separation region, the bandgap of $Mg_xZn_{1-x}O$ can be tuned again until the composition of $Mg_xZn_{1-x}O$ approaches to pure MgO, which has a bandgap of 7.8 eV. Note that the slope of each transmittance curve near the absorption edge, which serves as another indicator of the film crystalline quality, is sharp (~0.10/nm for hexagonal phase, and 0.12 for cubic phase).

The energy bandgap was derived from the UV-Vis transmission spectra using the $[\alpha(\lambda)h\nu]^2$ versus $h\nu$ plot, where $\alpha(\lambda)$ is the wavelength-dependent absorption coefficient, and $h\nu$ is the photon energy[32]. The bandgap of $Mg_xZn_{1-x}O$ increases almost linearly from 3.3 eV to 4.2 eV. The corresponding $Mg_xZn_{1-x}O$ film composition varies from $x=0$ to $x=0.37$. The hexagonal-cubic mixed phases are found between $x=0.37$ and

x=0.62, corresponding to a gap between E_g=4.2 eV to 5.3 eV. Above E_g=5.3 eV, the composition-bandgap curve bends upward for cubic phases.

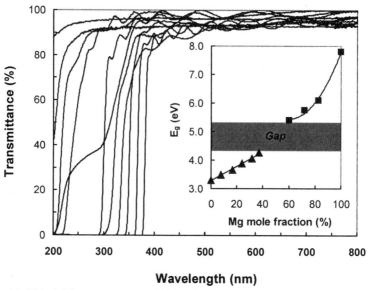

Figure 11: UV-visible spectroscopy of ZnMgO alloys of variable Mg mole fraction. Inset shows the relationship of $Mg_xZn_{1-x}O$ bandgap versus composition.

Figure 12 shows the XRD θ-2θ scans of the $Mg_xZn_{1-x}O$ thin films as a function of the composition. The hexagonal to cubic phase evolution is evident as indicated by the corresponding XRD peak intensity. The hexagonal phase is represented by the $Mg_xZn_{1-x}O$ (0002) diffraction peak at 2θ= 34.3°. Up to x= 0.37, the $Mg_xZn_{1-x}O$ remains in a *wurtzite* structure with almost identical lattice constants as ZnO. Only small lattice shrinking was found as indicated by the slight right shift of the $Mg_xZn_{1-x}O$ (0002) peak with the increase of the x value. It is worthwhile to note that there is no significant broadening of the XRD line widths within this composition range, implying that the substitution of Mg inside the ZnO lattice does not cause significant defects. For composition of x= 0.43, the peak at 2θ= 36.9° corresponds to $Mg_xZn_{1-x}O$ (111) representing a cubic phase alloy. The lattice constant is found to be a= 4.23 Å, which is almost identical to that of MgO[19]

10.3.2 Homo- and Heteroepitaxy of ZnMgO alloys

Homoepitaxial growth of oxide heterostructures has tremendous potential to achieve superior material quality that can result in extremely smooth surface morphology, narrow photoluminescence (PL) linewidths,

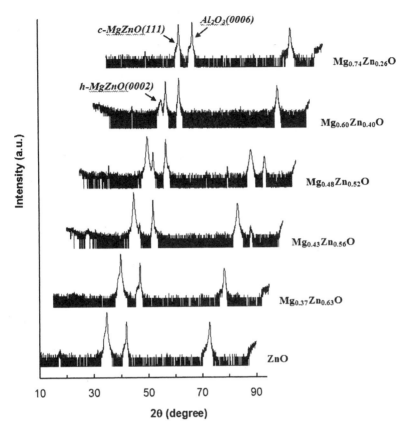

Figure 12: XRD θ-2θ Scans of $Mg_xZn_{1-x}O$ thin films with Mg composition, x = 0 to 0.74.

and a reduction of the dislocation densities by several orders of magnitude. These material qualities can only be attained using a substrate which is identical in crystal structure, lattice parameter and thermal expansion coefficient. Under those conditions, two-dimensional layer-by-layer growth can be obtained and the generation of dislocations can be inhibited. Additional process steps such as surface preparation and nucleation layers or buffer layers, mandatory in heteroepitaxy are no longer required, thus significantly simplifying the growth process for high quality device fabrication.

In the previous sections, we discussed that the wide band gap ZnMgO oxide alloys have two composition dependent crystal structures: hexagonal wurtzite structure and cubic MgO-type structure. For optoelectronic device applications it is imperative that these heterostructures (also quantum wells with sharp interfaces) to be grown on lattice-matched substrates to reduce the interface dislocation density as the quantum efficiency of many optoelectronic devices depends on the defect densities. It is thus important to study effect of substrate (lattice structures, orientation, lattice constants, and surface chemistry) on the growth properties of thin films. These results can be potentially useful for tailoring buffer layers on lattice mismatched substrates such as Si or GaAs for future integration of optoelectronic devices. In this context, we have studied homo and heteroepitaxy of ZnMgO alloys on single crystal substrates. Here we discuss mainly epitaxy of cubic ZnMgO alloys as homoepitaxy of hexagonal alloys on ZnO bulk substrate is very much similar to that of epitaxy of ZnO on GaN discussed in section 2.2.

For the growth of cubic ZnMgO alloys, MgO(100) (a case study for homoepitaxy) and SrTiO$_3$ (100) (a case study for heteroepitaxy) were used as lattice matched (lattice misfit of 0%) and mismatched (8%) substrates, respectively. The properties of the substrates and their lattice misfits with the alloy systems are given in Table 3 and depicted in Figure 13.

To study the characteristics of thin film growth and interfaces, RBS ion channeling and high resolution x-ray diffraction were performed for the homoepitaxially grown and heteroepitaxially grown layers. To note here that the optimum growth temperatures are lower by 150°C for homoepitaxy than that discussed in the earlier sections. Using high resolution x-ray diffraction studies (results not shown), we confirmed that the nature of epitaxy is cube-on-cube. Atomic force microscopy results indicated that the films grown on MgO substrate were smoother than films grown on sapphire.

Table 3: Substrates for homo and heteroepitaxy

Substrate	Lattice constant	Lattice Misfit between substrate and h-ZnMgO	Lattice Misfit between substrate and c-ZnMgO
c-ZnO	3.24 Å	0%	----
c-sapphire	2.75 Å	17%	---
MgO	4.22 Å	---	0%
SrTiO3	3.92 Å	---	8%

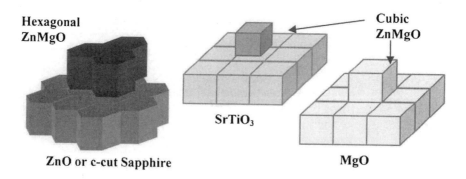

Figure 13: Schematics of homo and heteroepitaxy of ZnMgO alloys.

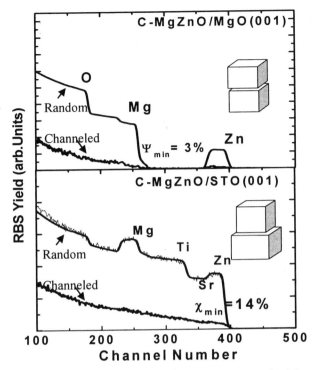

Figure 14: RBS and ion channeling results for homo and heteroepitaxial grown layers of $Mg_{0.8}Zn_{0.2}O$ on MgO and $SrTiO_3$ (100) substrates.

Figure 14 shows RBS and ion channeling for cubic phase $Zn_{0.2}Mg_{0.8}O$ alloy grown on single crystal MgO (100) and STO (100) substrates. The ion channeling results indicate that the surface of film grown on MgO is very ordered. The peak associated with the interface defects is relatively smaller than we observed for the large lattice mismatched Al_2O_3 substrate. However, ion channeling results for the cubic alloy grown on $SrTiO_3$ substrate indicate χ of 8% indicating a large concentration of defects. It is worthwhile to mention that oxide heteroepitaxy of novel ZnMgO alloys is still immature and systematic studies on the surface preparation and optimization of growth conditions are necessary.

10.3.3 Heterostructures on silicon

Epitaxy of oxide heterostructures on Si is challenging in view of the large lattice and thermal mismatch between the oxide films and the substrate[9]. For integration of MgZnO oxide heterostructures with silicon, we examined cubic phase $Mg_xZn_{1-x}O$ with implementation of variety of buffer layers including $SrTiO_3$ (STO) and MgO. Oxidation of Si gives rise to an amorphous SiO_x interfacial layer that might be detrimental especially for the epitaxial growth of the c-MgZnO layer. Thus, a proper buffer layer must be used and the right growth process must be followed to obtain high crystalline quality $Mg_xZn_{1-x}O$ epilayers. Based on our optimization studies, $SrTiO_3$ was found to be a suitable buffer layer between Si(100) and c-MgZnO[9]. The photograph of the epitaxial ZnMgO/STO grown on 4.0 inch diameter Si wafer is shown Fig 15. Right side of the figure shows the AFM morphology image for c-MgZnO film on Si (100). Smooth surface with root mean square (rms) roughness of only about 6 Å was obtained.

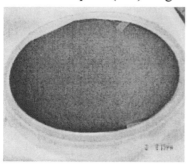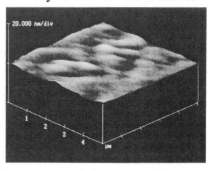

Figure 15: Photograph (left) of pulsed laser deposited c-ZnMgO alloy on a 4" dia. Si wafer with STO buffer layer. AFM morphology micrograph (right) of cubic $Mg_xZn_{1-x}O$ thin film.

Figure 16 shows XRD of c-MgZnO film grown on Si (100) with $SrTiO_3$ buffer layer. Note that the $SrTiO_3$ buffer layer is also epitaxially grown on Si(100). Figure 16 (b) is the ϕ scan of the (111) diffractions for c-MgZnO, $SrTiO_3$ and Si. Sharp diffraction peaks of c-MgZnO (average FWHM = 0.5^0) and $SrTiO_3$ (average FWHM = 0.7^0) are comparable with that of Si. As can be seen from Fig. 16 (b), the c-MgZnO and $SrTiO_3$ have the same peak position, which is 45^0 away from that of Si, implying a 45^0 in-plane rotation of $SrTiO_3$ unit cells with respect to Si. This lattice arrangement reduces lattice mismatch considerably from -28.2% without lattice rotation to +1.6% with lattice rotation. Without the use of buffer layer, ZnMgO on Si (100) tends to orient along growth direction only. RBS random and aligned backscattering spectra of c-MgZnO films on $SrTiO_3$/Si (100) showed χ is 4% for Zn at the surface and 12% for Sr at c-MgZnO/$SrTiO_3$ interface (figure not shown). Good crystalline quality epilayers having smooth surface morphology are the prerequisite for device applications.

10.4 MgZnO HETEROSTRUCTURES FOR UV PHOTODETECTORS

$Mg_xZn_{1-x}O$ system has the potential to produce photodetectors capable of accessing a broad UV spectrum from 400 nm out to 157 nm that can

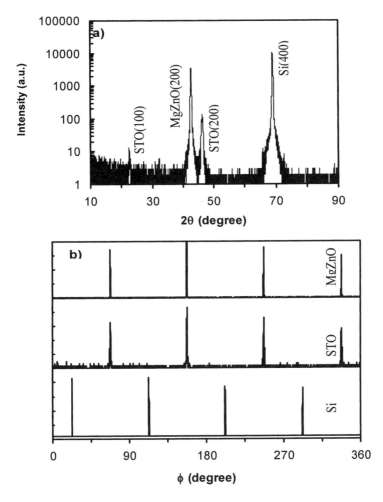

Figure 16: XRD (a) 2θ-scan and (b) φ-scan of cubic $Mg_xZn_{1-x}O$ thin film grown on $SrTiO_3$ buffered Si(100). The epitaxial relationship is $Mg_xZn_{1-x}O(100)//SrTiO_3(100)//Si(100)$ and $Mg_xZn_{1-x}O[110]//SrTiO_3[110]//Si[100]$.

enable a broad range of applications. For comparison, we begin this section with a brief discussion of the prior art in material technologies for UV detection including Si, GaN, and SiC.

Si-based *p-i-n* photodiodes are the most widely used UV photodetectors. As a result of well-established microfabrication techniques for Si, their economy and reliability are well-recognized. As a narrow bandgap semiconductor (E_g=1.14 eV) Si responds to a range of wavelengths: near-IR through visible and UV, unless certain coating or optical filters are applied. Within the UV region, the responsivity of Si is significantly lower than its peak photoresponse, which occurs between

λ_{MAX}= 700-1000 nm[33]. For UV photodetection, combinations of optical filters are applied. Other wide band gap materials recently emerged from rigorous research and development efforts are SiC and GaN. SiC-based UV detectors have a cutoff wavelength at $\lambda \sim 390$ nm. Typical responsivity of SiC-based detectors is 0.15 A/W[34]. However, material-related issues such as thin film growth quality, the ability to make ohmic contacts to *p*-type region[35], and the indirect bandgap currently constrain the range of detector applications for SiC. GaN and $Al_xGa_{1-x}N$-based photodiodes are among today's state-of-the-art UV photodetectors[36]. As a wide and direct bandgap semiconductor, GaN has peak photoresponse and cutoff wavelength at 370 nm and 385 nm respectively. The typical responsivity for a GaN *p-i-n* diode with an area of 500 μm^2 is 0.1 A/W[37]. By alloying it with Al, the $Al_xGa_{1-x}N$ has a wider bandgap than GaN alone and permits solar-blind operation[38]. Despite these successes, the cost of GaN and $Al_xGa_{1-x}N$ based UV detectors is high. The challenge of growing high quality films at relatively low temperatures[39], the difficulty to control dislocation density[40, 41] and lack of lattice matched single crystal substrates are the constrains on mass production of GaN and $Al_xGa_{1-x}N$ based optoelectronic devices.

As an alternative to the UV photodetector materials described above, $Mg_{1-x}Zn_xO$ have potential for visible and solar blind detection technologies. ZnO thin films are good for UV-A (320-400 nm) photon detection as its band gap (3.3 eV) lies in UV-A region. Tuning the band gap using $Mg_xZn_{1-x}O$ alloys above 3.9 eV enables the realization of UV-B (280-320 nm) and UV-C (200-280 nm) photodetectors, which have found both astronomical and terrestrial applications[42,43]. Here, we mainly emphasize on ZnMgO alloy heterostructures for UV detection application.

In this section, we will discuss the fabrication and characterization of $Mg_xZn_{1-x}O$ UV MSM photodetectors. Figure 17 is a scanning electron

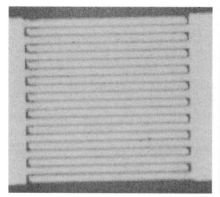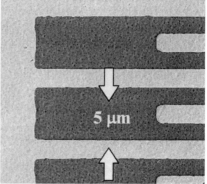

Figure 17: Left: Optical photograph of MSM MgZnO UV photodetector. Right: Scanning electron micrograph of MSM MgZnO UV photodetector.

micrograph of a $Mg_{0.34}Zn_{0.66}O$ UV detector with a size of 250x1000 μm^2. Interdigitated metal electrodes of Au/Cr bilayers (~1500Å thick) were defined by optical photolithography and subsequent ion milling. The features are approximately 250 μm long and 5 μm wide with a 5 μm pitch. Au/Cr produces an Ohmic contact after rapid thermal annealing at 350°C, 1 minute. As a reminder, samples were deposited on sapphire. Later in the section, we will compare these results with detectors grown on other substrates.

Spectral response of the ZnMgO UV detector is shown in Figure 18(a). Measurements were conducted with a Xenon source and 1200 lines/mm grating. Peak response is found at 308 nm and agreed with the cutoff wavelength from transmission data. The –3dB cutoff wavelength is 317 nm. Visible rejection ratio, (S(308nm)/S(400nm)) is more than four orders of magnitude, indicating a high degree of visible blindness. *Response time* was measured to be < 4 ns via pumping with an N_2 laser (337 nm). Detectors were subjected to a 3 V bias and 50Ω load. There was no observed persistent photoconductivity. The 10%-90% rise and fall times are 8 ns and ~1.4μs, respectively. The 8 ns rise time is limited by the excitation laser, which has a nominal pulse duration FWHM~ 4 ns. *I-V characteristics* at 5V bias showed an average dark current of ~ 40 nA. Figure 18(b) shows the current-voltage (I-V) curve of a $Mg_{0.35}Zn_{0.65}O$ UV detector. For this photoconductor, a typical dark current at 3V bias is below 1 μA. This value can be further quenched to 100 nA or lower if *Schottky* contacts[44,45] were applied. Direct deposition of Ag or Au metal layers on top of ZnO films leads to *Schottky*-type contacts[46]. However, the junction quality is highly processing dependent. A typical barrier height of ~0.7 eV with an ideal factor of 1.5 can be obtained.

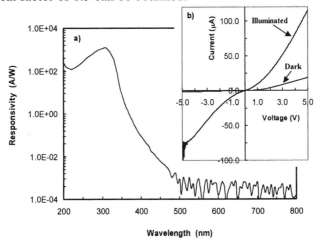

Figure 18: (a) Spectral response of MgZnO MSM photodetector as a function of wavelength (b) Current-voltage response of this photodetector under dark and illuminated conditions.

Figure 19 shows the relationship of the $Mg_xZn_{1-x}O$ photodetector peak response wavelength for various compositions. A chip with varying Mg composition was developed using combinatorial synthesis method[19] (also see the Chapter by I. Takeuchi in this book.) The linear shift of the response peaks is attributed to the increase of active material bandgap, which depends linearly on the film composition. The plot reveals that for UV-B applications, the Mg mole fraction of $Mg_xZn_{1-x}O$ films should be in the range of $0.30 \leq x \leq 0.37$. For *solar-blind* (a region in which the solar radiation background is negligible) detectors the cutoff wavelength is 280 nm.

We have also fabricated prototype UV detectors based on c-MgZnO on Si. The device shows a peak spectral response at 225 nm, which is in the solar blind region. The UV/visible rejection ratio was more than one order of magnitude (lower than hexagonal phase due to the high resistivity of the films and the weak UV output from the light source). Further improvement of the detector performance working in various wavelengths is currently in progress.

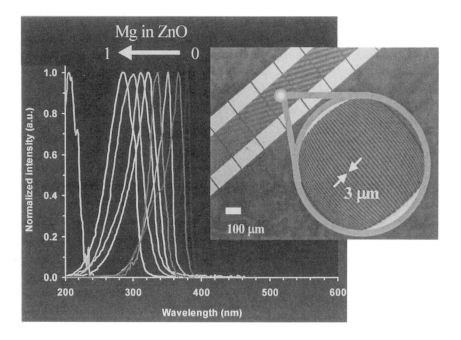

Figure 19: Peak response of MgZnO photodetector for Mg composition, x= 0 to o.74 (right to left on plot). Inset shows optical micrograph of photodetector.

10.5 DOPING IN OXIDE HETEROSTRUCTURES

Zinc oxide has been known to be an extrinsic, *n*-type semiconductor for over half a century[47]. By nature, ZnO is *n*-type due to the excess Zn in the lattice[48]. By controlling the density of defects from oxygen vacancies and zinc interstitials, one may control the resistance of ZnO, particularly for thin films. Group III elements such as aluminum[49-51] gallium[52] and indium are often selected to enhance *n*-type doping capabilities. Films with comparable mobilities to that of undoped ZnO, and carrier concentrations between $10^{20} - 10^{21}$ cm^3 have been grown by molecular beam epitaxy, laser ablation, chemical vapor deposition and sputtering techniques.

10.5.1 The search for p-type ZnO and ZnMgO

A vital step in any effort to widen the applications of ZnO and MgZnO in special applications like LED's and laser devices is realizing reliable and stable *p*-type dopant, which remains elusive due to the multifold challenges. In a recent report, Tsukazaki et al.[5] have convincingly demonstrated p-type doping by repeated temperature modulation epitaxy technique and the operation of homojunction ZnO based blue/violet LED.

Prior to this attempt diverse number of dopants and co-dopants have been attempted, primarily in Group V and Group I elements[53,54]. The ionic radii of these elements impose critical limits in the host material and one could conclude that P (2.18 Å), As (2.23 Å), K (2.42 Å), and Sb (2.45 Å) would be unlikely candidates for successful in *p*-type doping in ZnO. However, other dopants such as N[5], P, As[55], and Sb have shown the promising results in the current literature for *p*-type ZnO production. Homojunctions have been fabricated with As dopant, although resistances have been high, likely because of low doping levels and metal contact quality. In the case of phosphorous doping, thermal dissociation of P_2O_5 was considered to be the critical step ($T_{Sublimation} = 350^{\circ}C$) in which shallow donor level forms as a consequence of substitution of P on the Zn site[56]. Typically, *p*- type behavior in these films has been observed to be short-lived, although capping with polymers has been known to improve *p*-type conductivity lifetime. In general, issues related to the p-type doping in ZnO and ZnMgO films are on the forefront of the materials advancement. Improvement in materials through improved growth techniques, process, and stabilization of dopants is necessary for realization of optoelectronic devices.

328

10.6 CONCLUSIONS

In conclusion, recent advances in the wide band gap ZnO and ZnMgO heterostructures have been discussed addressing thin film growth, alloying, homo and heteroepitaxy, and the critical issues related to the surfaces and interfaces affecting optical and electronic properties of the heterostructures. Band gap tailoring in heterostructures has been demonstrated by alloying ZnO with MgO. Crystal structure of alloy heterostructure depends upon Mg concentration in ZnO lattice or Zn in MgO lattice. The band gap energy of MgZnO alloys has been shown to vary in a wide range from 3.3 eV to 4.2 eV for hexagonal phase and 5.3 to 7.8 eV for cubic phase, respectively. The band gap energy and the metastable structure of these films have been controlled by the growth temperature. We have implemented wider band gap ZnMgO alloys for the fabrication of UV photonic devices such as visible and solar blind UV detectors, and demonstrated UV detector (based on hexagonal $Mg_{0.34}Zn_{0.66}O$ thin film) with a high responsivity and a fast response (8ns rise and 1.4µs fall time). The detectors show peak responsivity at their band edge that depends on the composition. Visible rejection ratio is more than three orders of magnitude for hexagonal alloys. Cubic MgZnO alloy films were grown on silicon (100) with $SrTiO_3$ buffer layer for the fabrication of UV detectors. These UV sensors showed peak responsivity at 225 nm and could be promising for the integration with Si technology. Though, initial results are promising, there is a need for significant efforts toward further improvement of material technology with emphasis on doping and fabrication of stable homo and hetero p-n junctions, metal-semiconductor ohmic and Schottky contacts for realization of reliable LEDs, laser diodes, detectors, and transparent electronics based on FETs.

ACKNOWLEDGEMENTS

On of the authors (RDV) would like to thank National Science Foundation (SBIR DMI-0321465) and Maryland Industrial Partnerships for supporting this work. D.E. Pugel acknowledges support from NASA Goddard Space Flight Center. We would also like to thank Dr. S. Choopun, Dr. W. Yang, and Dr. S.N. Yedave for the technical support. We acknowledge Pelletron Shared Experimental Facility (SEF) supported under NSF-MRSEC grant number DMR-00-80008.

REFERENCES

1. S.J. Pearton, D.P. Norton, K. Ip, Y.W. Heo and T. Steiner, *Progress in Materials Science* **50**, 293 (2005).

2. Harvey E Brown, Zinc oxide: Properties and applications, *International Lead Zinc Research Organization* (1976).

3. D.C. Look, *Materials Science and Engineering B, Vol. 80, 383 (2001)*.

4. H. Ohta, K. Kawamura, M. Orita, N. Sarukura, M. Hirano, H. Hosono, *Materials Research Society symposia proceedings* **623**, 283 (2000).

5. A. Tsukazaki, A. Ohtomo, T. Onuma, M. Ohtani, T. Makino, M. Sumiya, K.Ohtani, S. F. Chichibu, S. Fuke, Y.Segawa, H. Ohno, H.Koinuma and M.Kawasaki, *Nature Materials* **4**, 42 (2005).

6. Ya. I. Alivov, E. V. Kalinina, A. E. Cherenkov, D. C. Look, B. M. Ataev, A. K. Omaev, M. V. Chukichev and D. M. Bagnall, *Appl. Phys. Lett.* **83**, 4719 (2003); *Material Research Society, Symposium Proceedings* **798**, 41 (2003).

7. Nuri W. Emanetoglu, Jun Zhu, Ying Chen, Jian Zhong, Yimin Chen and Yicheng Lu, *Appl. Phys. Lett.* **85**, 3702 (2004).

8. W. Yang, S. S. Hullavarad, B. Nagaraj, I. Takeuchi, R. P. Sharma, T. Venkatesan, R. D. Vispute and H. Shen, *Appl. Phys. Lett.* **82**, 3424 (2003).

9. W. Yang, R. D. Vispute, S. Choopun, R. P. Sharma, T. Venkatesan and H. Shen, *Appl. Phys. Lett.* **78**, 2787 (2001).

10. S J Pearton, W H Heo, M Ivill, D P Norton and T Steiner, *Semicond. Sci. Technol.* **19**, R59 (2004).

11. Darshan C. Kundaliya, S.B. Ogale, S.E. Lofland, S. Dhar, C.J. Metting, S.R.Shinde, Z. Ma, B. Varughese, K.V. Ramanjuchary, L. Salamanca-Riba and T. Venkatesan, *Nature Materials,* **3,** 709 (2004).

12. Lee Eun-Cheol and K.J. Chang, *Physical Review. B*, **69**, 85205 (2004).

13. Min Sik Park and B. I. Min, *Physical Review. B* **68**, 224436 (2003).

14. P. Sharma, A. Mansingh and K. Sreenivas, *Appl. Phys. Lett.* **80**, 553 (2002)

15. Agisilaos A. Iliadis, Soumya Krishnamoorthy, Wei Yang, Supab Choopun, R. D. Vispute and T. Venkatesan, *Proc. SPIE Int. Soc. Opt. Eng.* **4650**, 67 (2002).

16. Elvira M. C. Fortunato, Pedro M. C. Barquinha, Ana C. M. B. G. Pimentel, Alexandra M. F. Gonçalves, António J. S. Marques, Rodrigo F. P. Martins and Luis M.N. Pereira, *Appl.Phys.Lett.* **85**, 2541 (2004).

17. A. Ohtomo, M. Kawasaki, T. Koida, K. Masubuchi, H. Koinuma, Y. Sakurai, Y. Yoshida, T. Yasuda, Y. Segawa, *Appl. Phys. Lett.* **72**, 2466 (1998).

18. A.K. Sharma, J. Narayan, J.F. Muth, C.W. Teng, C. Jin, A. Kvit, R.M. Kolbas, and O.W. Holland, *Appl. Phys. Lett.* **75**, 3327 (1999).

19. I. Takeuchi, W. Yang, K.-S. Chang, M. A. Aronova, T. Venkatesan, R. D. Vispute, and L. A. Bendersky, *J. Appl. Phys.* **94**, 7336 (2003).

20. Toshihiro Miyata, Shingo Suzuki, Makoto Ishii, Tadatsugu Minami, *Thin Solid Films*, **411,** 76 (2002).

21. T. Makino, Y. Segawa, M. Kawasaki, A. Ohtomo, R. Shiroki, K. Tamura, T. Yasuda, and H. Koinuma, *Appl.Phys.Lett.* **78**, 1237 (2001).

22. M. Levinshtein, S. Rumyantsev, M. Shur, Handbook series on semiconductor parameters,. , Singapore ; New Jersey : World Scientific, Vol. **1** (1996).

23. J. Newey, *Compound Semiconductor* **8**, 2 (2002).

24. T. Minami, *MRS Bulletin* **25**, 38 (2000).

25. Toshihiro Miyata, Tadatsugu Minami, Hirotoshi Sato and Shinzo Takata *Jpn. J. Appl. Phys.* **31**, 932 (1992).

26. R. D. Vispute, V. Talyansky, Z. Trajanovic, S. Choopun, M. Downes, R. P. Sharma, T. Venkatesan, M. C. Woods, R. T. Lareau, K. A. Jones, A. A. Iliadis, *Appl. Phys. Lett.* **70**, 2735 (1997).

27. S. Choopun, R. D. Vispute, W. Noch, A. Balsamo, R. P. Sharma, T. Venkatesan, A. Iliadis and D. C. Look , *Appl. Phys. Lett.* **75**, 3947 (1999).

330

28. R. D. Vispute, V. Talyansky, S. Choopun, R. P. Sharma, T. Venkatesan, M. He, X. Tang, J. B. Halpern, M. G. Spencer,Y. X. Li, L. G. Salamanca-Riba, A. A. Iliadis and K. A. Jones, *Appl. Phys. Lett.* **73**, 348 (1998).

29. F. Hamdani, A. E. Botchkarev, H. Tang and W. Kim, *Appl. Phys. Lett.* **71**, 3111 (1997).

30. J. Narayan, A.K. Sharma, A. Kvit, C. Jin, J.F. Muth and O.W. Holland, *Solid State Comm.* **121**, 9 (2002).

31. E.M. Levin, C.R. Robbins and H.F. McMurdie, *Phase Diagrams for Ceramists*, The American Ceramic Society, Columbus, Ohio (1964).

32. U. Rossler, *Landolt-Bornstein Numerical Data and Functional Relationships in Science and Technology: Semiconductors*, Vol. **III/41**, Springer-Verlag, Berlin (1999).

33. D. Dragoman and M. Dragoman, *Optical Characterization of Solids*, Springer-Verlag, Berlin (2002).

34. Q. Chen, *Electronics Letters* **31**, 1781 (1995).

35. B.K. Ng, F. Yan, J.P.R. David, R.C. Tozer, G.J. Rees, C. Qin, J.H. Zhao, *Photonics Technology Letters, IEEE*, **14** , 1342 (2002).

36. M. Razeghi and A. Rogalski, *J.App.Phys.* **79** (1996) 7433; F. Moscatelli, A. Scorzoni, A. Poggi, G.C. Cardinali and R. Nipoti, *Semi. Sci. Tech.* **18**, 554 (2003).

37. M.S. Shur and M.A. Khan, "GaN and AlGaN Ultraviolet Detectors," in *Semiconductors and Semimetals*, Academic Press, **57**, 407 (1999).

38. G. Parish, S. Keller, P. Kozodoy, J. P. Ibbetson, H. Marchand, P. T. Fini, S. B. Fleischer, S. P. DenBaars, U. K. Mishra, and E. J. Tarsa *Appl. Phys. Lett.* **75**, 247 (1999).

39. F. Omnes and E. Monroy *GaN-Based UV Photodetectors* in *Nitride Semiconductors: Handbook on Materials and Devices*, Wiley, New York (2003).

40. M. Sumiya and S. Fuke, *J. Nitride Semicond. Res.* **9**, 1 (2004).

41. A. Usui, H. Sunakawa, A. Sakai and A. Yamaguchi, *Jpn. J. Appl. Phys,.* **36**, L899 (1997).

42. O.H. Nam, M.D. Bremser, T. Zheleva and R.F. Davis, *Appl. Phys. Lett,.* **71**, 2638 (1997).

43. G.R. Carruthers *Electro-optics Handbook*, McGraw-Hill, New York, (1994).

44. M.J. Eccles, M.E. Sim and K.P. Tritton, *Low Light Level Detectors in Astronomy*, Cambridge University Press, Cambridge, England (1987).

45. S. Liang, H. Sheng, Y. Liu, Z. Huo, Y. Lu and H. Shen, *J. Cryst. Growth*, **225**, 110 (2001).

46. H. Sheng, S. Muthukumar, N.W. Emanetoglu and Y. Lu, *Appl. Phys. Lett.* **80**, 2132 (2002).

47. A.R. Hutson, *Phys. Rev,* **108**, 222 (1957).

48. S.E. Harrison, *Phys. Rev.* **93**, 52 (1954).

49. T. Minami, H.Nanto and S.Takata, *Jpn. J. Appl. Phys.* **23**, L280 (1983).

50. R. J. Hong, X. Jiang, G. Heide, B. Szyszka, V. Sittinger and W. Werner *J. Cryst. Growth* **249**, 461 (2003).

51. H. Tanaka *et al.* H. Tanaka, K. Ihara, T. Miyata, H. Sato and T. Minami, *J.Vac.Sci. Technol. A,* **22**, 1757 (2004).

52. T. Minami, *MRS Bulletin* **25**, 38 (2000).

53. K. Minegishi,Y.Koiwai, Y.Kikuchi, K. Yano, M.Kasuga and A.Shimizu, *Jpn J. Appl. Phys.* **36**, L1453 (1997).

54. Y. R. Ryu, S. Zhu, D. C. Look, J. M. Wrobel, H. M. Jeong and H. W. White, *J. Cryst. Growth* **216**, 330 (2000).

55. D. C. Look, D. C. Reynolds, C. W. Litton, R. L. Jones, D. B. Eason and G. Cantwell *Appl. Phys. Lett.* **81**, 1830 (2002).

56. Kyoung-Kook Kim, Hyun-Sik Kim, Dae-Kue Hwang, Jae-Hong Lim and Seong-Ju Park, *Appl. Phys. Lett.* **83**, 63 (2003).

Combinatorial Synthesis and
Materials Discovery

CHAPTER 11

COMBINATORIAL SYNTHESIS OF FUNCTIONAL METAL OXIDE THIN FILMS

Ichiro Takeuchi

Department of Materials Science and Engineering and Center for Superconductivity Research, University of Maryland, College Park MD 20742

11.1 INTRODUCTION

The combinatorial approach to materials is a new wave of research methodology which aims to dramatically increase the rate at which new compounds are discovered and improved. In this approach, up to thousands of different compositions can be synthesized and screened in an individual experiment for desired physical properties.[1-4] This methodology initially started in biochemistry, and within the last twenty years, combinatorial chemistry and high-throughput screening for new drugs and biomolecules have already revolutionized the pharmaceutical and DNA-sequencing industries.[5]

The scope of the combinatorial approach is far-reaching, and it can be used to address materials issues at different levels in a wide spectrum of topics ranging from catalytic powders[6] and polymers[7] to electronic and bio-functional materials.[3,7,8] In solid-state applications, the concept and effectiveness of the approach have been demonstrated in successful discoveries of new compounds in a number of key technological areas including optical materials, dielectric materials and magnetic materials.[1-4,8,9]

Various thin film synthesis techniques incorporating spatially varying or selective deposition can be used to create combinatorial libraries and composition spreads. There have been significant advances in this area, and at the most sophisticated level, tools for lattice engineering such as laser

334

molecular beam epitaxy (LMBE), have been used to fabricate compositionally varying samples.[10]

Figure 1 is a cartoon depicting the experimental steps of the combinatorial approach to materials using one synthesis technique. In this particular scheme, the library is synthesized using the thin-film precursor technique (discussed below), but the overall picture is representative of combinatorial thin film techniques in general. In the first step, fabrication of a thin film library with a large number of compositionally varying sites is achieved by carrying out a series of thin film deposition in conjunction with precisely positioned shadow masks that allow spatially selective deposition. Using a series of masks in a predetermined sequence, different combinations of amorphous precursor multilayers are deposited at all different sites on the library chip. Following the deposition, the chip is thermally processed in order to diffuse the precursors and to form the desired (crystalline) phases.

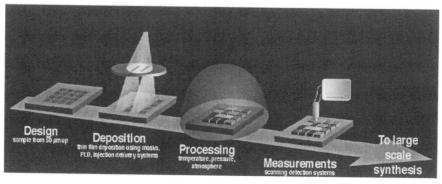

Figure 1: Cartoon showing experimental steps of a combinatorial experiment. In this scheme, a series of shadow masks are used to deliver different combinations of precursor thin films at different sites. Following the steps, one can take newly found compositions to large scale synthesis.

Various parallel measurements, scanning probe techniques, and/or sensor arrays are used to quickly characterize the specific physical properties of interest and screen the library.[1-4,11] Making quick and accurate measurements of a specific physical property from small volumes of materials in the libraries often represents a considerable feat of instrumentation. Researchers are taking on this challenge, which in some cases has led to invention of new measurement instruments.

Since the advent of high temperature superconductivity in cuprates, there has been an explosion of activities in transition metal oxides in general. A variety of functionalities observed in metal oxides provide a rich platform for exploring novel device applications. Strong correlation in many systems continues to dominate the focus of investigation in condensed matter physics. In magnetism alone, metal oxides exhibit a myriad of intriguing properties including colossal magnetoresistance and charge ordering in manganites to high spin polarization in Fe_3O_4 and long spin relaxation times in ferrites. In addition, there are many metal oxide systems which have traditionally been used as important device materials such as dielectric insulators, ferroelectric materials and phosphors. Because there are always needs to find better materials with improved physical properties, combinatorial strategy can play a significant role in development of new metal oxides.

This chapter will discuss the utility of the combinatorial approach in the field of functional metal oxide thin film materials. The emphasis is placed on work carried out in our group at the University of Maryland.

11. 2 FABRICATION OF COMBINATORIAL LIBRARIES OF METAL OXIDES

There are a number of thin film deposition techniques that can be used for fabrication of combinatorial samples of metal oxides. Essentially any physical vapor deposition technique can be used for achieving the spatially selective deposition facilitated by implementation of shadow masks as shown in the first step in Figure 1. Common techniques include pulsed laser deposition (PLD) and sputtering. In some instances, chemical vapor deposition techniques have also been demonstrated.[12,13]

PLD has several key advantages over other techniques in library fabrication. Laser ablation of materials off of bulk targets is a highly non-equilibrium process which allows stoichiometric transfer and delivery of target composition to the substrate. This aspect is crucial for metal oxides where materials systems of interest often comprise of multiple components with different vapor pressures. PLD can be performed in high vacuum or in low pressure of reactive gases. A single film can be made in a relatively short period of time with a quick turn-around. By monitoring the number of laser pulses, one can control the deposition of materials at an atomic layer level, and by incorporating a layer-by-layer deposition technique, one can design and explore novel materials systems that do not exist in bulk forms. One requirement of the deposition technique in performing fabrication of combinatorial libraries is that it needs to facilitate deposition of a number of different materials in a single series of deposition sequence. In PLD, target materials do not need to be biased, and a multiple-target carrousel allows sequential deposition of different materials in a single run. In general, PLD systems are relatively inexpensive, and they can be set-up quickly and easily with minimum maintenance requirements.

Another valuable synthesis method for creating compositionally varying thin films is co-sputtering. In this scheme, naturally non-uniform deposition thickness profiles from several sputtering targets are used to create composition spreads across wafers as discussed below.

11.3 SYNTHESIS TECHNIQUE USING THIN FILM PRECURSORS

The very first demonstration of the combinatorial approach to materials was performed using libraries fabricated by the thin-film precursor technique.[14] One goal library designs is to survey as diverse a compositional variation as possible in individual experiments using mathematically designed masking strategies. To this end, a series of high precision shadow masks have been used to define the layout of multilayers of amorphous precursors for all the sites. Elemental metals, simple metal oxides, flurides and carbonates are often used as precursors. For instance, in order to make $(Ba,Sr)TiO_3$, a multilayer consisting of amorphous layers of BaF_2, SrF_2 and TiO_2 are used (Fig. 2).[15,16] The substrate is usually held at room temperature or very low temperature (100~200 °C) during the deposition of precursor layers. Post-annealing and heat treatment are necessary for diffusion of amorphous precursors and phase formation.

There are different combinatorial mask configurations to effectively create and investigate large compositional phase spaces. Figure 3 illustrates the quaternary combinatorial masking scheme.[15,17] This scheme involves n different masks, which successively subdivide the substrate into a series of self-similar patterns of quadrants. The rth ($1 \leq r \leq n$) mask contains 4^{r-1} openings where each opening exposes one quarter of the area exposed in the preceding mask. Within each opening, there are an array of 4^{n-r} gridded sample sites. Each mask is used in up to four sequential depositions, where for each deposition the mask is rotated by 90°. This process results in up to 4^n different combinations of precursors created after $4 \times n$ precursor depositions. This can be effectively applied to survey a large number of different compositions each consisting of up to n elemental components, and each component is selected from a group of up to four precursors. In many ways, the use of the precursor technique and the masking strategy such as this brings out the most in the combinatorial approach since the number of compositionally varying samples is determined in a mathematically combinatorial way.

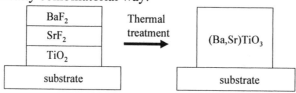

Figure 2: Scheme of phase formation in the precursor technique. An amorphous precursor multilayer deposited at room temperature is converted to a crystalline compound.

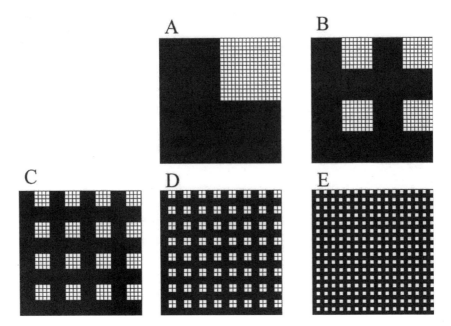

Figure 3: A set of physical shadow masks used in the quaternary masking strategy. Each mask is used in up to 4 depositions.

Implementation of masking schemes is accomplished using either photolithographic physical shadow masking or lift-off steps. Because of its high spatial resolution and alignment accuracy, the lift-off method is particularly suitable for generating chips containing a high density of sites. The trade-off is that after each deposition, a lift-off step and spinning and patterning of the next layer of photoresist needs to be performed. Figure 4 is a photograph (taken under daylight) of a 1024-member library chip designed to search for new luminescent materials. This library was laid out mostly using PLD together with lift-off of each layer using 5 quaternary masks on a 1 inch x 1 inch substrate. The photograph was taken following the deposition prior to annealing.[17] Different sites on the library all appear differently due to the varied thicknesses and different optical indexes of precursor multilayer films. The color variation and its distribution in the library seen here reflect the diversity one can achieve using this technique. This particular experiment yielded a number of leads in new luminescent materials and has led to the discovery of an efficient blue emitting photoluminescent composite material, $Gd_3Ga_5O_{12}/SiO_2$.[17]

338

Figure 4: A photograph of a luminescent materials library (1" x 1"). Library was laid out using the quaternary masking technique implemented using photolithography.[Ref. 17]

11.4 HIGH-THROUGHPUT THIN FILM DEPOSITION FLANGE DEVELOPED AT THE UNIVERSITY OF MARYLAND

In incorporating the combinatorial strategy for exploring metal oxide systems using PLD, our philosophy was to maintain and enhance the versatility and the relative simplicity of PLD. To this end, we have developed a compact 8" diameter combinatorial thin film deposition flange where all the necessary components for creating combinatorial libraries and composition spreads of different designs are integrated in a single vacuum flange.[18] Figure 5 is a photograph of the flange. Essentially, it is a 1.5-inch diameter substrate–mounting heater plate (which can go up to ~ 800 °C) integrated with a two-dimensional shuttering system. The shutters allow spatially selective shadow deposition of different layouts anywhere on mounted substrates up to 1-inch square in size. The main advantage of the flange is that it is not a permanent fixture to any one system, and it can be placed on any deposition system as long as the flange size is compatible. We have used one combinatorial flange on several different PLD chambers each dedicated to a specific materials system to demonstrate its portable nature. The current version of the flange is approximately 19" in height and weighs approximately 25 lbs. The heart of the flange is entirely modular, and it can, in principle, be placed on flanges of any sizes 6" and up. Although, up to now, it has been exclusively used for PLD, it is also applicable to other physical vapor deposition systems such as sputtering and evaporation processes where shadow deposition is readily achievable.

x-y movable
shutters/masks

drive chain   rotatable heater plate

Figure 5: A photograph of the top face of the combinatorial thin-film deposition flange. It can be mounted in any appropriate physical vapor deposition chamber to perform combinatorial library synthesis. X-Y shutters are replaceable, and any aperture shape or size can be cut into them.

Apertures of any shapes can be cut into the shutter plates depending on the required layout design of a particular combinatorial experiment. For instance, it can be a simple square opening used for fabricating composition spreads (described below) or a set of window patterns with the quaternary masking configuration discusses above. The shutter plates and the heater plates are replaceable, and they are exchanged from experiment to experiment in order to minimize cross-contamination through re-sputtering of materials deposited on them. We have found that the two overlapping shutters which move in two orthogonal directions combined with a rotatable heater allows us to accommodate virtually any type of library/spread layout designs. The shutters can move with variable speeds from 0.18 mm/sec to 1.2 mm/sec. The shutter-driving motors, the target

340

carousel, which fits up to 6 target materials, and the laser are all synchronized by a computer and controlled via a Labview compiled software.

Using the aperture patterns for the two shutters as shown in Figure 5, one can make a discrete library where thin film samples of different compositions are deposited on separated individual sites of square areas. Figure 6 is a photograph of a gas sensor library made in this manner.[19] The utility of combinatorial libraries for gas sensor applications is two-fold: one is to search and optimize the compositions for high sensitivity and selectivity of gases, and the other is to make use of the natural array geometry of the libraries for an electronic nose, where an array of sensors are multiplexed for gas identification by pattern recognition. Prior to the deposition of 16 sites on the library, a two-terminal Au electrode pattern was created by a lift-off process on a sapphire substrate. 500 Å thick sensor films of different compositions were deposited on selected areas (2 mm x 2 mm) at fixed positions on the Au electrode pattern. The array consisted of 16 compositions where SnO_2 was the host material and ZnO, WO_3, In_2O_3, Pt, and Pd were the dopants. The films were deposited at 550 °C in oxygen partial pressure of 2×10^{-3} Torr. These dopants were selected based on previous reports on single composition gas sensor studies.

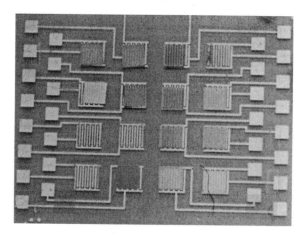

Figure 6: A photograph of an electronic nose: a combinatorial library of semiconductor based gas sensors (1" x 1"). [Ref. 18]

One major drawback of PLD is the limited area of uniformly deposited film thickness on substrates because of the finite size of the plasma plume. Under a typical deposition condition of 100 mTorr of O_2, this area is at most ~ 1 cm^2. Because the gas sensor library chip was 1" square, deposition was performed in four steps where a quarter of the substrate was exposed for deposition at a time. This was facilitated by 90° rotations of the substrate between steps. To create thin films of selected chemical compositions, we deposited materials in a layer-by-layer process where SnO_2 and other layers were deposited in an alternating manner. Each layer of SnO_2 and its dopant material was less than 4 Å thick so that dopants are intimately mixed at a sub-nanometer level. For deposition of the 16 sensors, the total deposition time of one library was ~ 3.5 hours, most of which was spent on moving and aligning the automated shutters. The testing of the chips was successfully performed in a gas flow chamber where all sensors were electrically connected to the outside electronics to monitor its resistance change. We were able to demonstrate the detection of different gas species by pattern recognition. In this manner, the entire library is used as one electronic nose device.[19]

11.5 COMBINATORIAL LASER MOLECULAR BEAM EPITAXY

While the precursor multilayer method described above has proven to be effective in creating diverse compositional variation across libraries, the process is very different from the sequence-controlled synthesis carried out in combinatorial chemistry of organic molecules where bonding of molecules can be sequentially regulated at each site.

If one were to extend the notion of sequence-controlled synthesis to inorganic synthesis, an atomically controlled layer-by-layer growth of a thin film at each reaction site needs to be achieved. In this manner, one can fabricate a library of artificially designed crystal structures, superlattices, or nano-structured devices. Every crystalline material can be viewed as being composed of molecular layers which are stacked periodically. State-of-the-art thin film deposition techniques such as molecular beam epitaxy and atomic layer epitaxy are used to control atomic layering sequences in thin films. In ceramic thin films, such technologies have undergone a considerable progress following the discovery of high T_c superconductors. Atomic layer controlled deposition of oxides was introduced in 1990[20,21] and another major breakthorugh took place in 1994 when an atomically finished substrate surface was employed for laser molecular beam epitaxy of $SrTiO_3$.[22]

In order to carry out parallel fabrication of layered structures by atomic layer-by-layer process, combinatorial laser molecular beam epitaxy (CLMBE) was developed by H. Koinuma et al. at Tokyo Institute of Technology[10]. CLMBE allows concurrent fabrication of a number of artificial lattices or heterojunctions where each layer composition, thickness, and the deposition sequence are manipulated at the atomic scale. This represents a significant improvement in one's ability to design novel materials systems: libraries can now be constructed where the arrangement of atoms is essentially controlled in three dimensions.[23,24]

Three key features of a CLMBE system make it a unique combinatorial synthesis tool: a set of physical masks for defining the film growth site, a scanning reflection high energy electron diffraction (RHEED) system for in-situ diagnostics of film growth mode at various sites during the growth, and a fiber guided Nd-YAG laser for substrate heating (Figure 7). By monitoring the growth mode with RHEED and by synchronizing the target switching with mask movements, synthesis of a number of atomic layer controlled materials can be coordinated on a single substrate. The lattice engineering achieved in this manner is best exemplified in simultaneous fabrication of perovskite oxide multilayers and superlattices. A clear intensity oscillation in scanning RHEED, each one of which corresponds to the growth of a unitcell layer, can be observed everywhere on the surfaces defined by the mask pattern. The RHEED beam scanning across the

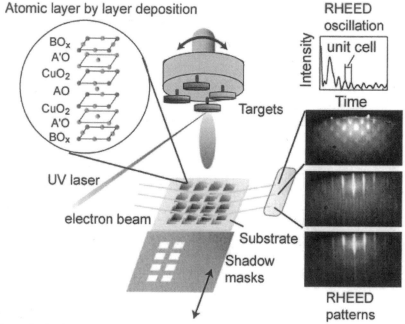

Figure 7: Combinatorial laser molecular beam epitaxy. Schematic setup for parallel growth of atomically controlled thin film materials. Reflection high-energy electron diffraction (RHEED) intensities versus time is obtained during the growth for a number of samples on the library simultaneously. The blow-up shows the schematic for "constructing" CuO_2 based layered material by atomic layer-by layer deposition.

substrate surface with a pair of coils and image acquisition can be synchronized with the moving of shadow deposition masks to control the growth of as many as ten thin film strips in parallel. The laser which heats substrates up to 1400 °C even in O_2 pressure as high as 1 atm can also be used to create a temperature gradient on a substrate.[25] Fine-tuning the optimum deposition temperature is a perennial problem in any thin-film research, and this capability allows one to exploit temperature as a combinatorial parameter. Thus, it is now possible to simultaneously optimize temperature and composition in one experiment. This technique has been successfully implemented in optimizing dielectric [26] and luminescent materials.[27,28]

In another mode of operation, by automating the motion of a shadow mask during depositions, atomically mixed epitaxially grown continuous composition spreads can be fabricated.[29] This technique can also be extended to ternary composition spreads.[30,31]

A good example of serendipitous discovery brought about by the combinatorial approach is the unexpected finding of optically transparent magnetism in TiO_2 anatase films doped with Co from a library made with CLMBE.[32] Ferromagnetic semiconductors formed by doping magnetic impurities into host semiconductors are key materials for spintronics, where the correlation between the spin and the charge of electrons gives rise to spin-dependent functionalities such as giant magnetoresistance and spin field-effect transistors. All known magnetic semiconductors until now had been based on non-oxides such as GaAs and ZnSe, and their highest Curie temperature was ~100 K. There was a theoretical prediction that ZnO would become ferromagnetic when doped with certain $3d$ transistion elements, but to date no sign of ferromatetism has been observed in such systems.

Titanium dioxide (TiO_2) is commonly used for photcatalysis for water cleavage and is well known for its unique properties including high refractive index, excellent optical transmittance in the visible and infrared regions, and high dielectric constant. Libraries of $3d$ transistion-metal-doped TiO_2 were originally fabricated by CLMBE in a search for photocatalysts. When their magnetic properties were checked with a scanning superconducting quantum interference device (SQUID) microscope, Co-doped TiO_2 was found, surprisingly, to exhibit ferromagnetism at room temperature while maintaining its transparency and single phase up to the composition of 8% Co. In order to systematically study the changes in film properties as a function of doping level, a library containing nine $Ti_{1-x}Co_xO_2$ films with different x values was fabricated. TEM of the films has indicated no sign of segregation of impurity phases in the compositional range of $x < 0.08$. This discovery has launched a new field of oxide-based dilute magnetic semiconductors, and researchers around the world are now competing to find other oxide materials with similar magnetic properties.[33]

11.6 COMPOSITION SPREADS AND COMBINATORIAL MATERIALS SCIENCE

In addition to serving the discovery/optimization needs of materials in key technology areas, the combinatorial approach can be used to address fundamental solid-state physics and chemistry questions as well. Composition spread experiments are highly effective in rapidly mapping compositional phase diagrams and systematically investigating physical properties which vary across compositional ranges.[34-36] They can be used to map the details of known materials systems or probe previously unexplored materials phase-space. This represents a novel and rapid way to construct composition-structure-property relationships.

Prior to the recent surge of activities in combinatorial materials sicnece, co-sputtering had been used as the main technique for creating thin-film composition spreads.[37] van Dover et al employed three magnetron sputtering guns in the 90° off-axis configuration to form a two-dimensional pseudoternary spread of oxide dielectric materials.[38] Over 30 chemical systems were explored in this study, resulting in the identification of amorphous $Zr_{.2}Sn_{.2}Ti_{.6}O_4$ as a particularly promising material. For each chemical system a codeposited composition spread was prepared on a 7×7 cm substrate coated with a metallic base electrode. The deposition was performed using metal targets and a reactive Ar-O_2 atmosphere. The capacitance and current-voltage characteristics of the resulting films were measured using a scanning Hg-probe counterelectrode, and a virtual array of about 4000 measurements was constructed by making measurements of the film on 1 mm centers. The product of the capacitance and breakdown voltage, normalized to the area of the Hg counterelectrode, is an important figure of merit for capacitor materials, as it represents the maximum charge per unit area that can be stored in the capacitor. It was found that compositions in the vicinity of $Zr_{.2}Sn_{.2}Ti_{.6}O_4$ have optimal properties, with a maximum stored charge of about 25 $\mu C/cm^2$. This value compares favorably with that obtained for deposited films of SiO_2 (3-5 $\mu C/cm^2$), Al_2O_3 (6-10 $\mu C/cm^2$), and Ta_2O_5 (10-12 $\mu C/cm^2$). Films made by conventional single-composition on-axis sputtering were subsequently shown to give identical results,[39] raising the prospect that this material might prove useful as an alternative dielectric for DRAM capacitors.

As mentioned above, using combinatorial PLD one can synthesize *in-situ* epitaxially grown composition spreads. In this technique, typically, two ceramic targets (whose compositions serve as the two end compositions of the spread) are ablated in an alternating manner. A linear compositional gradient across the spread can be created by performing a series of shadow depositions through a rectangular opening in a shutter, which moves back and forth over the substrate during the deposition. The motion of the shutter is synchronized with the firing of the laser in such a way so that for each deposition, a thickness gradient "wedge" is created on a chip. In order to ensure alloy-like intermixing of two compositions at atomic level at any position on the spread, less than a unit cell is deposited for each pair of deposition (for the two targets), and up to hundreds of pairs of gradient deposition are carried out for making relatively thick samples. This process is shown schematically in Figure 8. The resulting spread has the composition varying continuously from one component at one end to another at the other end. The samples are typically less than 1 cm long in the spread direction.

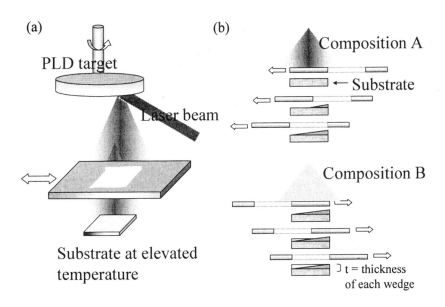

Figure 8: Fabrication scheme of a linear epitaxial composition spread. (a) Laser ablation is performed while the shutter with an edge moves across the substrate creating a thickness wedge. (b) Layers of alternating wedges are deposited for comspositions A and B. Each layer at the thick end is less than a unit cell. The thickness of each wedge is t. This process is repeated for as many as cycles as necessary.

This technique has the advantage that it provides specifically designed layouts of composition spread, unlike in co-sputtereing where composition distribution is dictated by the very nature of the co-sputtering process off of multiple targets. By ablating three targets in a repeating cycle, one can also use this method to construct ternary phase diagram chips on equilateral triangular shaped sample area.[30,31] This synthesis was first demonstrated for composition spreads of the manganite $La_{1-x}Sr_xMnO_3$ system where the composition was varied continuously from $LaMnO_3$ at one end to $SrMnO_3$ at the other end.[29] A variety of characterization techniques were used to obtain mapping of structure-composition-magnetic properties, and a magnetic phase diagram across the spread chip was obtained.

We have fabricated $Ba_{1-x}Sr_xTiO_3$ composition spreads were made using $BaTiO_3$ and $SrTiO_3$ targets on $LaAlO_3$ substrates, and a multimode scanning microwave microscope was used for their dielectric characterization at microwave frequencies.[40] Measurements were taken on the spread at different positions as the microscope tip was scanned over the spread, so that data are collected as a function of composition. Figure 9 shows the composition-spread profile of the dielectric constant at three different frequencies measured simultaneously at room temperature. The profile shows a continuous change with the expected compositional dependence: the peak in the profile is located near $Ba_{0.65}Sr_{0.35}TiO_3$, which has its Curie temperature near room temperature. On either side of this composition, the dielectric constant is lower.

From Figure 9, it is evident that the dielectric constant tends to decrease as the measurement frequency is increased for compositions with high Ba/Sr ratio. This frequency dispersion is minimal or close to zero near the $SrTiO_3$ end of the spread. It increases with increasing Ba content, reaches the highest values for x at around 0.2~0.4, and decreases again as x approaches zero. The observed dispersion here directly points to the presence of strong dielectric relaxation, and it is closely tied to the onset of ferroelectricity. It is largest for compositions experiencing the ferroelectric transitions. Above and below the transition temperature, the dispersion is reduced.

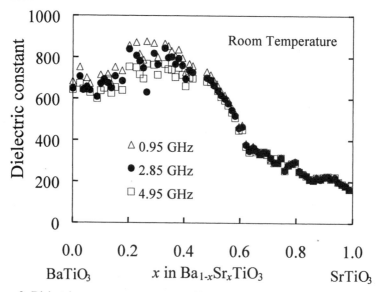

Figure 9: Dielectric constant versus composition on an epitaxially fabricated $Ba_{1-x}Sr_xTiO_3$ spread measured using a multimode microwave microscope at 0.95, 2.85 and 4.95 GHz simultaneously at room temperature.

This behavior is entirely consistent with the behavior of the soft mode near the paraelectric-ferroelectric transition, where the mode undergoes significant softening.[41] The dielectric frequency dispersion is maximum near the compositions undergoing transition, and above and below the transition range, it is markedly lower.[42] Thus, a single temperature measurement of the composition spread here allows an indirect probe of the temperature dependent behavior of the entire $(Ba,Sr)TiO_3$ system.

In another experiment, linear composition spreads of $BaTiO_3$ (BTO)-$CoFe_2O_4$ (CFO) were fabricated.[43] The idea behind this experiment is to place pure BTO which is ferroelectric at one end of the spread and place CFO which is ferromagnetic at the other end, and gradually mix the two toward the middle of the spread. The goal was to see if we can find thin film compositions which are simultaneously ferroelectric and ferromagnetic somewhere in the middle of the spread. Such multiferroic materials are of great interest from the point of view of developing novel devices which couple functionalities of ferroelectricity and ferromagnetism. Because BTO and CFO do not form solid solutions, in order to provide some "integrity" in the respective materials, spreads were made in such a way so that there is more than one unit cell of BTO and CFO at the thick end of each wedge layer (defined as t in Figure 8(b)). We made a series of composition spreads where t was approximately 0.84 nm, 2.52 nm, 8.4 nm, and 12.6 nm (see Fig. 1(b)). These numbers roughly correspond to 1x, 3x, 10x, and 15x the unit cell of CFO. BTO has a pseudo-cubic structure with the lattice constant a = 0.403 nm and CFO has a spinel cubic structure with the lattice constant a = 0.839 nm. Although the two materials have different structures, the mismatch between twice the lattice parameter of BTO and that of CFO is roughly 5 %, and we have found that they can be grown together in a hetero-epitaxial manner on (100) MgO (cubic with a = 4.2 nm). The typical size and the thickness of a composition spread sample was 6 mm (in the spread direction) x 10 mm and 300 nm (uniform across spreads), respectively. In each spread, the average composition continuously changes from pure BTO to pure CFO.

Structural changes of the BTO - CFO artificial multilayer spreads were studied by the ω scan mode of a scanning X-ray microdiffractometer (D8 DISCOVER with GADDS for combinatorial screening by Brucker-AXS). A 300 μm spot size was used to take twenty points across each spread. The X-ray intensity as functions of 2θ and the average composition for a spread in the 2θ range of 42 ° to 47 ° is shown in Figure 10. The peaks are integrated from -94.9° < χ < -78.4°. t was approximately 1.7 nm for this sample. It is seen that the (200) BTO peak and the (400) CFO peak continuously "evolve" toward the middle of the spread. The continuous shifts and the broadening of the peaks are indicative of significant and progressive lattice distortion caused by diffusion of cations (Ba^{2+} and Ti^{4+}

of $BaTiO_3$ and Co^{2+} and Fe^{3+} of $CoFe_2O_4$) at the multilayer interfaces. We have found that there is less lattice distortion toward the middle of the spread for spreads with larger thickness of the wedge layers: the X-ray peak positions do not shift in the middle of the spread for samples where t is 10x and 15x the unit cell of CFO. This is consistent with the fact that in comparing spreads with different wedge thicknesses, when the individual layer is thicker, there are fewer interfaces, and thus there is less overall fraction of materials affected by the interdiffusion at the interfaces.

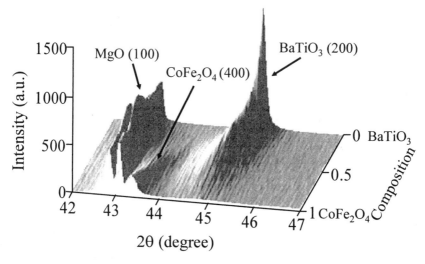

Figure 10: The scanning X-ray microdiffraction of a $BaTiO_3$-$CoFe_2O_4$ spread sample for which t is 1.68 nm. The intensity is plotted as a function of 2θ (from 42° to 47°) and as a function of average composition (from pure BTO to pure CFO).

In order to study the continuously changing properties of the spreads, microwave microscopy[40] and room temperature scanning superconducting quantum interface device (SQUID) microscopy were used. The linear dielectric constant measured at 1 GHz using the microwave microscope is shown in Fig. 11 for a spread where t was 8.4 nm. The linear dielectric constant systematically decreases from 450 for the pure BTO end to 40 for the pure CFO end. The decreasing trend across the spread was found to be similar for all spreads, and the general trend is consistent with the one obtained from calculating the effective dielectric constant using the series capacitance model where layers with different dielectric constants are alternating. The dielectric loss value at 1 GHz was found to vary continuously from ≈ 0.2 for the BTO end to ≈ 0.3 for the CFO end for all the spreads. These loss values are comparable to values obtained by other measurement methods for epitaxial $(Ba,Sr)TiO_3$ films at microwave frequencies.[44,45]

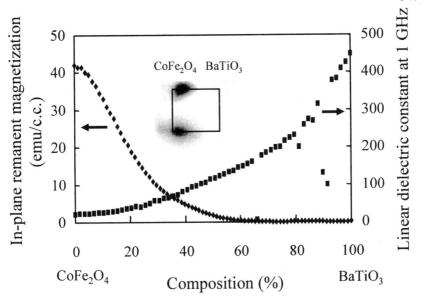

Figure 11: Dielectric constant (right) measured by the microwave microscope and the in-plane remanent magnetization (left) characterized by the scanning SQUID microscope as a function of average composition across a BaTiO$_3$-CoFe$_2$O$_4$ spread. t for this spread is 8.4 nm. The inset shows the false color scanning SQUID image of the spread. The black line outlines the spread sample. The magnetic dipole with varying magnetization across the spread can be seen.

For characterizing the ferromagnetic properties, mapping of magnetic field emanating from the sample due to remanent magnetization is obtained using a room temperature scanning SQUID microscope. We magnetize our samples in the in-plane direction even though the films studied here predominately showed the out-of-plane direction as the easy axis. This is because our present algorithm[46] for obtaining quantitative remanent magnetization uses an in-plane magnetized model. The inset of Fig. 11 shows a false color scanning SQUID image (field distribution) obtained for the spread where t was 8.4 nm. The trend of how the in-plane remanent magnetization changes as a function of various parameters is used as an indicator of the overall magnetic properties. The in-plane remanent magnetization value of the pure CFO end is consistent with those measured on individual composition samples measured with a SQUID magnetometer, and with reported values for CFO thin films. Despite the decaying of properties as one goes toward the middle of the spread, we see that there is a region that is magnetic and has a reasonably high dielectric constant ($\approx$ 120 at (BTO)$_{0.5}$ – (CFO)$_{0.5}$). High values of dielectric constant are often a good indicator of the presence of ferroelectricity.

In order to confirm the dielectric and ferromagnetic properties observed here, a single composition sample with a uniform average composition of $BTO_{0.45} - CFO_{0.55}$ where the periods of CFO and BTO layers were $\approx$ 1.4 nm and $\approx$ 1.1 nm respectively was fabricated. The dielectric constant of this sample was found to be $\approx$ 30, and a SQUID magnetometer measurement result indicated the presence of ferromagnetism in this sample.

11.7 CONCLUSION

I have discussed applications of the combinatorial strategy for exploring functional metal oxide thin films. I have given examples where thin film deposition techniques were used to fabricate combinatorial samples. Pulsed laser deposition techniques are extensively used to make libraries of metal oxide films. Thin film libraries have been effectively used to discover various new compounds. Composition spreads have been demonstrated as a powerful tool for quickly mapping the composition-structure-property relationship in different materials systems. The high-throughput methodology promotes fundamental changes in the practice of materials discovery and development. Because the notion of search and discover is central to materials science, it is expected that the role of the combinatorial approach will continue to expand in the field of functional metal oxides.

ACKNOWLEDGEMENT

The author is grateful to many colleagues for extensive collaboration and many insightful discussions over the years. I would particularly like to acknowledge the following colleagues for key contributions in the work described here: X.-D. Xiang, H. Chang, Y. Yoo, C. Gao, P. G. Schultz, K.-S. Chang, M. Murakami, M. A. Aronova, L. Knauss, L. A. Bendersky, and H. Koinuma. This work was supported by NSF DMR 0094265 (CAREER), NSF DMR 0231291, NSF DMR-00-80008 (MRSEC), ONR N000140110761, and ONR N000140410085.

REFERENCES

[1] H. Koinuma and I. Takeuchi, *Nature Materials* **3**, 429-438 (2004).

[2] I. Takeuchi, R. B. van Dover, and H. Koinuma, *MRS Bulletin*, **27**, 301-308 (2002).

[3] I. Takeuchi, J. M. Newsam, L. C. Wille, H. Koinuma, and E. J. Amis, ed., Combinatorial and artificial intelligence methods in materials science, MRS Symp., Proc., 700 (2002).

[4] X.-D. Xiang and I. Takeuchi, I, ed., Combinatorial Materials Synthesis, Marcel Dekker, (2003). ISBN: 0-8247-4119-6

[5] See for example, Michal Lebl, *J. Comb. Chem.* **1**, 3-24 (1999).

6   S. Senken, K. Krantz, S. Ozturk, V. Zengin, I. Onal, *Angew. Chem. Int. Ed.* **38**, 2794-2799 (1999).

7   U. S. Schubert and E. Amis, ed., *Macromol. Rapid Comm.*, Special issue, on "Combinatorial Research and High-Throughput Experimentation in Polymer and Materials Research" **24** (1) (2003).

8   Xiang, X.-D. Combinatorial materials synthesis and screening: an integrated materials chip approach to discovery and optimization of functional materials. Ann. Rev. Mater. Sci., **29**, 149-171 (1999).

9   Y. Matsumoto, H. Koinuma, T. Hasegawa, I. Takeuchi, F. Tsui, and Y. K. Yoo, *MRS Bulletin* **28**, 734-739 (2003).

10   H. Koinuma, *Solid State Ionics* **108**, 1-7 (1998)

11   R. A. Potyrailo and W. G. Morris, *Review of Scientific Instruments* **75**, 2177-2186 (2004).

12   Bin Xia, Ryan C. Smith, Tyler L. Moersch, Wayne Gladfelter, *Applied Surface Science* **223**, 14-19 (2004).

13   J. O. Choo, R. A. Adomaitis, G. W. Rubloff, L. Henn-Lecordier and Y. Liu, manuscript in preparation.

14   X.-D. Xiang, X.-D. Sun, G. Briceno, Y. Lou, K.-A. Wang, H. Chang, W. G. Wallace-Freedman, S.-W. Chen, and P. G. Schultz, *Science* **268**, 1738-1740 (1995).

15   H. Chang, C. Gao, I. Takeuchi, Y. Yoo, J. Wang, P. G. Schultz, X.-D. Xiang, R. P. Sharma, M. Downes, T. Venkatesan, *Appl. Phys. Lett.* **72**, 2185-2187 (1998).

16   I. Takeuchi, K. Chang, R. P. Sharma, L. A. Bendersky, H. Chang, X.-D. Xiang, E. A. Stach, and C.-Y. Song, *J. of Applied Physics* **90**, 2474-2478 (2001).

17   J. Wang, Y. Yoo, C. Gao, I. Takeuchi, X.-D. Sun, H. Chang, X.-D. Xiang, and P. G. Schultz, *Science* **279**, 1712-1714 (1998).

18   K.-S. Chang, M. A. Aronova, and I. Takeuchi, *Appl. Surface Sci.* in press.

19   M. A. Aronova, K. S. Chang, I. Takeuchi, H. Jabs, D. Westerheim, A. Gonzalez-Martin, J. Kim, B. Lewis, *Appl. Phys. Lett.* **83**, 1255-1257 (2003).

20   H. Koinuma, H. Nagata, T. Tsukahara, S. Gonda, and M. Yoshimoto, Ext. Abst. of 22nd Conf. on Solid State Devices and Materials (SSDM'90) Aug.22-24, 1990 (Sendai) (1990) pp.933-936.

21   T. Terashima, Y. Bando, K. Iijima, K. Yamamoto, K. Hirata, K. Hayashi, K. Kumagaki, H. Terauchi, *Phys. Rev. Lett.* **65**, 2684-2687 (1990).

22   M. Kawasaki, K. Takahashi, T. Maeda, R. Tsuchiya, M. Shinohara, O. Ishiyama, T. Yonezawa, M. Yoshimoto, H. Koinuma, H. *Science* **266**, 1540-1542 (1994).

23   H. Koinuma, N. H. Aiyer, and Y. Matsumoto, *Science and Technology of Advanced Materials* **1**, 1-10 (2000).

24   M. Lippmaa, M. Kawasaki, H. Koinuma. H. Chapter 5 in Xiang, X.-D. and Takeuchi, I, ed., Combinatorial Materials Synthesis, Marcel Dekker, (2003). ISBN: 0-8247-4119-6

25   T. Koida, D. Komiyama, H. Koinuma, M. Ohtani, M. Lippmaa, M. Kawasaki, *Appl. Phys. Lett.* **80**, 565-567 (2002).

26   H. Minami, K. Itaka, P. Ahmet, D. Komiyama, T. Chikyow, M. Lippmaa, M., and H. Koinuma, *Jpn. J. Appl. Phys.* **41**, L149-L151 (2002).

27   H. Kubota, R. Takahashi, T.-W. Kim, T. Kawazoe, M. Ohtsu, N. Arai, M. Yoshimura, H. Nakao, H. Furuya, Y. Mori, T. Sasaki, Y. Matsumoto, H. Koinuma, *Appl. Surf. Sci.* **223**, 241-244 (2004).

28   K. Tamura, T. Makino, A. Tsukazaki, M. Sumiya, S. Fuke, T. Furumochi, M. Lippmaa, C. H. Chia, Y. Segawa, H. Koinuma, M. Kawasaki, *Solid State Comm.* **127**, 265-269 (2003).

29   T. Fukumura, M. Ohtani, M. Kawasaki, Y. Okimoto, T. Kageyama, T. Koida, T. Hasegawa, Y. Tokuran and H. Koninuma, *Appl. Phys. Lett.* **77**, 3426-3428 (2000).

352

[30] K. Hasegawa, P. Ahmet, N. Okazaki, T. Hasegawa, K. Fujimoto, M. Watanabe, T. Chikyow, H. Koinuma, *Appl. Sur. Sci.* **223**, 229-232 (2004).

[31] R. Takahashi, H. Kubota, M. Murakami, Y. Yamamoto, Y. Matsumoto, Y., and H. Koinuma, *J. Comb. Chem.* **6**, 50 -53 (2004).

[32] Yuji Matsumoto, Makoto Murakami, Tomoji Shono, Tetsuya Hasegawa, Tomoteru Fukumura, Masashi Kawasaki, Parhat Ahmet, Toyohiro Chikyow, Shin-ya Koshihara, and Hideomi Koinuma, *Science* **291**, 854-856 (2001).

[33] See for example, MRS Bulletin 28, October 2003, special issue on spintronics

[34] H. Chang, I. Tacheuchi, and X.-D. Xiang, *Appl. Phys. Lett.* **74**, 1165-1167 (1999).

[35] Y. K. Yoo, F. W. Duewer, H. Yang, D. Yi, J.-W. Li, and X.-D. Xiang, *Nature* **406**, 704-706 (2000).

[36] I. Takeuchi, O.O. Famodu, J.C. Read, M.A. Aronova, K.-S. Chang, C. Craciunescu, S.E. Lofland, M. Wuttig, F.C. Wellstood, L. Knauss, A. Orozco, *Nature Materials* **2**, 180-184 (2003).

[37] J. J. Hanak, in *Combinatorial Materials Synthesis*, editors X.-D. Xiang and I. Takeuchi (Marcel Dekker, New York, 2003).

[38] Robert Bruce van Dover, L. F. Schneemeyer, and R. M. Fleming, *Nature* **392**, 162-164 (1998).

[39] R B van Dover, and L F Schneemeyer, IEEE Electron Device Letters **19**, 329- (1998) p.329.

[40] K.-S. Chang, M. A. Aronova, O. Famodu, I. Takeuchi, S. E. Lofland, J. Hattrick-Simpers, H. Chang, *Appl. Phys. Lett.* **79**, 4411-4413 (2001).

[41] M. E. Lines and A. M. Glass, *Principles and Applications of Ferroelectrics and Related Materials* (Oxford University Press, New York, 1977).

[42] James C. Booth, R. H. Ono, Ichiro Takeuchi, Kao-Shuo Chang, unpublished.

[43] K.-S. Chang, M. A. Aronova, C.-L. Lin. M. Murakami, M.-H. Yu, J. Hattrick-Simpers, O. O. Famodu, S. Y. Lee, R. Ramesh, M. Wuttig, I. Takeuchi, C. Gao, L. A. Bendersky, *Applied Physics Letters* **84**, 3091-3093 (2004).

[44] Y. G. Wang, M. E. Reeves, W. J. Kim, J. S. Horwitz, and F. J. Rachford, *Appl. Phys. Lett.* **78**, 3872-3874 (2001)

[45] J. C. Booth, L. R. Vale, and R. H. Ono, in Materials Issues for Tunable RF and Microwave Devices, edited by Q. X. Jia, F. A. Miranda, D. E. Oates, and X. X. Xi (Materials Research Society, Warrendale, PA, 2000), Vol. 606, pp. 253-264.

[46] E. F. Fleet, Ph.D. Thesis, University of Maryland, College Park (2000).

Film Growth and Real-Time *in situ* Characterization

CHAPTER 12

REAL-TIME GROWTH MONITORING BY HIGH-PRESSURE RHEED DURING PULSED LASER DEPOSITION

Guus Rijnders and Dave H.A. Blank

Faculty Science & Technology, MESA+ Institute for Nanotechnology, University of Twente

12.1 INTRODUCTION

The application of oxide thin films in electronic devices, relying on multilayer technology, requires (atomically) smooth film surfaces and interfaces. Understanding of the different mechanisms affecting the growth mode is, therefore, necessary to control the surface morphology during thin film growth.

Two independent processes, i.e., nucleation and growth, play an important role during vapor-phase epitaxial growth on an atomically flat surface. Here, nucleation causes the formation of surface steps and subsequent growth causes the lateral movement of these steps. Both processes are determined by kinetics, since they take place far from thermodynamic equilibrium. These kinetic processes affect the final surface morphology and are, therefore, extensively studied. Because of its high surface sensitivity, reflection high-energy electron diffraction (RHEED) has become an important tool for in-situ studies of these growth processes. RHEED, however, was considered as a high-vacuum analysis tool and was mainly utilized in low-pressure growth techniques like molecular beam epitaxy (MBE). The development of high-pressure RHEED enabled real-time growth monitoring and growth characterization at the high pressures used in PLD of oxides.

At the high temperatures necessary for the epitaxial growth of oxides, the very short deposition pulses in PLD cause a separation of the deposition and growth in time. Almost no nucleation and growth takes place during the deposition pulses. This enables measurement of the kinetic parameters, determining the growth of complex oxides in Pulsed Laser Deposition (PLD), at growth conditions by monitoring the decay of the adatom density between the deposition pulses. This information resulted in a novel growth technique, which we called Pulsed Laser Interval Deposition.

12.2 PULSED LASER DEPOSITION

Pulsed laser deposition refers to a thin film deposition technique using laser ablation from a target acting as the particle source. These particles are deposited on a substrate, placed opposite to the target, resulting in thin film growth. The most often mentioned reason to apply PLD is the easy stoichiometric transfer of material from target to substrate at high ambient pressure. PLD is, therefore, frequently used for the fabrication of complex oxide thin films, because a high oxygen pressure is favorable during growth.

PLD became widespread just after the discovery of the high-T_C superconductor $YBa_2Cu_3O_{7-\delta}$[1]. Here, stoichiometric transfer at high oxygen pressure is essential and first attempts of thin film fabrication of this complex oxide with PLD were very successful leading to the increased interest in PLD.

12.2.1 Basic principles

In PLD a pulsed high-energetic laser beam is focused on a target resulting in ablation of material. At the early stage of the laser pulse a dense layer of vapor is formed in front of the target. Energy absorption during the remainder of the laser pulse causes, both, pressure and temperature of this vapor to increase, resulting in partial ionization. This layer expands from the target surface due to the high pressure and forms the so-called "plasma plume". During this expansion, internal thermal and ionization energies are converted into the kinetic energy (several hundreds eV) of the ablated particles. Attenuation of the kinetic energy due to multiple collisions occurs during expansion into low-pressure background gas. Typically, a background pressure of 1 to 50 Pa is used for deposition of complex oxides. At these pressures, thermalization occurs at a penetration length comparable to the target to substrate distance. Several models, like the "drag force" and the "shock wave" model[2,3], have been proposed which describe this attenuation in low and high background pressure, respectively. Using an adiabatic thermalization model[4] a characteristic length scale is determined which defines the "plasma range". Plasma particles thermalize at distances larger than this characteristic length.

Ambient gas parameters, i.e., mass and pressure, determine the interaction with the ablated particles and, subsequently, the kinetic energy of the particles arriving the substrate. As a consequence, this kinetic energy can be varied from high initial energy in vacuum to low energies resulting from thermalization at sufficient large ambient pressure. This wide range is a unique feature of PLD and can be used to modify thin film growth. Here, the diffusivity[5,6] and, both, absorption and desorption probability[7] of the energetic particles at the film surface are the controllable parameters.

Another unique feature of PLD is the high deposition rate. In PLD of oxides, for instance RE123, typical deposition rates range from 10^{-2} to 10^{-1} nm/pulse with deposition pulse duration in the order of several μs to ~500 μs[8,9]. As a result, the deposition rate can be as high as 10^{2}-10^{5} nm/sec[10,11]. This value is orders of magnitude higher compared to other PVD techniques, like sputter deposition and MBE, which have typical deposition rates of 10^{-2}-10^{-1} nm/sec. The short duration of the intense deposition pulse has implications for the nucleation and growth processes. This is illustrated by comparing the pulse duration with the mean diffusion time t_D of deposited atoms, which is given by:

$$t_D = v^{-1} \exp\left(\frac{E_A}{k_B T}\right) \tag{1}$$

with v the attempt frequency for atomistic processes, E_A the activation energy for diffusion and k_B Boltzmann's constant. Here, the mean diffusion time t_D is of interest because it sets the time scale for the atomistic processes, including collision and nucleation. For complex oxides E_A range from a few tenths of eV to more than 2 eV and typical growth temperatures range from 800 to more then 1100 K. Using eq. (1) the mean diffusion time can be estimated. For a wide range of deposition conditions t_D exceeds the deposition pulse duration. As a consequence, the deposition can be regarded as instantaneous for every pulse. In PLD, this instantaneous deposition is followed by a relative long time interval, where no deposition takes place. During this time interval, which is determined by the pulse repetition rate, the adatoms rearrange on the surface by migration and subsequent incorporation through nucleation and growth. This rearrangement can be considered as an anneal process. Because of the instantaneous deposition, the two basic processes, i.e., random deposition and growth through rearrangement are separated in time, which is again unique for PLD.

Parameters which control the instantaneous deposition rate are the laser energy density at the target, pulse energy, distance between target and substrate and the ambient gas properties, i.e., pressure and mass. The average growth rate is determined by the pulse repetition rate and can be varied independently from the instantaneous deposition rate. The extremely high deposition rate leads to a very high degree of supersaturation $\Delta\mu$[12]:

$$\Delta\mu = k_B T \ln \frac{R}{R_0} \qquad (2)$$

where k_B is Boltzmann's constant, R the actual deposition rate and R_0 its equilibrium value at temperature T. The high degree of supersaturation $\Delta\mu$ causes two-dimensional (2D) nucleation of a high density of extremely small clusters. In case of PLD, the clusters can be as small as one atom to a few molecular blocks of the complex oxides. Because of the instantaneous deposition at typical PLD conditions, the nucleation takes place after the deposition pulse and can be considered as post-nucleation[13,14]. Sub-critical clusters are unstable in the time interval between the deposition pulses and dissociate into mobile atoms. These nucleate into new clusters or cause stable clusters to grow. The latter process is similar to Oswald ripening where larger islands grow at the expense of small islands or clusters. The separated random deposition and subsequent growth is, furthermore, advantageous for the study of growth kinetics.

12.2.2 PLD set-up

Figure 1 shows the schematic drawing of the PLD system equipped with high-pressure RHEED. In the experiments described in this chapter, a KrF excimer laser (Lambda Physic Compex 105, wavelength $\lambda = 248$ nm) with maximum pulse repetition rate of 50 Hz is used. The maximum pulse energy is 650 mJ with pulse duration ~25 nsec. A mask is used to select the homogeneous part of the laser beam, resulting in a spatial energy variation is ~5%. The mask is projected at an inclination of 45° on the target by means of a focusing lens (focal length ~450 mm). The energy density on the target is controlled by adjustment of, both, mask size and demagnification.

The multi-target holder and substrate holder including heater are mounted on a computer controlled XYZ-rotation stage and can be inserted via a load-lock system without breaking the vacuum (base pressure ~10^{-5} Pa).

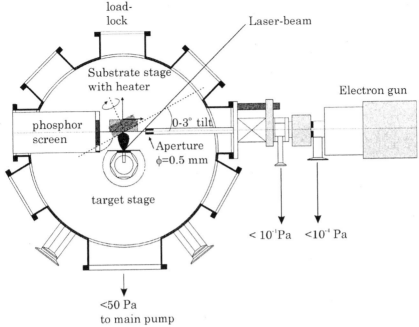

Figure 1: A schematic view of the deposition chamber, including the electron source assembly. The system is fully computer controlled including selection of the targets allowing for automated multi-layer deposition.

12.3 HIGH-PRESSURE REFLECTION HIGH-ENERGY ELECTRON DIFFRACTION (RHEED)

Reflection high-energy electron diffraction (RHEED) has become an important tool in surface science because of its high surface sensitivity. It utilizes diffraction of electrons by surface atoms and provides information of the periodic arrangement of the surface atoms. Because of the compatibility with (ultra) high vacuum deposition techniques, it is often used for the investigation of the surface morphology during thin film growth. It requires minimal hardware and thus a relative low-cost analysis technique.

12.3.1 Geometry and basic principles of RHEED

A schematic drawing of a typical RHEED geometry is sketched in figure 2. The incident electron beam (e-beam) strikes the sample surface at a grazing angle θ_i. The electrons are mono-energetic with a typical energy of $E \sim 10\text{-}50$ keV. The corresponding amplitude of the wavevector k_0 for these high-energy electrons can be estimated using:

$$E = \sqrt{\frac{\hbar^2 |k_0|^2}{m^*}} \qquad (3)$$

where m^* is the effective mass of the electron. Without relativistic correction the electron wavelength λ can be estimated by:

$$\lambda(\overset{\circ}{A}) = \sqrt{\frac{150}{E}} \qquad (4)$$

with E given in eV. At the typical energies used in RHEED, the electron wavelength λ is ~0.05-0.1 Å, which is an order of magnitude smaller than the thickness of an atomic layer. The angle of incidence is typically set to a few degrees (0.1-5°). At these grazing angles the penetration depth is only as small as a few atomic layers, which makes RHEED a surface sensitive diffraction technique; the electrons are easily scattered by surface steps and terraces. The coherence length[15], which is the maximum distance between reflected electrons that are able to interfere, is determined by the beam convergence and the energy spread of the electrons. This coherence length is typically of the order of several hundreds nm.

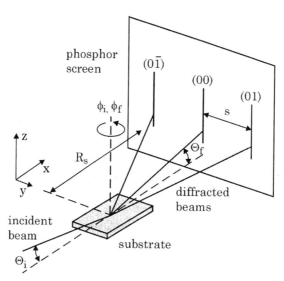

Figure 2: Schematic view of the RHEED geometry. θ_i (θ_f) and ϕ_i (ϕ_f) are the incident and azimuthal angles of the incident (diffracted) beam. R_S is the distance between substrate and phosphor screen and S the distance between the diffraction spots or streaks.

The electron gun (e-gun) and phosphor screen (acting as detector) are located far from the sample to avoid interference with the deposition process. In this geometry, electrons are scattered from the crystal surface resulting in a characteristic diffraction pattern on the phosphor screen. This pattern is instantaneously displayed and can be used to define the crystallographic surface structure of, for instance, a growing thin film.

The kinematical scattering theory is used to describe weakly interacting diffraction techniques, like x-ray or neutron diffraction. RHEED, however, involves strong interaction[16] of the electrons with the periodic potential of the crystal surface and can, therefore, not be described quantitatively by the kinematic approach. Nevertheless, the kinematic approach is used for, both, physical understanding and qualitative description of RHEED.

RHEED diffraction spots are produced when the momentum of the incident beam and that of the diffracted beam differ by a reciprocal lattice vector G:

$$k_S - k_0 = G \qquad (5)$$

where k_S and k_0 are the wavevectors of the diffracted and incident beams.

A useful geometrical representation of the conditions for diffraction in elastic scattering, i.e., $|k_S| = |k_0|$, is provided by the Ewald sphere construction, depicted in figure 3. Here, the reciprocal lattice of a 2-dimensional surface is a lattice of infinitely thin rods, perpendicular to the surface. The tip of the incident wavevector is attached to a reciprocal lattice rod. The Ewald sphere is defined by the sphere around the origin of k_0 with radius $|k_0|$ (equals $2\pi/\lambda$ for elastic scattering). The condition for diffraction is satisfied for all k_S connecting the origin of k_0 and a reciprocal lattice rod. For atomically flat and single domain crystalline surfaces, this result in diffraction spots lying on concentric circles, called Laue circles. Due to the high electron energy, used in RHEED, the Ewald sphere is very large compared to the reciprocal lattice spacing of oxide crystals. As a result, only a few reciprocal lattice rods are intersected at the small grazing incident angle. Essentially, a one-dimensional map of the reciprocal space is obtained, see figure 3 (b). Other areas of the reciprocal space are mapped by rotation of the sample about the incident angle or, alternatively, the azimuthal angle ϕ_i.

Figure 4 shows a diffraction pattern of a perfect $SrTiO_3$ surface[17]. It exhibits sharp diffraction spots lying on the 0th order Laue circle, i.e., intersections of the (0k) rods with the Ewald sphere. Due to the substrate size and RHEED geometry, i.e., e-beam diameter and grazing angle, part of the e-beam is not blocked by the substrate and is visible in the diffraction pattern. The specular reflected beam is the diffraction spot "lying" on the same rod as the incident beam. It is, therefore, never forbidden, unlike other spots.

362

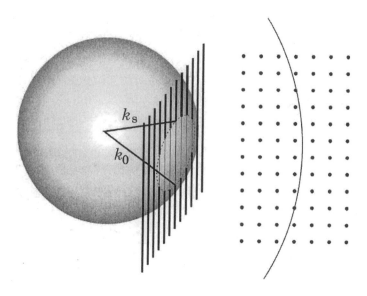

Figure 3: Ewald sphere construction in three dimensions (a) and a section of the horizontal z=0 plane (b).

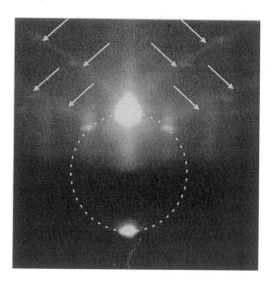

Figure 4: Typical RHEED pattern as recorded from a perfect SrTiO₃ surface.

As mentioned previously, RHEED is a strong interaction diffraction technique and dynamical scattering has to be taken into account. As a result non-linear effects like multiple scattering occur leading to anomalies in the RHEED intensity. One of these anomalies is a phenomenon often observed in RHEED, the Kikuchi lines. They originate from diffraction of diffuse scattered electrons and appear as curved lines on the phosphor screen, indicated by arrows in figure 4. The intensity of a diffraction spot or streak is affected at intersections with Kikuchi lines, since diffusive scattered electrons contribute to the diffracted intensity. Intensity measurements during growth near such intersections should be avoided.

Kikuchi lines move rigidly fixed to the crystal and are, therefore, often used to determine the crystal orientation and for alignment of the e-beam. The occurrence of clear and sharp Kikuchi lines is an indication of a flat and crystalline surface.

12.3.2 Utility of RHEED: monitoring thin film growth

In the early days of RHEED, the surface sensitivity has been exploited mainly for the study of cleaved crystals and surface reconstruction. Nowadays, the main applications are thin film growth monitoring and the study of growth kinetics. The existence of intensity oscillations[18] corresponding to the 2-dimensional (2D) growth of atomic or molecular layers is probably the most important reason to use RHEED. Here, the surface is periodically roughened and smoothened during the 2D nucleation and growth, resulting in a periodically varying density of surface steps. Electrons are easily scattered out of the specular beam by the step edges since the layer thickness is much larger than de wavelength of the electrons. As a result, periodic intensity variations are expected during 2D growth, which can be used to determine the growth rate. A detailed description of RHEED intensity variations during 2D growth of complex oxides in PLD will be given later in this chapter.

12.3.3 High-pressure RHEED

12.3.3.1. Electron scattering

The pressure in a RHEED set-up has to be sufficiently low to avoid electron scattering by the ambient gas. Attenuation of the electron beam (e-beam) intensity is to be expected at high ambient pressure due to elastic and inelastic electron scattering[19]. Furthermore, a low pressure is required near the filament of the e-gun to avoid breakage or a short lifetime. With the introduction of the differentially pumped e-gun, the maximum operating pressure was raised to ~1 Pa[20]. To increase the operating pressure in RHEED to the high deposition pressures used in PLD of oxides, both mentioned requirements have to be fulfilled: a low pressure in the e-gun should be maintained and the attenuation of the e-beam intensity has to be minimized. The latter be expressed as[21]:

$$\frac{I}{I_0} = \exp\left(-\frac{l}{L_E}\right) \tag{6a}$$

where L_E is the mean free path of the electrons, l is the distance of the traveling path of the electrons and I_0 is the intensity of the beam for $l \ll L_E$, i.e., at sufficiently low pressure. The mean free path L_E is defined as:

$$L_E = \frac{1}{\sigma_T n} \tag{6b}$$

where σ_T is the total cross-section for, both, elastic and inelastic scattering and n is the molecular density given by:

$$n = \frac{P}{k_B T} \tag{6c}$$

with P the pressure, k_B Boltzman's constant and T the temperature.

At the high electron energies used in RHEED the total cross-section σ_T is decreasing with increasing electron energy[21]. Unfortunately, no experimental data for electron scattering by O_2 at the high electron energies (>10 keV) used in RHEED have been reported. By extrapolating experimental data[21] a value for $\sigma_T \sim 3 \times 10^{-21}$ m^2 at 10 keV is determined. Using this value, the attenuation I/I_0 can be estimated. Figure 5 shows the distance l for a 10 keV e-beam, calculated using eq. (6) as function of the oxygen pressure for several values of I/I_0. From this figure it becomes clear that the traveling distance has to be reduced to minimize electron scattering losses at the high oxygen pressures used in PLD.

12.3.3.2 High pressure RHEED set-up

To satisfy the two requirements, i.e., low pressure in the e-gun and short traveling distance at high pressure, a two-stage, differentially pumped RHEED system has been developed[1]. A schematic view of this high-pressure RHEED system is depicted in figure 1. The e-gun (EK-2035-R, STAIB Instrumente) has a minimum beam size of ~ 250 μm (FWHM), even at the working distance of 500 mm. It is mounted on a flange

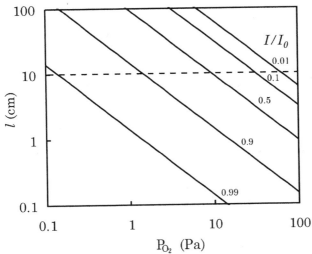

Figure 5: Attenuation I/I_0 of a 10 keV electron beam as function of the oxygen pressure P_{O_2} and penetration length l. The dashed line represents the traveling distance in our high pressure RHEED set-up.

connected to a stainless steel extension tube with an inner diameter of 8 mm. A differential pumping unit is used to maintain a vacuum of better than 10^{-4} Pa in the e-gun. An aperture (diameter 0.5 mm) separates the tube from the deposition chamber. The pressure in the tube, which depends on the pump speed and the size of the aperture, is kept below 10^{-1} Pa. Using this two-stage pumping system, the pressure in the deposition chamber can be increased up to 100 Pa maintaining the low pressure in the electron source. The e-beam, which passes through the apertures inside the differential pumping unit and the tube, enters the deposition chamber near the substrate at a distance of 50 mm. The XY deflection capability of the electron source is used to direct the e-beam through the aperture at the end of the tube.

The fluorescent phosphor screen (diameter 50 mm) is mounted on a flange located near the substrate. The distance between the screen and substrate is 50 mm. The screen is shielded from the laser plasma in order to minimize deposition. The electron source, including the extension tube, is mounted on a XYZ-stage allowing adjustment of the distance between substrate and end of the tube.

Using the extension tube, the traveling distance in the high-pressure regime is reduced to 100 mm. Intensity losses due to electron scattering inside the extension tube are negligible since the pressure is kept well below 10^{-1} Pa. In the high-pressure regime, however, a significant decrease in the intensity ($I/I_0 \sim 0.01$) is expected at oxygen pressure P_{O_2} of 100 Pa.

366

However, compensation of the scattering losses is possible by adjusting the e-beam current.

At large working distance, the e-beam is deflected over several millimeters by small magnetic fields, like the earth magnetic field. Therefore, special care has been taken to shield the e-beam from magnetic fields using µ-metal. Furthermore, the substrate can be rotated in order to adjust the angle of incidence and azimuthal angle of the e-beam on the substrate. The diffraction pattern is monitored using a Peltier cooled CCD camera and acquisition software (K-Space Associates). The software allows for time-resolved spot intensity measurement. The minimum acquisition time for each data point is 33 ms.

Figure 6 (a)-(e) shows RHEED patterns recorded from perfect $SrTiO_3$ surfaces at several oxygen pressures from 10^{-5} to 50 Pa. As expected, the specular reflected intensity is a clear function of the ambient pressure. At pressures above 10 Pa, features like the sharp Kikuchi lines and a clear shadow edge disappear whereas the background intensity is increased due to the collection of the forward scattered electrons. At an oxygen pressure of 50 Pa, the specular reflected spot is still clearly visible. Consequently, growth monitoring is feasible by measurement of the intensity variations of the specular reflected beam.

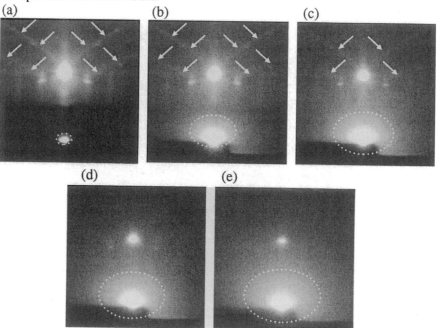

Figure 6: RHEED patterns from a perfect $SrTiO_3$ surface recorded at different oxygen pressure: 10^{-5} Pa (a), 10^{-1} Pa (b), 7 Pa (c), 15 Pa (d), 50 (Pa) (e). The arrows indicate the Kikuchi lines, the additional intensity due to the diffuse forward scattered incident electrons is indicated by the dotted lines.

This high-pressure RHEED system is used for the study as well as the in-situ monitoring of the growth of complex oxides and multilayered structures[22,23,24,25,26,27,28,29,30]. Before showing the results on the homoepitaxial growth of SrTiO$_3$ as a model system for complex oxide growth, a short overview is given of the possible growth modes during PLD.

12.4 GROWTH MODES IN PULSED LASER DEPOSITION

The thermodynamic approach to crystal growth is used to describe crystal growth close to equilibrium, i.e., for a thermodynamic stable system. Local fluctuations from equilibrium lead to nucleation, which give rise to a phase transition from, for instance, the gas to the solid phase. A supersaturated[31] gas phase is a prerequisite for the formation of these nuclei, whereas the formation probability is determined by the activation energy. Nuclei will be formed until a critical density is reached. From this point onwards the nuclei will grow and crystallization is in progress.

This thermodynamic approach has been used to determine growth modes of thin films close to equilibrium[32], i.e., only at small or moderate supersaturation. In this approach the balance between the free energies of film surface (γ_F), substrate surface (γ_S) and the interface between film and substrate (γ_I) is used to determine the film morphology. Three different growth modes, schematically depicted in figure 7, can be distinguished. In the case of layer-by-layer growth (figure 7 (a); Frank-van der Merwe growth mode), the total surface energy, i.e., $\gamma_F + \gamma_I$, of the wetted substrate is lower than the surface energy of the bare substrate γ_S. That is, the strong bonding between film and substrate reduces γ_I such that $\gamma_F + \gamma_I < \gamma_S$.

On the other hand, when there is no bonding between film and substrate, three-dimensional (3D) islands are being formed. The film does not wet the substrate because this would lead to an increase of the total surface energy. This growth mode is referred to as the Volmer-Weber growth mode, see figure 7 (b). In heteroepitaxial growth, the so-called Stranski-Krastanov growth mode can occur, see figure 7 (c). Here, the growth mode changes from layer by layer to island growth. During heteroepitaxial growth, the lattice mismatch between substrate and film gives rise to biaxial strain, resulting in an elastic energy that grows with increasing layer thickness. Misfit dislocations at or near the substrate-film interface will be formed if the layer thickness exceeds a critical thickness h_c. At this thickness it is thermodynamically favorable to introduce dislocations because the elastic energy, relieved by the dislocations, becomes comparable to the increase in the interfacial energy. In other words, misfit dislocations are necessarily introduced to relieve the mismatch strain and, therefore, equilibrium defects.

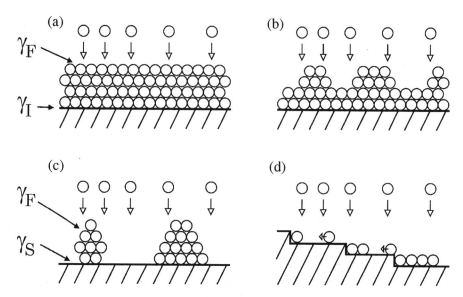

Figure 7: Film growth modes: layer-by-layer; Frank-Van der Merwe (a), island;Volmer-Weber (b), Stranski-Krastanov (c) and step flow (d).

In many vapor-phase deposition techniques, e.g., PLD, the growing film is usually not in thermodynamic equilibrium and kinetic effects have to be considered. Because of the limited surface diffusion, the deposited material cannot rearrange itself to minimize the surface energy. The high supersaturation of the vapor leads to a large nucleation rate, and kinetic effects will lead to the occurrence of different growth modes.

12.4.1 Homoepitaxial growth modes

In homoepitaxial growth the deposited material is identical to the substrate material and the crystalline structure of the substrate is extended into the growing film. Complicating effects like lattice parameter misfit and thermal expansion coefficients do not play a role. Consequently, only 2D growth modes are expected, depending on the behavior of the deposited species[33]. This behavior is determined by a number of kinetic parameters. Among them are the surface diffusion coefficient (D_S) of the adatoms, the sticking probability of an adatom arriving the edge of a terrace and the additional energy barrier (E_S) for adatoms to descend the edge to a lower terrace. From the mentioned kinetic parameters the diffusion coefficient is probably the most important one. It determines the average distance an atom can travel on a flat surface before being trapped. This distance is the surface diffusion length l_D and can be defined by:

$$l_D = \sqrt{D_S \tau} \tag{7}$$

where τ is the residence time before re-evaporation. The surface diffusion coefficient D_S is generally expressed as:

$$D_S = v a^2 \exp\left(-\frac{E_A}{k_B T} \right) \tag{8}$$

where E_A is the activation energy for diffusion, v the attempt frequency and a the characteristic jump distance. From eq. (8) it is clear that the deposition temperature is important because it controls the diffusivity of the adatoms. In order to understand the possible 2D growth modes on, both, singular and vicinal substrates, two diffusion processes have to be considered. First, the diffusion of atoms on a terrace (intralayer mass transport) and second the diffusion of an atom to a lower terrace (interlayer mass transport). Both processes are determined by the kinetic parameters.

A fast intralayer mass transport will lead to step flow growth on a vicinal surface, see figure 7 (d). In this case l_D is sufficiently larger than the average terrace width l_T. The mobility of adatoms is high enough to enable atoms to reach the edges of the substrate steps. Here, steps act as a sink for the deposited atoms diffusing towards the steps and nucleation on the terraces is prevented. As a result, the steps will propagate leading to step flow growth. A growing vicinal surface will be stable if, both, the terraces keep the same width and the step ledges remain straight. Otherwise, step meandering or step bunching can occur and, subsequently, the distribution in l_T broadens.

If the intralayer mass transport is not fast enough, nucleation on the terraces occur. Initially, nuclei will be formed until a saturation density is reached. After the saturation, the probability for atoms to attach to an existing nucleus exceeds the probability to form a new nucleus and islands will start to grow. The interlayer mass transport will have a large effect on the growth mode in this case. Two extreme growth modes can be distinguished, i.e., ideal layer-by-layer growth and ideal multilayer growth. To obtain a layer-by-layer growth mode a steady interlayer mass transport must be present; atoms deposited on top of a growing island must, first, reach the island edge and, second, diffuse to the lower layer. In the ideal case, layer-by-layer growth is obtained if nucleation starts after completion of a layer. When there is no or very limited interlayer mass transport, nucleation will occur on top of islands before these islands have coalesced.

This so-called "second layer nucleation" will lead to multilayer growth. The probability for second layer nucleation is related to a critical island size R_C. It is the mean island radius at a critical time during sub-monolayer growth where stable clusters nucleate on top of the islands. If R_C is small compared to the mean distance between islands in the first layer, islands will nucleate a second layer and multilayer growth will occur. An important parameter, which influences R_C, is the energy barrier for atoms to descend across the step edge to a lower terrace. An additional energy step edge barrier E_S has to be overcome if this energy barrier is large compared to E_A. Large values of E_S will lead to accumulation of adatoms on top of islands, leading to an increased second layer nucleation rate and, therefore, to a smaller value of R_C. The growth mode in real systems far from equilibrium will be in between the two extreme growth modes.

12.4.2 Homoepitaxial growth study of SrTiO$_3$

The specular RHEED intensity, monitored during homoepitaxial growth of SrTiO$_3$ at the deposition temperature of 650, 750 and 850 °C is depicted in figure 8 (a), (b) and (c), respectively. The oxygen deposition pressure was set to 3 Pa, the distance between target and substrate to 58 mm and the laser energy density on target to 1.3 J/cm^2. At these conditions, 2D nucleation and growth is observed, indicated by the clear intensity oscillations. The oscillation periods correspond to the deposition time of one unit cell layer. At the highest temperature, full recovery is obtained at completion of every unit cell layer. Here, the growth mode can be described as ideal unit cell layer-by-layer.

The angle of incidence of the electron beam (e-beam) was set to ~1°, resulting in a diffraction condition corresponding to the in-phase condition. Reflections from unit cell layers at different levels add constructively and a constant maximum intensity is predicted by the kinematic theory at this condition. Still, RHEED intensity variations are observed at the in-phase diffraction condition, caused by diffuse scattering at surface step edges. As already mentioned, electrons are easily scattered out of the specular beam by the step edges since the unit cell layer thickness is much larger than de wavelength of the electrons. The specular RHEED intensity is, therefore, a measure of the surface step density. During 2D nucleation and growth, oscillation of the step density is expected caused by the periodic formation, i.e., nucleation, and coalescence of growing islands. Note that a minimum specular RHEED intensity will be observed at the maximum step density. Because of the high pulse deposition rate and, subsequently, a high supersaturation within the deposition pulse an instantaneous nucleation can be assumed. Subsequently, 2D growth of equally spaced islands by step propagation is obtained. The step density S can therefore be determined by the widely used step density model[34,35]. Without second layer nucleation, S is given by:

$$S = 2\sqrt{\pi N_S}\,(1-\theta)\sqrt{-\ln(1-\theta)} \tag{9}$$

where N_S is the number of nuclei per unit area. Both, the diffusivity of the deposited material and deposition rate determine the quantity N_S and, therefore, the amplitude of the step density oscillation. The shape of the RHEED intensity, measured at the in-phase diffraction condition during homoepitaxial growth of SrTiO₃, supports the applicability of the step density model oscillations. A maximum step density $S_{MAX} = \sqrt{2\pi N_S / e}$ is reached at a surface coverage $\theta \sim 0.39$. This corresponds to the observed RHEED intensity minimum at $\theta \sim 0.4$ in figure 8.

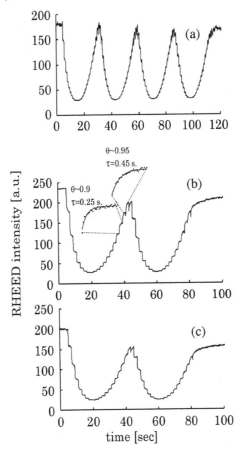

Figure 8: Specular RHEED Intensity oscillations during homoepitaxial growth of SrTiO₃ at 3 Pa, 850 °C (a), 750 °C (b) and 650 °C (c). The insets in (b) show the enlarged intensities after a laser pulse at different coverage.

12.4.2.1 Determination of kinetic growth parameters on singular surfaces

The RHEED intensities in figure 8 are clearly modulated by the laser pulse, giving rise to the typical RHEED intensity relaxations[36,37]. Following the model described above, i.e., instantaneous nucleation just at the start of every monolayer, this relaxation behavior is attributed to particles diffusing towards the step edges of the islands. Incorporation of these particles at the step edges causes the islands to grow resulting in a change in the step density. The evolution of the step density after every pulse is, therefore, a direct result from the decay of the density of the diffusing particles, which depends on the diffusivity and the average travel distance, see eq. (7). This distance is determined by the nucleation density N_S and the average island size. The latter depends on the coverage and gives rise to the coverage dependent characteristic relaxation times, see figure 8 (b). This dependence is illustrated by the following proposed model. Without nucleation on top of islands, the density of diffusing particles on top of a circular, 2D island can be found by solving the time dependent diffusion equation[38] given by:

$$\frac{\partial^2 n_S}{\partial r^2} + \frac{1}{r}\frac{\partial n_S}{\partial r} = \frac{1}{D_S}\frac{\partial n_S}{\partial s} \tag{10a}$$

with the initial condition $n_S(r,0)=n_0$, the density of instantaneously deposited particles due to one laser pulse. The boundary conditions are given by:

$$n_s\left(r = r_0\right) = n_{se} \tag{10b}$$

and

$$\left(\frac{\partial n_S}{\partial r}\right)_{r=0} = 0$$

(10c)

where n_{se} is the equilibrium density at the island edge; $n_{se} \sim 0$ assuming the edge acting as a perfect sink. The solution of eq.'s (10) is of the form:

$$n_S\left(r,t\right) = n_0 \sum_{m=1}^{\infty} A_m\left(r;r_0\right)\exp\left(-\frac{t}{\tau_m}\right) \tag{11}$$

where τ_m is given by:

$$\tau_m = \frac{r_0^2}{D_S \mu^{(0)^2}_m} \tag{12}$$

A_m are pre-factors depending only weakly on r and r_0. Eq. (11) converges rapidly and, for long enough time, only the first term needs to be considered. Without nucleation on top of the 2D island, an exponential decrease in the density of diffusing particles is expected after the deposition pulse. This exponential decrease depends on the size of the growing islands, which depend on the coverage by:

$$\pi r \rho_2{}^2(t) = \frac{\theta(t)}{N_S} \tag{13}$$

where $\pi \rho_2{}^2(t)$ is the area of the islands, see figure 9 Substituting eq. (13) into eq. (12) gives the exponential decay time of diffusing particles on top of a 2D island as function of the coverage θ:

$$\tau_2 = \frac{\theta}{D_S \mu^{(0)^2}_1 \pi N_S} \tag{14}$$

The index 2 indicates diffusing particles at the 2^e level, i.e., on top of the islands. The same approach can be followed for diffusing particles deposited between the islands. Here, the maximum travel distance of a diffusing particle is also determined by the coverage dependent size of the islands (see figure 9). The area between the islands can be approximated by:

$$\pi \rho_1{}^2(t) = \frac{1 - \theta(t)}{N_S} \tag{15}$$

Substituting eq. (15) into eq. (12) gives the exponential decay time (now indexed 1) of diffusing particles between 2D islands as function of the coverage θ:

$$\tau_1 = \frac{1 - \theta}{D_S \mu^{(0)^2}_1 \pi N_S} \tag{16}$$

From the equations above it follows that, with increasing coverage the

374

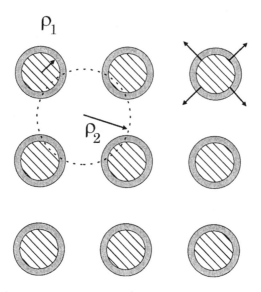

Figure 9: Schematic view of the coverage θ, given by the (hatched) area of the 2D islands. The density is determined by the nucleation density N_S. ρ_1 represent the radius of the islands at given coverage θ, whereas ρ_2 represents the radius of the area between the growing islands, given by 1-θ. The gray area represents the change in coverage due to one deposition pulse.

decay time in the density of diffusing particles on top of a growing island is increasing. On the other hand, the decay time in the density of diffusing particles between islands is decreasing with increasing coverage.

The change in the coverage $\Delta\theta(t)$ after a deposition pulse can be approximated by:

$$\Delta\theta_n(t) = \frac{\theta_{n-1}}{n_P}\left(1-\exp\left(-\frac{t}{\tau_2}\right)\right) + \frac{(1-\theta_{n-1})}{n_P}\left(1-\exp\left(-\frac{t}{\tau_1}\right)\right) \quad (17)$$

where θ_{n-1} is the coverage before applying the pulse and n_P the number of pulses needed for completion of one monolayer, i.e., $\theta = 1$. Here, n_P^{-1} represents the density of deposited particles. Multiplication of this density with θ_{n-1} and $(1-\theta_{n-1})$ gives the total number of particles, expressed in θ, deposited on top and in between the islands, respectively.

Figure 10 (a) shows $\theta(t)$ for $N_S = 2\times10^{11}/\text{cm}^2$, T = 850 °C, $E_A = 2.2$ eV and $n_P = 20$, calculated using the model described above. The parameters are chosen such that the density of diffusing particles is approximately zero before the next pulse is applied. Using eq. (9) the evolution of the step density $S(t)$ can be calculated, see figure 10 (b). Assuming a direct coupling between $S(t)$ and the diffuse scattering of electrons, the reflected RHEED intensity is approximated by:

$$I(t) \propto 1 - \frac{S(t)}{S_{max}} \tag{18}$$

which is shown in figure 10 (c) for the parameters given above. The simple model describes the observed RHEED intensity variations in figure 8 qualitatively, as shown by the coverage dependent characteristic relaxation time after every pulse. However, the observed RHEED intensity shows a pronounced decrease in the intensity just after each pulse is applied.

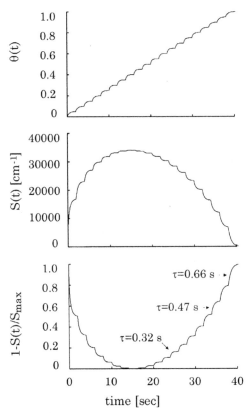

Figure 10: Calculated coverage $\theta(t)$ during mono layer growth (a), the corresponding surface step density $S(t)$ (b) and corresponding RHEED intensity, given by $1-S(t)/S_{max}$ (c). Here, the nucleation density N_S is $2 \times 10^{11}/cm^2$, the temperature is 850 °C, the energy barrier of diffusion E_A is 2.2 eV and the number of pulses for one monolayer n_p is 20.

This effect is especially visible in figure 8 (b) at coverage close to unity. This difference can be explained by the contribution of the diffusing particles. Not only surface steps but also the diffusing particles themselves act as diffuse scatterers[39] for electrons. The decay in the particle density n_s on top and between the islands is described by eq. (11) using the time constants given by eq. (14) and eq. (16), respectively. Assuming a direct coupling of the averaged particle density and the diffusive scattered intensity, an exponential increase of the intensity is expected, given by:

$$I \sim I_0 \left(1 - \exp\left(-\frac{t}{\tau} \right) \right) \tag{19}$$

where I_0 is determined by the particle density just after the deposition pulse and τ given by eq. (14) and (16) for particles deposited on top and in between the island. Here, only the first term of eq. (11), i.e., $m=1$, is used.

The evolution of the step density and the density of diffusing particles determine the relaxation behavior of the RHEED intensity. The time dependence of both is expected to be comparable since the two contributions are coupled. At coverage close to unity ($\theta = 0.7$-1), most of the deposited material is deposited on top of the growing islands and only one time constant has to be considered. At these values of the coverage, the characteristic relaxation times could be used to estimate the activation energy for diffusion E_A. However, reliable determination of the activation energy for diffusion E_A from the temperature dependent characteristic relaxation times is only possible assuming the average travel distance for diffusing particles to be approximately constant in the temperature range of interest. That is, only small variations in the step density S are allowed and, as a consequence, only small variations of the coverage θ and the nucleation density N_S are allowed, see eq. (14) and (16). The first requirement is easily obtained by depositing a sufficiently small amount of material in one pulse. The latter, however, is affected by the supersaturation $\Delta\mu$ and diffusivity D_S of the material. Both depend on the substrate temperature; with increasing temperature $\Delta\mu$ and D_S are decreasing. Consequently, the nucleation density N_S is not expected to be constant in a wide temperature range. A higher nucleation density N_S at lower temperature decreases the average travel length of the deposited material and the corresponding relaxation times will decrease, see eq. (14) and (16).

In conclusion, varying the temperature not only changes the diffusivity D_S but also N_S. As a result, determination of E_A not reliable from temperature dependent measurements during 2D growth.

To overcome the temperature dependent nucleation density and corresponding change in the average travel distance, step flow growth can be employed. No nucleation takes place during this growth mode, resulting in a steady average travel distance. Extraction of the energy barrier for diffusion during step flow growth is, therefore, more reliable, as will be discussed in the following section.

12.4.2.2 Determination of kinetic growth parameters on vicinal surfaces

The transition from 2D to step flow growth on a vicinal substrate is often used to determine the adatom diffusion length. Either the substrate temperature, determining the surface diffusion length l_D, or the vicinal angle, determining the terrace width l_T, is changed. A transition occurs when the diffusion length of adatoms becomes comparable to the terrace width, i.e., $l_D \approx l_T$. It has been proposed[40] to use RHEED to determine this transition as a function of the growth conditions. This technique has been used by several groups[41,42,43] to estimate the surface migration parameters. The RHEED intensity is expected to be constant during step flow growth, whereas RHEED intensity oscillations occur during 2D nucleation and growth.

However, nucleation, being a stochastic process, can take place on a terrace smaller than the diffusion length. Furthermore, the absence of RHEED intensity oscillations is not always a clear signature of pure step flow growth. Even when a steady RHEED intensity is observed, nucleation on the terraces may occur. Such a "mixed" growth mode causes a constant step density and, therefore, a constant RHEED intensity. This growth behavior is observed during growth of $SrRuO_3$ on vicinal TiO_2 terminated $SrTiO_3$[44].

During step flow growth, the step ledges of the vicinal substrates act as a sink for diffusing particles and nucleation on the terraces is prevented, resulting in a steady step density, determined by the vicinal angle of the substrate. The coverage dependent average island size in 2D nucleation and growth is replaced by the constant average terrace width of the vicinal substrate. Measurements of RHEED characteristic relaxation times during step flow growth at different temperature can, therefore, be used to determine the energy barrier for diffusion more reliable.

The relaxation time is inversely proportional with the diffusivity:

$$\tau \sim \frac{C}{D_S} \tag{20}$$

where C is a constant, determined by the vicinal angle of the substrate.

Figure 11 (a) and (b) show the RHEED intensity variations during homoepitaxial step flow growth of SrTiO$_3$ at different oxygen deposition pressure and temperature, using a substrate with vicinal angle of ~2°, corresponding to an average terrace width of ~10 nm. Full intensity recovery is observed in these cases, indicating step flow growth. The characteristic relaxation time, obtained by a fit using eq. (19), versus oxygen deposition pressure for different temperatures is given in figure 11 (c). A clear dependence of the relaxation times on the oxygen deposition pressure is observed. The sharp transition observed during 2D growth is, however, not observed at step flow conditions. In figure 11 (d), the data is presented in an Arrhenius plot. From the slopes a value for the energy barrier of diffusion E$_A$ of 2.6 ± 0.3, 3.0 ± 0.3 and 3.5 ± 0.3 eV have been derived for oxygen deposition pressure of 3, 12 and 20 Pa, respectively. As expected, these values are larger than the values derived during 2D growth. Still a lower value for E$_A$ is found at the lowest oxygen deposition pressure, which can be attributed to the kinetic energy of the particles arriving the substrates.

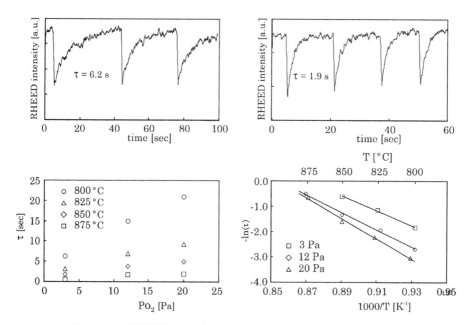

Figure 11: Specular RHEED intensity variations during homoepitaxial step flow growth of SrTiO$_3$ at oxygen deposition pressure of 3 Pa and 800 °C (a) and at 20 Pa and 875 °C (b). Relaxation times obtained from a fit with eq. (13) for different temperature and oxygen deposition pressure (c) and the same values in the Arrhenius form (d).

The activation energy E$_A$ for the diffusing particles is expected to be smaller than the estimated values[45]. Due to the high supersaturation in the deposition pulse, nucleation of a high density of 2D clusters does occur.

These clusters are, however, unstable after the deposition pulse and dissociate into the mobile particles. The absolute values of E_A are, therefore, related to the dissociation of the clusters as well as the diffusion of the particles.

12.5 GROWTH MANIPULATION

The possibility to control the growth during PLD using high-pressure RHEED allows for atomic engineering of oxide materials and growth of hetero structures with atomically smooth interfaces. Layer-by-layer growth is a prerequisite: nucleation of each next layer may only occur after the previous layer is completed. Occasionally, the deposition conditions such as the substrate temperature and oxygen pressure can be optimized for true 2D layer-by-layer growth, e.g., homoepitaxy on $SrTiO_3$ (001), see figure 8 (a). The relatively high temperature in combination with a low oxygen pressure enhances the diffusivity of the adatoms on the surface. As a result, the probability of nucleation on top of a 2D island is minimized, i.e., the adatoms can migrate to the step edges of the 2D islands and nucleation only takes place on fully completed unit cell layers.

The probability of second layer nucleation, however, increases at lower substrate temperature and/or higher deposition pressure, leading to multilevel growth. This is indicated by the damping of RHEED intensity oscillations. See for example figure 12 (a), where $SrTiO_3$ is depositing at a temperature of 800 °C and oxygen pressure of 10 Pa with a continuous pulse repetition rate of 1 Hz.

In general, roughening of the surface is observed during deposition of different kinds of materials, i.e., metals, semiconductors and insulators. Assuming only 2D nucleation, determined by the super saturation, limited interlayer mass transport results in nucleation on top of 2D islands before completion of a unit cell layer. Still, one can speak of a 2D growth mode. However, nucleation and incorporation of adatoms at step edges is proceeding on an increasing number of unit cell levels, which is seen by damping of the RHEED intensity oscillations.

Several groups[46,47,48] investigated the possibility of growth manipulation by enhancement of the interlayer mass transport. Two different temperatures, two different growth rates or periodical ion bombardment was applied to increase the number of nucleation sites and thus to decrease the average island size. This will enhance the transport of material from an island to a lower level. Usually, for epitaxy of complex oxide materials, the regime of temperatures and pressures is limited by the stability of the desired phases, e.g., Y123 can only be grown in a specific temperature and pressure regime[49]. At low temperatures a-axis oriented films are formed whereas at high temperatures the material decomposes.

380

Periodically ion bombardment is very difficult to realize in view of the stoichiometry of oxide materials[50]. Improvement of the interlayer mass transport by applying temperature variations or periodic ion bombardment during growth of complex oxides is, therefore, almost not feasible.

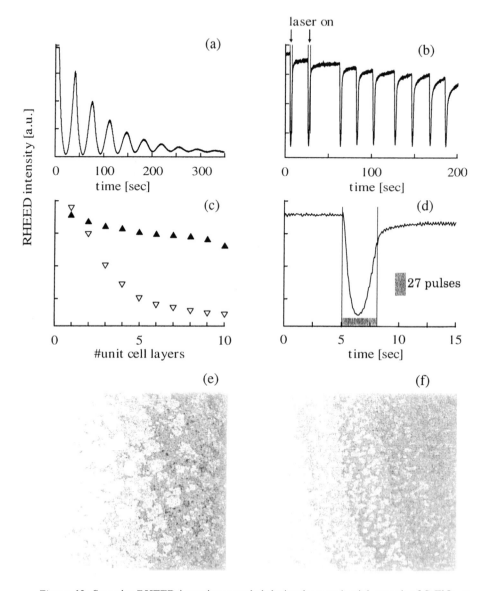

Figure 12: Specular RHEED intensity recorded during homoepitaxial growth of SrTiO₃ at 10 Pa and 800 °C using "standard" PLD (a) and interval PLD (b). Intensity maxima using "standard" PLD:▽ and interval PLD: ▲ (c). Intensity variation during one deposition interval (d). The surface morphologies of ~30 nm thick SrTiO₃ films are depicted in the AFM micrographs (1 μm²): "standard" PLD (e) and interval PLD (f).

12.5.1 Pulsed laser interval deposition

Manipulation of the growth rate, i.e., both the instantaneous and average growth rate, can be used to impose layer-by-layer growth. Here, the unique features of PLD are employed. As discussed in previous sections, a high density of small 2D nuclei is formed due to the high supersaturation just after the deposition pulse. Subsequently, larger islands are formed through recrystallization in between deposition pulses, exhibited by the typical relaxation of the RHEED intensity of the specular spot during PLD. As a result, the probability of second layer nucleation on the growing islands increases. Hence, to decrease this probability, coarsening should be avoided. This can be achieved by maintaining the high super saturation for a longer time by decreasing the time between the deposition pulses. The latter is easily obtained at high repetition rates. For that reason we introduced a growth method[51], based on a periodic sequence: fast deposition of the amount of material needed to complete one unit cell layer in a short[52] interval followed by a long interval in which no deposition takes place and the deposit can reorganize by recrystallization. The deposition interval should be in the order of the characteristic relaxation times (typically 0.5 s). This makes it possible to grow in a unit cell layer-by-layer fashion in a growth regime (temperature, pressure) where otherwise island formation would dominate the growth.

Figure 12 (b) shows the RHEED intensity during 10 cycles of SrTiO$_3$ deposition (at 10 Hz) and subsequent a period of no deposition, using the same oxygen pressure and substrate temperature as in figure 12 (a), following the new approach. In this case the number of pulses needed per unit cell layer was estimated to be about 27 pulses. Formation of a multilayer growth front is suppressed, indicated by the small decay of the intensity after each unit cell layer; in figure 12 (c) the intensities at each maximum of both methods are compared.

The small decrease of the maximum intensity is ascribed to the fact that only an integer number of pulses can be given to complete one unit cell layer, resulting in small deviations between the deposited and the required amount of material. A higher number of pulses decreases this deviation and is, therefore, favorable. The deposition time for one unit cell layer, however, should be smaller or comparable to the characteristic relaxation time. This is visualized in figure 12 (d). Here, the intensity change during deposition of one unit cell layer at 10 Hz is shown. The shape of the intensity curve at 10 Hz strongly resembles the parabola when calculating the intensity change of a two-level growth front with random distributed island and island sizes using eq. (9).

From the shape of the curve it can be seen that the time needed to deposit one unit cell layer is still too long. This is because the deposition

382

time interval of 2.7 sec. is longer than the characteristic relaxation time (~0.5 sec.). However, a significant suppression of the formation of a multilevel system has already been achieved. This suppression is clearly noticed by comparison of the surface morphologies after deposition using "standard" PLD and the new approach, see figure 12 (e) and (f), respectively. At least four levels are observed on the terraces in figure 12 (e), compared to only two levels in figure 12 (f).

Using the above-presented method[53,54] it is possible to impose a single level 2D growth mode or layer-by-layer growth mode with PLD despite unfavorable deposition conditions for SrTiO$_3$ with respect to mobility. The value of this method is demonstrated in the epitaxial growth and the infinite layer structure SrCuO$_2$/BaCuO$_2$[8,55], for which the choice of temperature and pressure is more critical, proving the importance of this growth method.

12.6 CONCLUSIONS

The growth of complex oxides is far from thermal equilibrium and, therefore, dominated by kinetic processes. High-pressure RHEED can be used to extract the energy barrier for diffusion E_A and showed a dependence of E_A on the oxygen deposition pressure attributed to the kinetic energy of the deposited particles. Reliable determination of E_A is feasible at step flow growth condition. Here, a steady step density causes a constant average travel distance for the diffusing particles. It is shown that this constant distance is essential for the reliable determination of E_A.

A single level 2D growth mode or layer-by-layer growth mode for SrTiO$_3$ can be imposed with PLD despite unfavorable deposition conditions with respect to mobility. Depositing every unit cell layer at a very high deposition rate followed by a relaxation interval, the typical high supersaturation in PLD is extended keeping the average island size as small as possible. Therefore, the interlayer mass transport is strongly enhanced and the formation of a multi-level growth front does not occur. This technique, which we call Pulsed Laser *interval* Deposition, is unique for PLD, no other technique has the possibility to combine very high deposition rates with intervals of no deposition in a fast periodic sequence.

REFERENCES

[1] M.K. Wu, J.R. Ashburn, C.J. Torng, P.H. Hor, R.L. Meng, L. Gao, Z.J. Huang, Y.Q. Wang, C.W. Chu, *Phys. Rev. Lett.* **58**, 908 (1987).
[2] D.B Geohegan, *Thin Solid Films* **220**, 138 (1992).
[3] D.B. Geohegan in *Laser Ablation of Electronic Materials: Basic Mechanisms and Applications*, E. Fogarassy, S. Lazare Eds. (North Holland, Amsterdam, 1992), p. 73.
[4] M. Strikovski and J.H. Miller, *Appl. Phys. Lett.* **73**, 1733 (1998).
[5] B. Dam, B. Stäuble-Pümpin, *J. Mat. Sci.: Mat. in Elec.* **9**, 217 (1998).

6   D.H.A. Blank, G.J.H.M. Rijnders, G. Koster, H. Rogalla, *Appl. Surf. Sci.* **138-139**, 17 (1999).

7   M. Tyunina, J. Levoska, S. Leppävuori, *J. Appl. Phys.* **83**, 5489 (1998).

8   T. Okada, Y. Nakata, M. Maeda, W.K.A. Kumuduni, *J. Appl. Phys.* **82**, 3543 (1997).

9   J. Cheing, J. Horwitz, *MRS Bulletin* **XVII**, 30 (1992).

10 D.B. Geohegan, A.A. Puretzky, *Appl. Phys. Lett.* **67**, 197 (1995).

11 M.Y. Chern, A. Gupta, B.W. Hussey, *Appl. Phys. Lett.* **60**, 3045 (1992).

12 Markov, *Crystal growth for beginners*, chapter 1 (World Scientific, 1994).

13 V.M. Fuenzalida, *J. of Cryst. Growth* **183**, 497 (1998).

14 V.M. Fuenzalida, *J. of Cryst. Growth* **213**, 157 (2000).

15 M.G. Lagally, D.E. Savage, M.C. Tringides in *Reflection high energy electron diffraction and reflection electron imaging of surfaces*, P.K. Larsen and P.J. Dobson Eds. (Plenum Press, London, 1989), pp. 139–174.

16 T. Kawamura in *Reflection high energy electron diffraction and reflection electron imaging of surfaces*, P.K. Larsen, P.J. Dobson Eds. (Plenum Press, London, 1989), pp. 501-522.

17 G. Koster, B.L. Kropman, G.J.H.M. Rijnders, D.H.A. Blank, H. Rogalla, *Appl. Phys. Lett.* **73**, 2920 (1998).

18 J.H. Neave, B.A. Joyce, P.J. Dobson, N. Norton, *Appl. Phys. A* **31**, 1 (1983).

19 G. García, F. Blanco, A. Williart, *Chem. Phys. Lett.* **335**, 227 (2001).

20 H. Karl, B. Stritzker, Phys. Rev. Lett. 69, 2939 (1992).

21 G. García, M. Roteta, F. Manero, *Chem. Phys. Lett.* **264**, 589 (1997).

22 S. Bals, G. Rijnders, D.H.A. Blank, G. Van Tendeloo, *Physica C* **355**, 225 (2001).

23 S. Bals, G. Van Tendeloo, G. Rijnders, M. Huijben, V. Leca, D.H.A. Blank, *IEEE Trans. Appl. Superconductivity* **13**, 2834 (2003).

24 G. Rijnders, S. Curras, M. Huijben M, D.H.A. Blank, H. Rogalla, *Appl.Phys. Lett.* **84**, 1150 (2004).

25 G. Rijnders, D.H.A. Blank, J. Choi, C-.B. Eom, *Appl. Phys. Lett.* **84**, 505 (2004).

26 J. Choi, C-.B. Eom, G. Rijnders, H. Rogalla, D.H.A. Blank, *Appl.Phys. Lett.* **79**, 1447 (2001).

27 X. Ke, M.S. Rzchowski, L.J. Belenky, C-.B. Eom, *Appl. Phys. Lett.* **84**, 5458 (2004).

28 G.Y. Logvenov, A. Sawa, C.W. Schneider, J. Mannhart, *Appl. Phys. Lett.* **83**, 3528 (2003).

29 Y.F. Chen, W. Peng, J. Li, K. Chen, X.H. Zhu, P. Wang, G. Zeng, D.N. Zheng, L Li, *Acta Physica Sinica* **52**, 2601 (2003).

30 J. Klein, C. Hofener, L. Alff , R. Gross, *J. of Magnetism and Magnetic Mat.* **211**, 9 (2000).

31 I. V. Markov, *Crystal growth for beginners* (World Scientific, London, 1995), pp. 81-86.

32 E. Bauer, *Z. Kristallographie* **110**, 372 (1958).

33 G. Rosenfeld, B. Poelsema and G. Comsa in *Growth and Properties of Ultrathin Epitaxial Layers,* D. A. King and D.P. Woodruff Eds. (Elsevier Science B.V., 1997) chapter 3

34 S. Stoyanov, *Surface Science*, **199**, 226 (1988).

35 S. Stoyanov, M. Michailov, *Surface Science*, **202**, 109 (1988).

36 H. Karl and B. Stritzker, *Phys. Rev. Lett.* **69**, 2939 (1992).

37 V.S. Achutharaman, N. Chandresekhar, O.T. Valls and A.M. Goldman, Phys. Rev. B 50, 8122 (1994).

38 G. Koster, *Artificially layered oxides by pulsed laser deposition*, PhD thesis, University of Twente, The Netherlands, (1999).

39 J.P.A. Van der Wagt, Reflection High-Energy Electron Diffraction during Molecular-Beam Epitaxy, PhD thesis, Stanford University, USA, (1994).

384

40 J.H. Neave, P.J. Dobson, B.A. Joyce, J. Zhang, *Appl. Phys. Lett.*, **47**, 100 (1985).

41 T.Shitara, D.D. Vvedensky, M.R. Wilby, J.Zhang, J.H. Neave and B.A. Joyce, *Phys. Rev. B* **46**, 6825 (1992).

42 T. Shitara, J. Zhang, J.H. Neave and B. A. Joyce, *J. Appl. Phys.* **71**, 4299 (1992).

43 H.J.W. Zandvliet, H.B. Elswijk, D. Dijkkamp, E.J. van Loenen and J. Dieleman, *J. Appl. Phys.* **70**, 2614 (1991).

44 G. Koster, B.L. Kropman, G.J.H.M Rijnder, D.H.A. Blank and H. Rogalla, *Appl. Phys. Lett.*, **73**, 2920 (1998).

45 M. Lippmaa, N. Nakagawa, M. Kawasaki, S. Ohashi and H. Koinuma, *Appl. Phys. Lett.* **76**, 2439 (2000).

46 G. Rosenfeld, R. Servaty, C. Teichert, B. Poelsema and G. Comsa, *Phys. Rev. Lett.* **71**, 895 (1993).

47 G. Rosenfeld, B. Poelsema and G. Comsa, *Journ. of Cryst. Growth* **151**, 230 (1995).

48 V.A. Markov, O.P. Pchelyakov, L.V. Sokolov, S.I. Stenin ans S.Stoyanov, *Surf. Sci.* **250**, 229 (1991).

49 R.H. Hammond and R. Bormann, *Physica C* **162-164,** 703 (1989).

50 A different approach is the use of surfactants, see for instance J. Vrijmoeth, H.A. van der Vegt, J.A. Meyer, E.Vlieg and R.J. Behm, *Phys. Rev. Lett.* **72**, 3843 (1994). However, a suitable candidate for complex oxides has to our knowledge not yet been found..

51 G. Koster, G.J.H.M. Rijnders, D.H.A. Blank, and H. Rogalla, *Appl. Phys. Lett.* **74**, 3729 (1999).

52 G. Koster, A.J.H.M. Rijnders, D.H.A. Blank, H. Rogalla, *Mater. Res.Soc. Symp. Proceedings* **526,** 33 (1998).

53 G. Rijnders, G. Koster, V. Leca, D. Blank and H. Rogalla, *Appl. Surf. Sci.* **168**, 223 (2000).

54 D.H.A. Blank, G. Koster, G.A.J.H.M. Rijnders, E. van Setten, P. Slycke and H. Rogalla, *J. Cryst. Growth* **211**, 98 (2000).

55 G. Koster, K. Verbist, G. Rijnders, H. Rogalla, G. van Tendeloo and D.H.A. Blank, *Physica C* **353**, 167 (2001).

CHAPTER 13

RECENT ADVANCES IN THE DEPOSITION OF MULTI-COMPONENT OXIDE FILMS BY PULSED ENERGY DEPOSITION

T. Venkatesan[*], K. S. Harshavardhan, M. Strikovski and J. Kim

Neocera, Inc.,
10,000 Virginia Manor Road,
Beltsville MD 20705, USA.

13.1 INTRODUCTION

The potential of Pulsed Laser Deposition (PLD) was realized for the first time with the deposition of high temperature superconducting films of YBCO nearly 17 years ago.[1] It was shown that the non-equilibrium thermal evaporation process preserves the stochiometry of the target material in the film under material-specific window of laser energy density and collection angle of the evaporant.[2,3] Preservation of target stochiometry, a unique feature of PLD, is considered to be a result of the following two important factors.

1. The time scale of the evaporation process and the heat transfer to the target is short compared to the time scale in which the surface and bulk atomic species intermix.

2. Due to high evaporant density at the target surface, collisional processes dictate the angular distribution of the ejected species (atoms, neutrals and ions) and up to a specific distance from the target, the trajectories are species independent. This distance is specific to a target material, laser energy density and the deposition pressure.

Besides preserving the target stochiometry in the deposited film, some of the attractive features of PLD include the following.

1. Since the energy source (pulsed laser) is external to the deposition chamber, film depositions can be undertaken over a wide dynamic range of

chamber pressures and substrate temperatures. In RF sputtering for example, the chamber can not be at arbitrary pressures during the film deposition. Similarly, in Chemical Vapor Deposition (CVD), the substrate temperature is governed for example, by the precursor cracking temperature.

2.   Multi-layers can be prepared by sequentially locating a target in the laser beam and this can be accomplished easily by using a multi-target carrousel and simple target manipulation schemes. Either manual or automated approaches can be used.

3.   For exploring new materials, a target typically an inch or less in diameter, a few millimeters thick is adequate and such targets can be synthesized with relatively low investment and in short turn around times using standard ceramic processing routes.

As a result,   PLD has emerged as the technique of choice among researchers working in the area of multi-component oxide thin films, and is today the fastest route to prototyping films of novel materials. While the early PLD systems were simple, consisting of a target, substrate heater and a pumping stack, today's systems have evolved considerably in their functionality and features. In this review we hope to capture the salient progress made to date and also speculate on what the future holds for this technique.

13.2 ADVANCED PLD TECHNIQUES

13.2.1 PLD-MBE

A large number of groups have demonstrated, what has been termed as PLD-MBE (or laser MBE), where individual layers of materials can be deposited with an exceptional control at the growth interface.[4-7] Unlike the typical MBE (Molecular Beam Epitaxy) systems used in the deposition of semiconductors, the PLD-MBE chambers may not have a liquid nitrogen jacket around the chamber but yet can achieve fairly low vacuum levels typical of most MBE systems. This feature, coupled with the non-thermal evaporation process (without any bulk heating) facilitates an exceptionally 'clean' deposition environment. The use of Reflection High Energy Electron Diffraction (RHEED) to monitor the in-situ film growth and particularly the ability to use RHEED under high growth pressures[5] has been an advancement over the conventional semiconductor based MBE systems. As an example, a layer of Strontium Ruthenate ($SrRuO_3$ or SRO), grown on a Strontium Titanate ($SrTiO_3$ or STO) substrate is illustrated in Figure 1.[7] Notable is the preservation of atomically smooth STO surface steps in a relatively thick film of SRO. Also shown in the figure are the sharp interfaces of Barium Titanate layers grown on the SRO layer. The

RHEED oscillations observed in the growth of an alternating superlattices of Barium Titanate (BaTiO$_3$ or BTO), Strontium Titanate (STO) and Calcium Titanate (CaTiO$_3$ or CTO), establishing the capability of this process for atomic layer growth control over micron thick deposition[7] are shown in Figure 2. The TEM cross sections with atomically

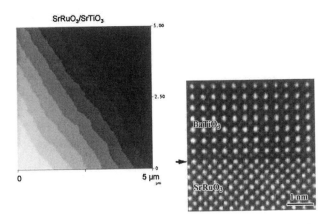

Figure 1: Atomically flat surfaces of SrRuO$_3$ films grown on SrTiO$_3$ substrate as measured by AFM (left); Epitaxy of BaTiO$_3$ film on SrRuO$_3$ layer seen in a TEM cross section (right). [Ref. 7]

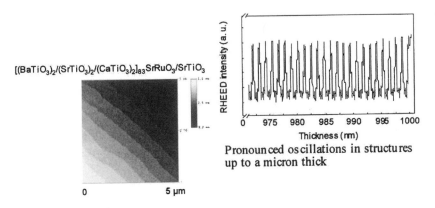

AFM confirms surface flatness
(single unit-cell steps after hundreds of
perovskite layers – identical to substrate)

Figure 2: RHEED Oscillations observed over a growth of a micron thick superlattice layer consisting of BaTiO$_3$, SrTiO$_3$ and CaTiO$_3$ layers. The preservation of the surface atomic steps as seen in the AFM signal confirms the layer-by-layer growth mode. [Ref. 7]

abrupt interfaces shown in Figure 3 further establish the level of control achievable by PLD in a multilayer film process. Thus despite the original

388

fears about PLD being a supersaturated growth process with a large arrival rate of vapor species and its potential negative impact on the growth of films, very high quality films have been demonstrated to date by many groups, lending strongly to the validation of PLD process in its ability to accomplish thin film growth with exceptional control at the atomic level.

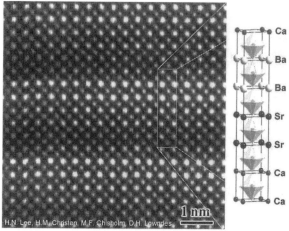

Figure 3: TEM cross section of the superlattice showing atomically abrupt interfaces of three different layers of $BaTiO_3$, $SrTiO_3$ and $CaTiO_3$. [Ref 7]

PLD-MBE systems with integrated RHEED diagnostics are now available commercially. A Typical Neocera PLD-MBE system with an integrated high pressure RHEED diagnostics is shown in Figure 4.

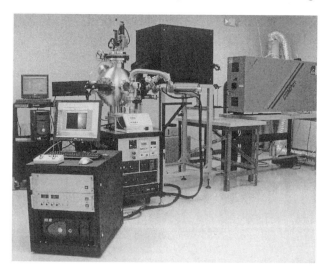

Figure 4: A PLD-MBE system with integrated high-pressure RHEED diagnostics for process control.

13.2.2 PLD - Combinatorial Systems

On account of their simplicity and ease of implementation, PLD based combinatorial systems for the growth of material libraries are becoming increasingly popular.[8-10] Two approaches are currently in practice - one using a set of shadow masks and the other without masks and termed continuous composition spread PLD (CCS-PLD). Figure 5 shows a typical mask-based arrangement used by Chang et al [10]. In Figure 6, a ternary phase diagram of a materials library consisting of Ba-Ca-SrTiO$_3$ [ref. 10] obtained with such a system is presented. The material alloys which exhibit high dielectric constant also exhibit relatively low microwave losses over a large region of the phase spread.

x-y movable
shutters/masks   modular and compact 8" flange

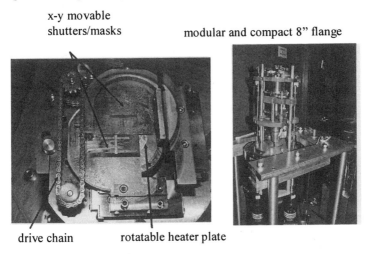

drive chain   rotatable heater plate

Figure 5: PLD Combinatorial system using a sequence of masks.

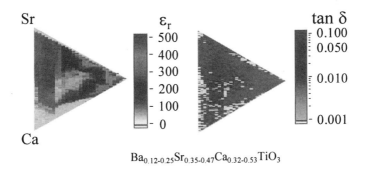

$$Ba_{0.12-0.25}Sr_{0.35-0.47}Ca_{0.32-0.53}TiO_3$$

Figure 6: Ternary Phase diagram of a material library of Ba-Sr-CaTiO$_3$ for their microwave dielectric properties. [Ref.10]

390

Continuous Composition Spread PLD (CCS-PLD) is a novel deposition technique[11-12] and is based on the deposition rate profiles naturally occurring in PLD as a result of the $\cos^n\theta$ ($5 \leq n \leq 11$) dependence.[13] Using rapid sequential deposition of the phase spreads constituents, intermixing at the atomic level is accomplished without resorting to any further post annealing treatments. Pseudo-binary, pseudo-ternary or ternary phase spreads can easily be accomplished as a result. No masks are used in this approach and this technique can operate in a wide dynamic range of pressures (up to about 500 mTorr) which are typically not possible in a mask-based approach. In this approach, combinatorial synthesis is carried out under optimal film growth conditions (no post anneal) representing the 'true' deposition parameter space. Composition determination across the sample and mapping of physical properties onto the ternary phase diagram can be achieved using a simple algorithm that incorporates the deposition rate profiles of the individual constituents.

Figure 7 illustrates the approach of CCS-PLD. Sub-monolayer amounts of each constituents are carried out sequentially through an automated process, coupled with a mechanical and synchronized movement of both the target and the substrate stages. Ternary or pseudo-ternary phase spreads can be readily accomplished by a synchronized $120°$ rotation of both the substrate and the target, and controlling the number of laser pulses on each target.

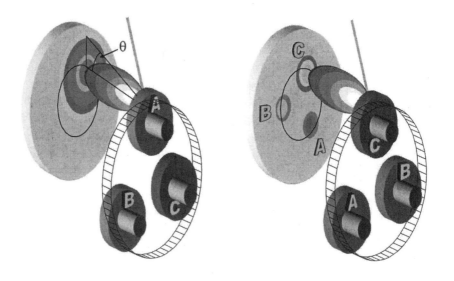

Figure 7: Schematic of the ternary phase library generation using the CCS-PLD process. [Ref. 11]

To illustrate the applicability of CCS-PLD to ternary or pseudo-ternary systems, a phase diagram consisting of SnO_2, In_2O_3 and ZnO is chosen.[11] Since for this system the compositions with relatively small amounts of SnO_2 are of interest (transparent conductors) and are therefore useful to be able to 'zoom-in' to that range. Two of the targets have small amounts of SnO_2 (ITO 0.5%: $In_{99.5}Sn_{0.5}O_x$ and ITO 10%: $In_{90}Sn_{10}O_x$ respectively) rather than a true In_2O_3 and SnO_2. The compositional spread film was deposited onto a 2 inch x 3 inch glass plate and the sheet resistance was measured by a standard 4-probe method at 368 evenly spaced points. The results are shown in Figure 8. The observed variations in sheet resistance are due largely to the varying composition, with some contribution from non-uniform thickness. The resistivity calculated from the thickness and resistance data is then plotted as a function of composition.

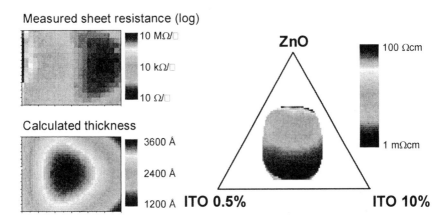

Figure 8: A pseudo-ternary phase diagram of In, Sn and Zn Oxides for a library of transparent conductors. [Ref. 11]

13.2.3 Pilot-scale Production PLD Systems

From the inception of the PLD process, there has been a great deal of skepticism regarding the potential for the commercial scale up of the process. Two important issues that have contributed to this notion are the following.

1. Deposition over large areas with acceptable thickness and phase uniformity.

2. Ability to produce films with acceptably low particulate densities.

Significant progress has occurred in this area and some important milestones have been accomplished. As an example of these developments, a PLD cluster module built by Neocera is shown in Figure 9. This module

392

is designed and optimized for the deposition of ferromagnetic oxide films such as Fe_3O_4 and other novel oxides for magnetic storage and spintronics applications. This system is an integral part of a larger cluster consisting of other deposition and processing modules (deposition and etch stages), all designed for 6-inch diameter wafers handled robotically. Thickness uniformity data and phase uniformity data obtained using this PLD module is presented in Figures 10 and 11. Both within wafer and wafer-to-wafer measurements are presented. Within wafer thickness data for a batch of 10 wafers is presented in Figure 10(a). The data show that within wafer

Figure 9: A Neocera- production PLD cluster module that docks with a cluster tool used in magnetic storage industry.

Thickness uniformity data of 10 wafers
Within wafer thickness uniformity: ($\pm$) 1%

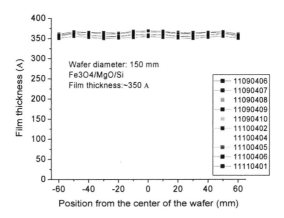

*Figure 10(*a*):* Within wafer thickness uniformity data for a batch of 10 wafers.

thickness variations can be limited to $\pm1\%$ and further improvements are possible. In Figure 10(b), wafer-to-wafer thickness variation for a batch of 12 wafers is shown indicating $\pm1\%$ variations in thickness in all the wafers.

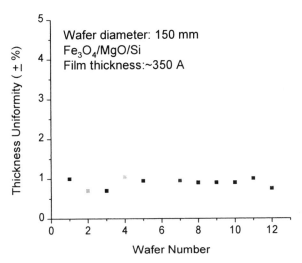

Figure 10(b): Wafer-to-Wafer thickness uniformity for a batch of 12 wafers

Variation of the electrical property of the film (resistivity) in a batch of six wafers showing better than $\pm5\%$ variation within the entire batch is presented in Figures 11 (a) and 11(b) respectively for within wafer and

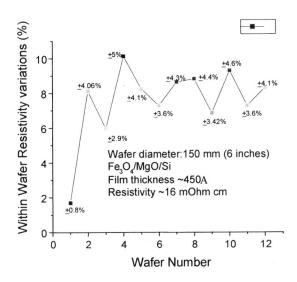

Figure 11(a):. Within wafer resistivity uniformity for a batch of wafers. The non-uniformities are within $\pm5\%$.

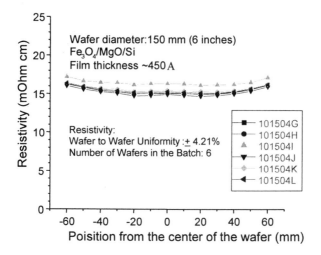

Figure 11(b): Wafer-wafer resistivity uniformity data for a batch of 6 wafers. The data show a high degree of phase uniformity and reproducibility of the Fe_3O_4 phase over the entire wafer area (diameter of 6 inches or 150 mm).

wafer-to-wafer measurements. Since iron oxide is a multi-phase oxide (Fe_2O_3, FeO and Fe_3O_4) and the desirable Fe_3O_4 phase can be stabilized only in a narrow deposition parameter space, the results presented clearly depict the true potential of PLD in its ability to stabilize 'difficult to form' phases such as Fe_3O_4 over a 6-inch diameter area or larger. In addition, for the films deposited, the particle density (for particle size exceeding 0.1 microns) was measured below 2000 per six inch wafer. While this number is still two orders of magnitude higher than acceptable by the magnetic storage and semiconductor industries, it is a significant leap for the PLD process towards large scale production platforms. With the adaptation of other active processes the particle density can be brought down by another two orders of magnitude, and this constitutes a future research direction for PLD.

13.2.4 RF-assisted PLD Systems

In recent times, nanotechnology with its large applications potential has become the driving force in contributing towards the development of several novel thin film growth processes. Preparation of nano-wires and nano-particles with unique functionalities has been in the forefront of this exciting research arena. In PLD, since the energy source (pulsed laser) is external to the deposition chamber, a wide range of operating pressures can be realized during film growth. Atomically controlled nanostructures in ultra-high vacuum conditions as well as nanoparticle growth where the size of the particles can be controlled by the ambient gas pressure can be realized rather easily due to this feature.

Unique nanostructures can also be realized by RF excitation during film growth.[14] As an example, a novel PLD-RF hybrid approach is presented. In this approach, an RF plasma is maintained near the target surface as well as near the substrate during ablation. By controlling the deposition conditions as well as the RF plasma parameters, oriented nanotubes can be grown with unique functionalities. Figures 12 shows the schematic of the experimental geometry and Figure 13 shows the results obtained in such an approach.

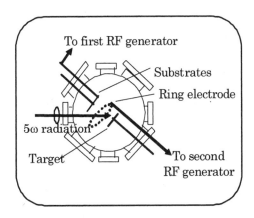

Figure 12: RF - assisted PLD Technique for the growth of oriented nanotubes. [Ref. 14]

Figure 13: Oriented multi-walled carbon nanotubes grown by a hybrid RF assisted-PLD technique. [Ref. 14]

13.2.5 Ion-assisted PLD Systems

Ion beam-assisted deposition (IBAD) has recently emerged as an innovative deposition method for in-plane aligned structural templates on polycrystalline substrates.[15-17] In this method, an energetic beam striking the growing film at an angle corresponding to the angle of a channeling direction for a particular crystalline orientation can enhance film growth in this orientation, while inhibiting grain growth in the other non-channeled orientations. A schematic representation of the experimental arrangement is shown in Figure 14 for the typical case of growing (100) oriented in-plane textured Yttria Stabilized Zirconia (YSZ) films. Under optimum deposition conditions, selective growth of (100) YSZ for example, occurred over (111) or (110) YSZ orientations. This is interpreted in the following manner. The [111] channeling in cubic YSZ is 54.7° from the [001] direction and hence ions directed near this angle to the substrate normal during growth, will enhance the required (100) orientation, through

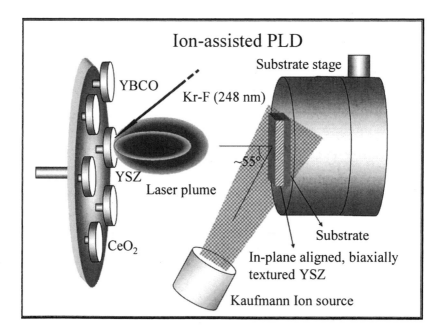

Figure 14: Schematic of experimental arrangement for ion-assisted PLD of YSZ films. [Ref. 17]

a preferential growth and selective etching mechanism. It is therefore possible to achieve an in-plane aligned film growth, which could become an excellent growth template for subsequent film depositions. Critical experimental parameters required to achieve this preferred orientation are the deposition rate of YSZ (determined by the laser energy density, pulse repetition rate and target-substrate distance) and the selective etch rate by the incident ion beam (determined by the ion beam parameters such as ion beam current density, accelerating voltage and ambient gas pressure in the chamber).

Figure 15 shows the 4-circle x-ray diffraction Phi-scan data of in-plane textured YSZ films as well as the subsequently grown in-plane textured CeO_2 and high temperature superconducting YBCO films. The (103) reflections of the YBCO film are seen at the top of the figure. The in-plane texture is visible in the Phi-angle data by the 4 peaks separated by 90°. The FWHMs of these peaks are ~ 7° indicating good in-plane texture. The remarkable feature of this data is that the polycrystalline substrate on

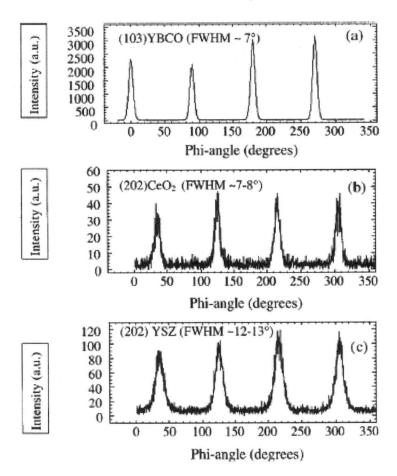

Figure 15: 4-circle x-ray diffraction data for biaxially textured YSZ, CeO_2 and YBCO films. [Ref. 17]

which the in-plane textured YSZ, CeO_2 and YBCO films are deposited is randomly oriented with no in-plane texture. Such remarkable modifications in the crystallographic texture of the films are possible only with ion assisted deposition and opens up several application possibilities. One such possibility is in the development of coated conductors based on superconducting YBCO films deposited on polycrystalline or randomly oriented substrates. As an example, high critical current densities (in excess of $10^6 A/cm^2$) realized in YBCO films grown on polycrystalline substrates with ion-assisted PLD is shown in Figure 16.

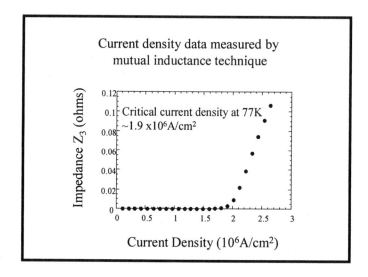

Figure 16: Critical current data of in-plane textured YBCO films prepared by ion-assisted PLD, showing excellent current carrying capability. [Ref. 17]

13.3 PULSED ELECTRON DEPOSITION (PED)

One of the exciting developments to occur over the last decade has been the emergence of a novel pulsed electron source based on a pseudo spark capillary source.[18-23] The schematic of this source is shown Figure 17. A discharge is allowed to occur inside the hollow cathode using the resident gas in the chamber and the cascading plasma generates a large number of ionized species and electrons. The capillary, appropriately biased at 10-20 KeV extracts the electrons from the plasma producing a pulsed current in the range of a kA over a time duration of the order of 80-100 nsec. The pulsed electron beam, forming a self-pinched plasma in the capillary, emerge at the end of a ceramic tube with a typical pulse energy of the order of a joule, comparable to the pulse energy produced in a typical excimer laser.

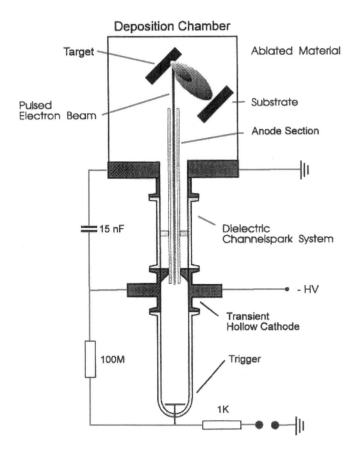

Figure17: Schematic of pulsed electron-beam source. [Ref. 20]

This section has two main components. Pulsed Electron-beam Deposition (PED) technique is introduced and described in the first part to the extent that it provides the basic understanding of the working principle of the technique. In the second part, data obtained on films are presented establishing the effectiveness of PED as a complimentary technique to PLD in extending the range of materials that can be prepared as thin films using pulsed energy techniques. Some basic differences and similarities between PED and PLD are also discussed.

Pulsed Electron Deposition is a process in which a pulsed high power electron beam is incident on the surface of a target, penetrates approximately 1-2 μm resulting in a rapid evaporation of target material.[19] The ablation process lasts about 100 ns leading to non-equilibrium heating, which in turn facilitates stochiometric preservation of the target composition in the deposited film. Most solid state materials can be deposited as thin films with PED. The energetic beam of pulsed electrons is created in a low pressure gas discharge known as "Pseudo-spark" or "Channel-spark" in the literature [18-21] and is briefly discussed below.

13.3.1 Pulsed Electron Beams from Pseudo-Spark and Channel-Spark Discharges

The transient, low-pressure gas discharge called "pseudo-spark" was first reported in 1979 [18] and the mechanism of the phenomenon has not been fully understood yet. The low-pressure gas discharge occurs between a planar anode and a hollow cathode with unique charged particle emission characteristics.[20] The resulting electron beam is generated in a plasma inside a hollow cavity and travels through a background gas, forcing the beam to be magnetically pinched tightly as it propagates. Pseudo-Spark and Channel Spark discharges are quite similar in nature even though certain differences exist in the source design. The main difference however, is the conversion efficiency of stored electrical energy into beam energy at the target in these two cases which is about 4 % in the case of pseudo-spark discharge and about 30% in the case of channel-spark discharge. This improvement in the electrical energy conversion efficiency in channel spark discharge is a direct consequence of modifications carried out in the accelerator part of the discharge source. The term 'Pulsed Electron Beams' used in the following text represents electron beams generated in a channel spark discharge. Schultheiss and Hoffman discovered that such pseudo-spark/channel spark discharges producing well-pinched electron beams with beam diameters as small as several millimeters in size leading to high beam brightness $\sim 2 \times 10^{11}$ A/m^2-rad^2 (current densities up to 10^6 A/cm^2) are capable of producing power densities up to 10^9 W/cm^2 with a pulse duration of ~ 100 ns.[23] This power density is comparable to the power densities obtained with pulsed lasers and can therefore be used for stochiometric thin film deposition.

13.3.2(a) Pulsed electron-beam source

A special type of hollow cathode is used for generating high currents. The hollow cathode is a metal tube, positioned in front of a planar anode.[20] The electrons are generated at the inner walls of the hallow cathode either by means of ion impact or by photoelectric effect. Due to the reduced electric field inside the hollow cathode the electrons move slowly, increasing the probability for ion impact ionization. Special cathode geometries facilitate oscillations around the cathode axis, ionizing the gas in the cavity effectively, before escaping the cathode along the axis towards the anode. The current density of the axially focused electron beam are of the order of 1 A/cm^2.

For very high currents the hollow cathode is modified in such a way that it operates only in a 'pulsed' mode. This is called transient hollow cathode, consisting of a similar cathode but only with a narrow exit.[20] The transient hollow cathode is a low pressure gas discharge electron source, which develops a well focused electron flow with currents of even up to a few kilo-amperes with pulse widths ~100 nanoseconds. Large current densities in the range of 10-100 kA/cm^2 are attainable in this case. A pre-discharge trigger circuit, which affectively controls the discharge voltage, is used to ignite the plasma.

13.3.2(b) Acceleration: Pseudo-spark discharge

In the pseudo-spark discharge occurring in the high voltage low pressure region of the Paschen curve,[20] electron acceleration is affected by a set of parallel electrodes separated by insulators. The special arrangement of hollow cathode and electrodes leads to an electric field gradient that focuses electrons along the central axis. At a given breakdown voltage (pressure dependent) the low pressure gas discharge escalates into[19] a very fast spark-like discharge characterized by an over exponential current rise. The discharge leads to the formation of intense pulsed electron beam along the central axis of the accelerator which can be extracted out of the cavity through an opening. The pulse width of the electron beam is ~ 100 ns and the current density in the beam is about 5000A/cm^2. The electrical energy of the pseudo-spark chamber is stored in a variable number of high voltage ceramic capacitors which can deliver up to about 3-5 J/pulse. For a more detailed description of Psuedo-spark discharges the reader is referred to excellent papers available in the literature.[19,22]

In the case of psuedo-spark discharge the transfer efficiency of electrically stored energy to beam energy at the target is rather low and is about 4%, as seen in the film deposition experiments of Hobel et al.,[19] and Jiang et al.,[22] on YBCO and is similar to the transfer efficiency (~3%) of electrical to optical power by the excimer lasers.[25]

13.3.2(c) Channel-spark discharge

Even though the psuedo-spark discharge based pulsed electron beams have been able to produce good quality films, there have been some problems associated with the stack of metal disks used in the accelerator. The metal discs were subject to oxidation, changing properties of the discharge over a period of time, thereby calling for a frequent cleaning of the discs as a routine maintenance requirement.[22] To overcome these difficulties, Jiang et al., have developed a novel scheme in which the a dielectric 'channel' replaces acceleration section consisting of alternating metal and insulator discs.The channel-spark discharge is quite similar to pseudo-spark discharge. The electron beams are generated from similar hallow cathodes and the beams are also magnetically self pinched and accelerated through specially designed electrodes providing the required electric field gradients. The major advantage of channel-spark is realized in generated electron beams which are found to be much more stable relative to those generated in the pseudo-sparks. More importantly, the efficiency of energy transfer from beam to target is about 30% which is about 7-8 times higher than in the case of pseudo-spark discharges.[22] Table I presents typical electron beam parameter generated in a channel spark discharge.

Table I: Typical Pulsed electron-beam parameters

Maximum discharge voltage	20 kV
Electrical efficiency	30%
Stored Energy	3J
Gas pressure	4-30 mTorr
Discharge time	~80-100 ns
Repetition rate	1-100 Hz
Electron energy in the beam	$\leq$ 15 keV
Electron current	1.5 kA
Electron beam diameter at the target	~2-3 mm
Beam Current density	$\leq 10^5 A/cm^2$
Power in the beam	$\leq$15 MW
Power density in the beam	$\leq$500 MW/cm^2
Maximum Range of electrons	0.4 μm

The above parameters of the pulsed electron beam are very attractive for thin film applications. The pulsed nature of the energy source (electron beam) facilitates the ability to deposit (i) complex metal-oxides and (ii) very large band gap (optically transparent) materials such as SiO_2, MgF_2 etc., which are difficult to process by PLD. The technique also can be used to deposit organic thin films such as PTFE and therefore is poised to become a complimentary thin film deposition technique to PLD extending the range of materials that can be prepared as thin films using pulsed energy techniques.

13.3.3 Film depositions by PED and evaluations

In this section, data obtained on thin films of representative materials systems is presented

13.3.3.1 Multi-component films: High-temperature super conducting YBCO and GdBCO

Figures 18(a) and 18(b) show the AC susceptibility data of superconducting YBCO and GdBCO films on LaAlO$_3$ substrates respectively. These films were deposited at a substrate temperature of 800 C and an oxygen partial pressure of about 12 mTorr. The films show superconducting transition temperatures (T_c's) around 89-90K with transition widths of ~1K.

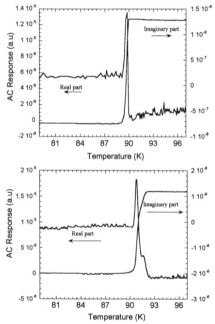

Figure 18: (a). AC susceptibility data of YBCO on LaAlO$_3$ substrate. (b). AC susceptibility data of GdBCO on LaAlO$_3$ substrate.

The critical current densities of the films were measured by standard I-V measurements. YBCO and GdBCO films on single crystalline substrates were patterned to 40 micron wide bridges by standard photolithographic patterning and wet chemical etching in 0.5 % phosphoric acid. Figure 19 shows the critical current data for GdBCO films. From these measurements the critical current density for GdBCO films was estimated to be 1.1 x 10^{6}A/cm^2 at 77K, in self field.

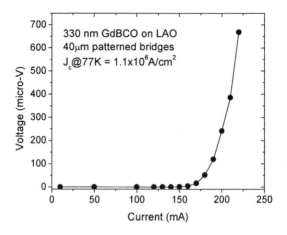

Figure 19: Critical current data of GdBCO on LaAlO₃ substrates.

The critical current density obtained on YBCO deposited on LaAlO₃ substrates is also measured and is greater than $1 \times 10^6 A/cm^2$ at 77K and zero field. These values compare well with state-of-the-art films deposited by PLD and other deposition techniques. Rutherford Backscattering data indicates that the composition of the target is reproduced in the film (Gd:Ba:Cu = 1:2:3)

13.3.3.2 Transparent conducting oxides (SnO₂)

Recently, much attention has been devoted to the deposition of n-type semiconductor tin oxide (SnO_2) from the standpoint of a variety of applications. Given its high transparency, good electrical conductivity, and superior chemical and mechanical stability[26] as compared to other wide band gap semiconductors, thin films of SnO_2 have found widespread applicability in fields such as transparent conducting electrodes in flat-panel displays and solar cells, opto-electronic devices, gas sensing applications (even for flammable and toxic gases) and IR detectors.[27-30]

A typical XRD pattern for the PED-SnO_2 film grown in 8 mTorr of forming gas (95% Ar + 5% H₂) is shown in Figure. 20. The data obtained are similar to that for a film grown by PLD in optimized condition of background O_2 pressure of 10^{-4} Torr. The XRD data show the presence of only (101) family of the rutile phase in the film. The rocking curve (inset to Figure. 20) full width at half maximum (FWHM) for the (101) peak is ~0.5°, comparable to high quality PLD films. Given the substrate rocking curve FWHM of 0.3°, this result signifies excellent (101) orientational quality of the films. Figure 21(a) shows temperature dependence of resistivity for the films grown by PED at 8 mTorr pressure of different

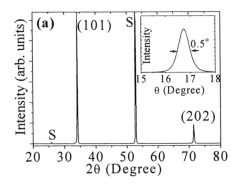

Figure 20: X-ray diffraction data of SnO$_2$ films deposited by PED. [Ref. 31]

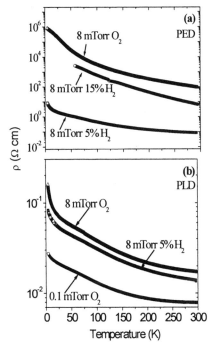

Figure 21: Resistivity data of SnO$_2$ films deposited by PED(a) and compared to films deposited by PLD (b). [Ref. 31]

gases. The films grown in oxygen are highly resistive with a room temperature resistivity ~ 100 ohm-cm which increases to about 10^6 ohm-cm at 5 K. The films grown in forming gas are more conducting with resistivity ~ 0.08 ohm-cm at 300 K. The resistivity of the most conducting film grown by PED is about an order of magnitude higher than that grown by PLD under optimized conditions of 10^{-4} Torr O$_2$ (Figure 21 (b)). The film grown by PLD on the other hand, either in 8 mTorr forming gas or 8 mTorr O_2 also had similar room temperature resistivity values ($\sim 1 \times 10^{-2}$ ohm-cm).This suggests that in the PLD process, the background gas may not have as drastic an effect on resistivity of the SnO_2 films as in the PED case. Clearly, SnO_2 films grown by PED and PLD have subtle differences in their properties and more understanding about the film growth process is needed.

Figure 22 shows the optical transmission spectra of the films grown by PED (top) and PLD (bottom) respectively. Very high transmission ($\sim$ 80 %) is observed in the visible and infrared regions for all the films grown by PED or PLD. Despite the resistivity differences in these films, the optical spectra are quite similar. The AFM images of the films grown by PED and PLD are shown as insets. The surface features of the films are comparable and the surface roughness values obtained from these scans are 4.5 and 3 nm for the PED and PLD films, respectively.

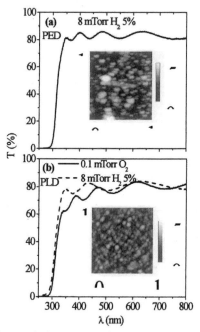

Figure 22: Optical transmission spectra of PED (top) and PLD (bottom) deposited SnO_2 and associated AFM images. [Ref. 31]

13.3.3.3 Films of other materials by PED
13.3.3.3.1 ZnO and Co-doped TiO₂

ZnO and Co-doped TiO_2 have also been successfully grown using PED and the results obtained have been found to be comparable to those obtained for PLD. Figure 23 shows x-ray diffraction rocking curve full width at half maximum (FWHM) and Figure 24 presents the optical transmission spectra as a function of the deposition temperature for ZnO films deposited by PED.

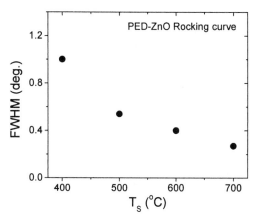

Figure 23: X-ray rocking angle FWHM as a function of deposition temperature for ZnO films deposited by Pulsed electron Deposition. [Ref. 32]

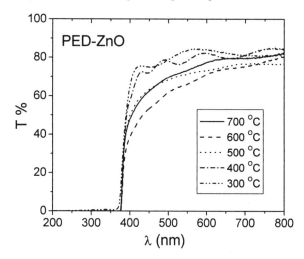

Figure 24:. Optical transmission spectra of ZnO as a function of deposition temperature for films deposited by Pulsed Electron Deposition. [Ref. 32]

13.3.3.3.2 Polymer films (PTFE and PET)

The most striking example of the versatility of PED has been the ability to deposit polymer films such as PTFE (Teflon) and PET using solid targets. The mechanism leading to polymer film formation is not understood at the present time. The IR spectra of a PTFE film deposited by

408

PED are presented in Figure 25. The film is deposited using a bulk PTFE target. The IR spectrum for the bulk target is also presented (at the top of figure) for comparison. The preliminary data indicate that the films retain their molecular structure of the bulk target to a large degree. The PTFE films are smooth, adherent, transparent, and are highly hydrophobic. The ability to

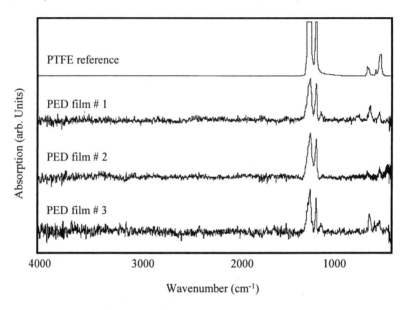

Figure 25: IR spectra of bulk Teflon target (top spectrum) and PED deposited films under various conditions (bottom three spectra). [Ref. 33]

form polymer films by PED opens up new possibilities where certain organic films need to be deposited in conjunction with inorganic layers all requiring vacuum processing. It may be mentioned here that even though it is possible to deposit polymer films by PLD, special approaches such as those used in the so-called MAPLE process[34] are required for obtaining the desired film properties. Also, in case of PLD of transparent polymers, a specific UV absorbing binder is required to facilitate ablation. With PED, a polymer target may be used directly without any additives since the ablation process does not depend on the optical absorption. Some of the possibilities demonstrated here may lead to interesting research directions for PED.

13.3.4 PLD and PED: A comparison

- The variation of the absorption depth for electrons over the material space is a much smaller range compared to the absorption coefficient of materials for UV photons, giving a larger phase space of materials accessible to the PED process. Wide band gap materials such as MgF_2, SiO_2, semiconductors such as Ge. Si, a variety of high reflectivity metals such Al and even organic films can be deposited relatively easily by PED compared to PLD.

• The capital and operating costs are likely to be significantly lower for PED than for PLD. This is a direct consequence of the energy conversion efficiency (electrical to electrical) of the PED process (~30%) relative to PLD with a conversion efficiency of only ~ 3% (electrical to optical).

• PED technology is readily scalable to large areas and high throughputs. From a cost effectiveness point of view therefore, PED will have advantages over PLD where the capital and running costs of industrial excimer lasers can be prohibitively large.

• In a simple R&D environment, the small foot print of PED (as well no use of halogen based corrosive gases required for operating an excimer laser) might be advantageous. The entire deposition unit can be plugged into standard power outlets with no special utility requirements. As an example, a typical PED system manufactured and marketed by Neocera is shown in Figure 26. The compact PED source is attached to the vacuum system on the lower right hand side.

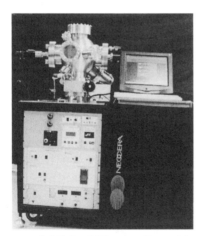

Figure 26: A commercial Pulsed Electron Deposition System.

In spite of several advantages mentioned above, PED, however, suffers from some disadvantages compared to PLD.

• The dynamic range of pressures for materials synthesis is somewhat limited in the case of PED since the operating pressure of the source and the deposition pressure are not decoupled as in the case of a PLD system. However, this could be overcome by adopting a differentially pumped scheme where the source and the deposition chamber operate at different pressures.

• The surface roughness of PED deposited films has been found to be higher than for PLD films on account of a larger range of electrons in target materials (a few microns) compared to UV photons ($\sim$ 1000 angstroms)

• The density of the evaporant over the target is significantly lower for PED than it is for PLD and hence the angular distribution of the evaporant is also lower ($Cos^{5-6}\theta$ dependence) compared to PLD ($Cos^{11}\theta$ dependence). As a result, for PED, the film deposition phase space for stochiometric film deposition in the case of a multi-component materials is much narrower relative to PLD.

• Unlike the case of PLD where a single laser can be shared among many deposition systems, each PED system requires a dedicated source.

In summary, PED and PLD are complimentary techniques and a hybrid system incorporating the two processes would stand to gain from the advantages of both the techniques.

13.3.5   PED for large scale applications

As previously mentioned, PED is scalable to large areas and high throughputs cost effectively. As an illustration of a continuous process based on PED, a reel-reel system[35] for the deposition of a superconducting precursor on a flexible Ni tape is shown in Figure 27. The BaF_2 based precursor after a subsequent anneal shows a high critical current density J_c ($\sim$1.7MA/cm^2) in the films validating the potential of PED technology for what is now known as the second generation coated conductors based on high T_c superconducting films.[36]

In Figure 28, a prototype reel-to-reel web coater incorporating 26 sources (2 x 13 arrays), for depositing SiO_2-based glass films on a continuously moving substrate (sheet)[37] is presented. Notable is the large number of sources integrated into the deposition system and this feature is quite attractive for high volume manufacturing and is an intrinsic advantage of PED over PLD.Continuous progress is being made in PED source technology and its reliability is beginning to approach that of the excimer lasers, which have enjoyed a significant head start.

Figure 27: Photograph of a PED based reel-to-reel system for continuous deposition of high T_c precursors for second generation coated conductor development. [Ref. 35]. The reel handling mechanism is shown in the enlarged image on the left.

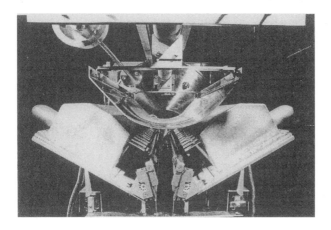

Figure 28: A 2x13 source PED array for continuous coating of one- meter wide plastic sheet [Ref. 37]. The arrays are seen in the middle of the figure.

13.4 SUMMARY

Over the last 17 years pulsed energy deposition (PLD/PED) has grown from its infancy to a mature R&D technology and is on its way to being accepted in the commercial production environment. Interest in multi-component materials stimulated the development of the PLD process and the commercial application of multi-component materials is also likely to be the engine that propels this technology forward. There are still challenges to be overcome to make this technology realize its full potential.

ACKNOWLEDGEMENTS

The authors would like to thank all the researchers that have contributed to the growth of Pulsed Laser Deposition and Pulsed Electron Deposition technologies. In particular, thanks are due to Dr. H. M Christen of Oak Ridge National Laboratory, Dr. Y. K. Yap of Michigan Technological University, Dr. I. Takeuchi, Dr. R. J. Choudhary, Dr. S. B. Ogale, Dr. S. R. Shinde and Dr. V. N. Kulkarni of University of Maryland and Dr. R. K. Singh of University of Florida. The authors also thank Dr. C. Schultheiss, Dr. G. Muller and Dr. W. Frey of Forschungszentrum Karlsrue for useful discussions.

REFERENCES

*   Also at Physics Department, University of Maryland, College Park MD 20742.

1. D. Dijkkamp, T. Venkatesan, X. D. Wu, S. A. Shaheen, N. Jasrawi, Y. H. Min-Lee, W. L. McLean and M. Croft, *Appl. Phys. Lett.* **51**, 619 (1987).

2. T. Venkatesan, X. D. Wu, A. Inam and J. B. Wachtman, *Appl. Phys. Lett.* **52**, 1193 (1988).

3. X. D. Wu, D. Dijkkamp, S.B. Ogale, A. Inam, E.W. Chase, P.F. Miceli, C. C. Chang, J. M. Tarascon and T. Venkatesan, *Appl. Phys. Lett.* **51**, 861 (1987).

4. J. T. Cheung, *History and Fundamentals of Pulsed Laser Deposition* (Pulsed Laser Deposition of Thin Films, edited by D. B. Chrisey and G. K. Hubler), John Wiley & Sons, Inc, New York (1994).

5. G. J. H. M. Rijnders, G. Kostner, D.H.A. Blank and H. Rogalla, *Materials Science and Engineering* B **56**, 223 (1998).

6. G. H. Lee, M. Yoshimoto, T. Ohnishi, K. Sasaki and H. Koinuma, *Materials Science and Engineering B* **56**, 213 (1998).

7. H. N. Lee, H. M. Christen, M. F. Chisholm, D. H. Lowndes, *Nature Materials* (In print).

8. X. D. Xiang, X. SunG. Briceno, Y. Lou, K. Wang, H. Chang, W. G. Wallace-Freedman, S. Chen and P. G. Schultz, *Science* **268**, 1738 (1995).

9. J. Wang, Y. Yoo, C. Gao, I. Takeuchi, X. Sun, H. Chang, X. D. Xiang and P. G. Schultz, *Science* **279**, 1712 (1998).

10. H. Chang, I Takeuchi and X. D. Xiang, *App. Phys. Lett.* **74**, 1165 (1999).

11. H. M. Christen, S. D. Silliman and K. S. Harshavardhan, *Review of Scientific Instruments* **72**, 2673 (2000).

12. H. M. Christen, S. D. Silliman and K. S. Harshavardhan, *Applied Surface Science* **189**, 216 (2002).

13. K. L. Saenger, *Angular Distribution of Ablated Material* (Pulsed Laser Deposition of Thin Films, edited by D. B. Chrisey and G. K. Hubler), John Wiley & Sons, Inc, New York, (1994).

14. Y. K. Yap et al., (*Private Communication*).

15. Y. Iijima, N. Tanabe, O. Kohno and Y. Ikeno, *Appl. Phys. Lett.* **60**, 769 (1992).

16. S. R. Foltyn, P. N. Arendt, P. C. Dowden, R. F. DePaula, J. R. Groves, J. Y. Coulter, Q. Jia, M. P. Maley and D. E. Peterson, *IEEE Trans. Appl. Supercond.* **9**, 1519 (1999).

17. K. S. Harshavardhan, H. M. Christen, S. D. Silliman, V. V. Talanov, S. M. Anlage, M. Rajeswari and J. Claassen, *Appl. Phys. Lett.* **78**, 1888 (2001).

18. J. Christiansen and C. Schultheiss, *Z. Physik A* **290**, 35 (1979).

19. M Hoebel, J. Geerk, G. Linker and C. Schultheiss, *Appl. Phys. Lett.* **56**, 973 (1990).

20. G. Muller, M. Konijnenberg, G. Kraft and C. Schultheiss, *'Thin film Deposition by means of Pulsed Electron Beam Ablation'* in Science and Technology of Thin Films, Eds. F.C. Matacotta, G. Ottaviani, World Scientific, p89, 1995.

21. V. I. Dediu, Q. D. Jiang, F. C. Matacotta, P. Scardi, M. Lazzarino, G. Nieva and L. Civale, *Superconductor Science and Technology* **8**, 160 (1995).

22. Q. D. Jiang, F.C. Matacotta, M. C. Konijnenberg, G. Mueller and C. Schultheiss, *Thin Solid Films* **241**, 100 (1994).

23. C. Schultheiss and F. Hoffman, *Nuclear Instruments and Methods in Physics Research B* **51**, 187 (1990).

24. Q. D. Jiang, F. C. Matacotta, G. Masciarelli, F. Fuso, E. Arimondo, M. C. Konijnenberg, G. Muller and C. Schultheiss, *Supercond. Science and Technol.* **6**, 567, (1993).

25. D. Basting, in *Industrial Excimer Lasers, Fundamentals, Technology and Maintenance,* 2 nd edition, Lambda Physik GmBH, Gottingen (1991).

26. K. L. Chopra, S. Major and D. K. Pandya, *Thin Solid Films* **102**, 1 (1983).

27. M. W. J. Prins, K. O. Grosse Holz, G. Muller, J. F. M. Cillessen, J. B. Giesbers and R. P. Weening and R. M. Wolf, *Appl. Phys. Lett.* **68**, 3650 (1996).

28. A.V. Tadeev, G. Delabouglise and M. Labeau, *Thin Solid Films* **337**, 163 (1999).

29. E. Comini, G. Faglia, G. Sberveglieri, Z. W. Pan and Z. L. Wang, *Appl. Phys. Lett.* **81**, 1869 (2002).

30. J. Y. Kim, E. R. Kim, Y. K. Han, K. H. Nam and D. W. Ihm, *Jpn. J. Appl. Phys.* **41**, 237 (2002).

31. R. J. Choudhary, S. B. Ogale, S. R. Shinde, V. N. Kulkarni, T. Venkatesan, K. S. Harshavardhan, M. Strikovski nad B. Hannoyer, *Appl. Phys. Lett.* **84**, 1483 (2004).

32. R. J. Choudhary et al., *(Private Communication)*.

33. K. S. Harshavardhan, M. Strikovski and J. Kim *(Private communication)*.

34. P. K. Wu, B. R.Ringeisen, D. B. Krizman, C. G. Frondoza, M. Brooks, D. M. Bubb, R. C. Y. Auyeung, A. Piqué, B. Spargo, R. A. McGill and D. B. Chrisey, *Review of Scientific Instruments* **74**, 2546 (2003).

35. H. M. Christen *(Private Communication)*

36. Coated Conductor Technology Development Roadmap, Prepared by Energetics Inc. for US Department of Energy, Superconductivity for Electric Systems Program, (2001).

37. H. C. G. Kebler, Forschungszentrum Karlsruhe GmBH, Technik und Umwelt *(Private communication)*.

Index

418

Index of Materials